The Electrician's Toolbox Manual

Rex Miller

ARCO

New York

ARCO

Simon & Schuster, Inc.
15 Columbus Circle
New York, NY 10023

DISTRIBUTED BY PRENTICE HALL TRADE SALES

Manufactured in the United States of America

2 3 4 5 6 7 8 9 10

Library of Congress Cataloging-in-Publication Data

Miller, Rex.
The electrician's toolbox manual.
Includes index.
1. Electric wiring—Handbooks, manuals, etc.
I. Title.
TK3201.M55 1989 621.319'24 88-34977
ISBN 0-13-247701-7

CONTENTS

FIGURE CREDITS

Figures 1-2, 1-3, 1-5, 1-12, 1-15, 1-17, 1-19, 1-23, 2-20, 2-21, 2-35, 10-3: courtesy of Stanley Tools, a division of The Stanley Works.

Figures 2-2, 2-3, 2-5, 2-10b: courtesy of ITW Linx.

Figure 2-4: courtesy of Star Industries.

Figure 2-10a: courtesy of Telecom Products Division/3M.

Figures 2-11, 2-12: courtesy of M.M. Rhodes & Sons Company.

Figure 2-24: courtesy of Wiremold Tools.

Figures 4-5, 4-6, 4-8: courtesy of Bryant Electric Company.

Figures 4-9, 5-32: courtesy of Bodine Electric Company, Chicago, IL.

Figures 5-1, 5-2, 5-3, 5-4, 5-5, 5-6, 5-7, 5-8, 5-9, 5-10, 5-11, 5-12, 5-13, 5-14, 5-15: courtesy of NuTone.

Figures 5-42, 5-43, 5-44, 5-45, 5-46: courtesy of Majara-Mohawk Power Corp.

Figure 5-51: courtesy of Automatic Switching Company.

Figures 5-71, 5-72: courtesy of NEMA.

Figures 5-73, 5-78, 5-79, 5-80: courtesy of Crouse-Hinds Electrical Construction Materials.

Figure 5-83: courtesy of Arrow Fastener Company, Inc.

Figures 8-6, 8-7: courtesy of Motorola.

Figure A1-3: courtesy of the National Electrical Code®.

Figure A2-1: courtesy of Underwriters Laboratories, Inc.

Figure A2-2: courtesy of Canadian Standards Association.

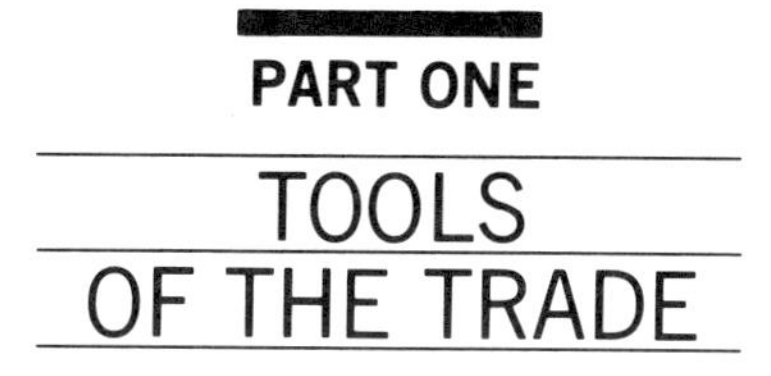

Every electrician needs tools to get a job done quickly and efficiently. The tools become an extension of the person, increasing productivity.

Every electrician apprentice should have a basic knowledge of the tools most useful for electrical work and of how to use and care for these tools. In that way, the apprentice will be able to choose which tools to use on a particular job, and perhaps also be creative, using a tool in an unusual way suited to the specific task at hand.

This chapter presents basic information on those tools commonly used for general residential and light commercial wiring.

AUGER BIT

"Auger bit" is a general term for bits used on a bit brace. (See Figure 1-1.) *See* Bit Brace; Bits, Specialty.

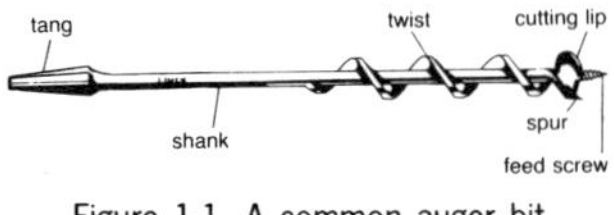

Figure 1-1. A common auger bit.

AWL (SCRATCH AWL)

A scratch awl is a versatile tool that can be used to

- scratch a line or mark a surface,
- make a pilot hole for starting a screw,
- punch a hole in sheet metal,
- hold a piece of equipment in place while it is being secured, and
- pry staples loose.

Scratch awls come with a large pear-shaped handle or with a screwdriver-type handle. The blade, which is usually very pointed, comes in sizes from 2 3/4 inches to 3 1/2 inches. (See Figure 1-2.)

Figure 1-2. Scratch awl.

BIT BRACE

A hand-operated drill, the bit brace uses the principle of the wheel and axle to create a driving force. The jaws of the chuck hold the bit in the brace. Pressure is applied to the head of the brace, and the handle in the middle of the brace is rotated. The drill bits have a screw tip to catch the material and pull the cutting edge forward. The twist of the bit causes a smooth flow of chips to the outside of the hole.

Some bit braces have ratchet action that allows the handle to be turned in one direction and then backed off and turned again. This aids in drilling in tight spots where a full swing of the handle is not possible. (See Figure 1-3.)

The bits used in the bit brace are known as *auger bits.* The size of the auger bit is stamped on the tang, or shaft, of the bit and is given as a whole number, indicating the number in sixteenths of an inch. (*See* Bits, Specialty for a discussion of different types of bits that can be used in a bit brace.)

Figure 1-3. Bit brace.

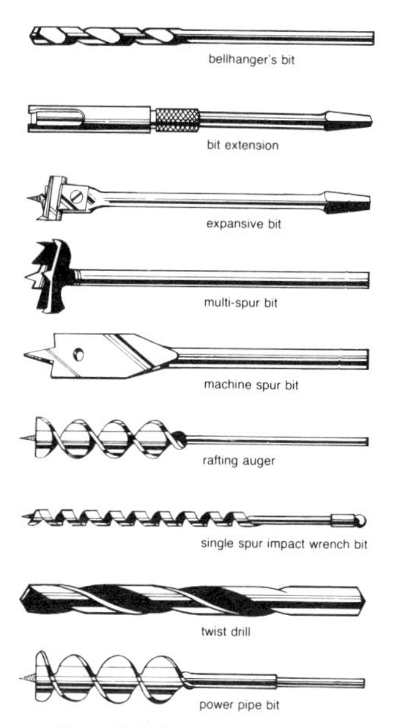

Figure 1-4. Special types of bits frequently used by electricians.

BITS, SPECIALTY

A wide variety of bits are available for use with the bit brace discussed above and also with hand drills and electric drills. Some bits are designed for use with particular materials, others for use in inaccessible locations, still others for use on only certain types of drills. (See Figure 1-4.)

Bellhanger's Drill Bit. The bellhanger's drill bit is an auger bit that is handy for drilling small-diameter holes such as those needed for telephone-wire installation and signaling circuits. This bit is standard equipment for telephone installers. The bit comes in diameters from 3/16 inch to 3/4 inch in 1/16-inch increments.

Brace Bit Extension. This device extends the bit for the person using the brace, a convenience when the operator cannot reach or extend the bit to the point where the drilling is needed. For example, a bit brace extension allows an operator to stand on the floor underneath and drill through rafters. It can also be used to drill through concealed areas for feeding wire through without having to get inside the area. For example, when the air return for a furnace has already been installed and Romex (branch hook-up wire) has to be pulled through it, a bit extension allows the operator to drill straight through and then feed the Romex through. Using the extension is better than trying to drill through from two sides and not having the holes align. Bit extensions are also available for power drills.

Expansive Bit. An expansive bit fits into a brace and has a screw adjustment that allows the blade to be adjusted to drill holes of various diameters. The bit can be obtained with 1/2-inch to 1 1/2-inch cutters, 7/8-inch to 1 3/4-inch cutters and 1 3/4-inch to 3-inch cutters.

Machine Spur Bit. A wood drilling bit designed for use in electric drills, the machine spur bit, also known as a spade bit, is very popular. It is inexpensive and provides good service for drilling holes for Romex and other wiring. It can be used at any angle. A machine spur bit is a little over 6 inches long and comes in diameters from 1/4 inch through 1 1/2 inches.

Masonry Bit. A precision-made, high-quality, carbide-tipped rotary masonry drill bit consists of a single carbide insert and a steel drill body. The carbide insert forms two cutting edges and the grade of carbide is chosen for its resistance to wear and shock. The insert is electronically brazed, ground, and inspected to ensure concentricity and is positioned in a recess in the forward end of the drill. The drill body retains and supports the carbide insert and by its design removes dust and cuttings from the hole as it is being drilled.

There are two types of drill bodies; both spiral-fluted. The standard type has a fast double-lead spiral and resembles a common twist drill. It removes dust and cuttings fast and also resists compacting of masonry dust and cuttings in the drill flutes. The F-type drill has an acme-type, double-lead worm, or slow spiral. This type has a slight advantage over the standard type in that the carbide insert has a little more steel backing and support.

Remember that these masonry drill bits are designed for use in a rotary-type drill motor. They should *not* be used with a rotary percussion-type (impact wrench) or hammer-type drill motor. (There are special drill bits designed to take the hammering of the percussion-type drill, and, in fact, percussion-type drill bits cut the time required to drill through masonry by two thirds.)

Multi-Spur Pipe Bit. Designed for use in portable electric or pneumatic drills and stationary boring machines, the multi-spur pipe bit has three milled flats that are sharpened to give them an edge. This bit is used to cut large-diameter holes through wood—for example, rafters or studs—to provide space for large sizes of conduit.

Power Bit Extension. This device extends the bit on power drills, making it possible to drill holes in spots otherwise inaccessible. The extension comes in 18-inch and 24-inch lengths. Two set screws allow for easy attachment of the drill bit.

Power Pipe Bit. The power pipe bit makes a very large hole and is designed for rough, rugged boring. It is very handy for drilling holes for rigid conduit, EMT, and tubing. Lengths run from 14 to 17 inches overall. Several cutter sizes are available, including 1 1/4 inches for 3/4-inch pipe; 1 1/2-inch diameter for 1-inch pipe; 1 3/4 inches for 1 1/4-inch pipe; 2-inch diameter for 1 1/2-inch pipe; and 2 1/2-inch diameter for 2-inch pipe.

Rafting Auger. Also known as a *boom auger*, the rafting auger has a double twist with a flat-cut pattern head. It provides a smooth cut and is a good bit for a slow-moving power drill. Rafting augers come in lengths from 14 to 40 inches with hole diameters from 1 1/4 inches to 4 inches. They are available for bit braces and for power drills.

Single-Spur Impact Wrench Bit. Designed for use in an impact wrench (not in a bit brace), the single-spur impact wrench bit can be identified by a grooved portion near the end of the bit. The single spur—spur means the cutting edge is near the screw tip—bit bores 3/16-inch to 1 1/2-inch holes in telephone or power line poles quickly, easily, and safely. Electricians often need to drill through power line poles to mount boxes and other equipment. Two sizes of the bits, used with 7/16-inch or 5/8-inch hexagonal adaptors, fit all impact wrenches. Overall length of the bit is 12,

16, or 24 inches, depending on the twist length (8, 12, or 18 inches) needed. (*See also* Roto-Hammer.)

Single-Twist Power Bit. Designed for use in power drills, the single-twist power bit is a smooth boring bit that can be used for drilling metal and wood as well as plastic. Chip clearance is excellent, since this bit has a single-twist design. It may be used to bore holes from 5/8 inch to 1 inch. Overall length is from 5 to 7 inches, depending on the size of the drill. With an extension, the single-twist power bit will drill through floor joists, studs, and other existing construction—important when installing Romex.

Twist Drill. Twist drill bits are available for both electric and hand drills. The drill bits come in a variety of sizes for almost any use, from 1/16 inch to over 1 inch. They are also available in special designs. Every electrician should have an assortment of twist drill bits available for a variety of tasks.

CHALK LINE AND REEL

A chalk line is used to mark a straight guideline and is useful when you need to install fixtures in a straight line. Pull the ring at the end of the string and, with a nail or similar anchoring device, attach it to the surface you will be working on. Pull the string taut and then snap it with two fingers so it will leave a straight chalk line on the surface. Snap it only once; repeated snappings will cause several lines to appear and lead to confusion. Although most chalk lines are coated with white dust, some are available with red or blue dust. A handle on the reel allows you to wind the string back into its protective case, where it is rechalked. (See Figure 1-5.)

Humidity can cause problems with the chalk. Therefore, it is good practice to hit the case against a solid object before pulling the string. This will cause the chalk to spread evenly on the line as it is pulled from the case.

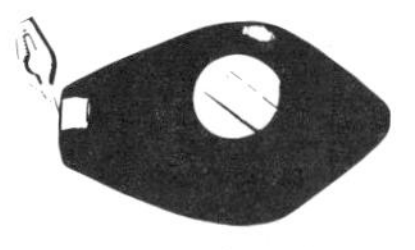

Figure 1-5. Chalk line.

CHISELS

Chisels are used to shave material such as wood, plastic, or metal. Electricians use them when material must be removed to allow the installation of electrical equipment. (See Figure 1-6.)

Firmer Chisel. A firmer chisel is used to cut wood away from unwanted locations so that electrical boxes can be installed with proper clearance. Firmer chisels are made of high-carbon steel and come in a variety of widths. The blade is usually thick and has beveled edges.

Framing Chisel. Larger than a firmer chisel, a framing chisel is used for heavy framing work, as a wrecking tool or, like a firmer chisel, to chip away wood. The blade does not have beveled edges.

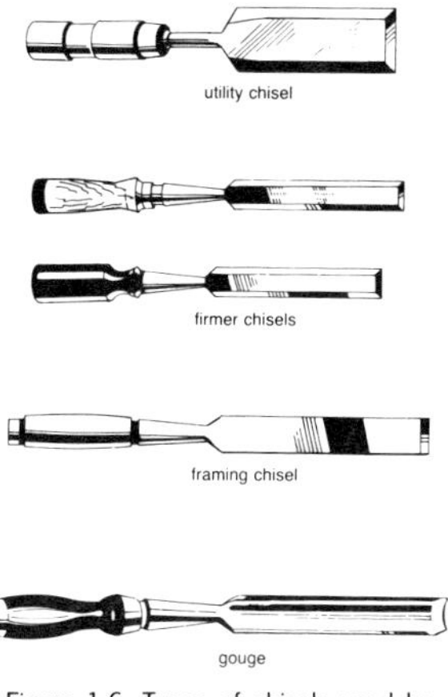

Figure 1-6. Types of chisels used by electricians.

Gouge. Although a gouge is primarily a woodworking tool, electricians use it when they need to remove excess wood when installing conduit. It has a hollow blade and is especially useful when a groove is needed.

Utility Chisel. The utility chisel is just what its name implies—a tool with many uses. However, it is so named because utility and telephone line workers commonly use it to cut daps in poles for crossarms. The 2-inch-wide beveled blade is heat-treated to withstand hard wear.

PROPER USE OF A CHISEL

- Always keep your attention focused on the work at hand.
- Keep the chisel sharp. Sharpen by using a grinder and then touch up the edge on an oilstone.

- Protect the sharp edges of the chisel—to prevent injuries and damage to the edge.
- Be careful not to drop a chisel on concrete.
- Before using a chisel, examine any wooden object to be worked on for nails, screws, or metal of any kind and remove.
- Do not use a chisel as a screwdriver or prybar.

CIRCUIT TESTER

A circuit tester is used to determine whether a circuit is energized, or live, and with some instruments, to determine exact voltage, current, or resistance. (See Figure 1-7.)

Clamp-on Ammeter. The clamp-on ammeter is used to locate devices that are drawing too much current but not enough to cause a short, to determine actual load, and to balance polyphase loads. The clamp-on feature opens when the lever on the side is pressed and then closes to encircle a single wire to determine if current is flowing and, if it is, to measure how much. One major advantage of the clamp-on ammeter is that it can be used without having to disconnect the lines.

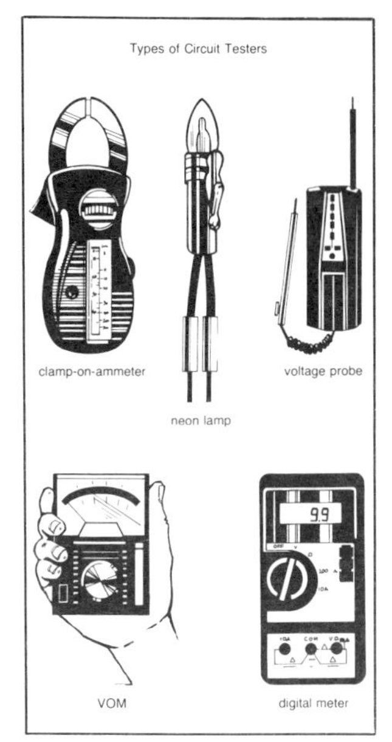

Figure 1-7. Circuit testers: clamp-on ammeter, neon lamp, voltage probe, VOM, and digital meter.

Neon Lamp. A very inexpensive tool, the neon lamp is an instrument that every electrician and apprentice should have in the toolbox. It is not accurate in determining exact voltages, but it is valuable in preventing shock. Just plug it in to determine whether a circuit is energized. With 120 volts in the circuit, the neon lamp will glow; with 240 volts, the lamp will glow twice as brightly.

Voltage Probe. A voltage probe allows the electrician to determine the exact voltage in a line—120 volts, 240 volts, 480 volts, or whatever the upper limit is on the probe. A voltage probe can be used to measure voltage in either alternating current (AC) or direct current (DC) lines. This instrument is valuable in quick-check situations when the electrician needs to know the line voltage for troubleshooting purposes.

VOM. A VOM is a rugged, self-contained, battery-operated unit used to measure voltage, current, and resistance. VOM stands for volts-ohms-milliamperes, signifying that the unit is capable of reading voltage, checking resistance, and measuring current, in most cases up to 10 amperes. The instrument's ability to measure resistance is particularly valuable when testing for shorts in a new system that has not yet been energized.

The VOM is an analog, not a digital, meter; in some instances an analog type is better, since it shows rapid up-and-down movement rather than the changing of numbers every second. In troubleshooting situations, many electricians prefer the VOM over the digital meter. However, the electrician must become familiar with the VOM scale to be able to read it quickly and accurately.

Digital Meter. A digital meter is also a battery-operated self-contained unit used to measure voltage, current, and resistance. It is more accurate—and more expensive—than the VOM. In fact, the digital meter may be too accurate for most electrical problems. Most digital meters can provide voltage readings to within 0.01 volt, resistance (ohmmeter) readings to 0.01 ohm, and current (ammeter) readings in microamperes (0.000001 amp.) The readouts resemble those of a digital watch or calculator. Electronics troubleshooting relies heavily on the digital meter.

DRILL, ELECTRIC

Hand-held, portable electric drills are inexpensive and found in many homes. They are very useful in drilling holes from 1/16 inch to several inches in diameter. In many models, variable speeds and special features, such as reverse motion, allow the drills to be used for drilling many different types of materials, as well as for driving and removing screws. (See Figure 1-8.)

The chuck on the electric portable drill is different from that on the bit brace; it has a clamping action that fits over smooth, round drill bits. Some drill bits have two or four flat surfaces to allow for a better chuck hold on the bit. Battery-powered drills and right-angle "hole hog" drills are also available and are useful in certain situations.

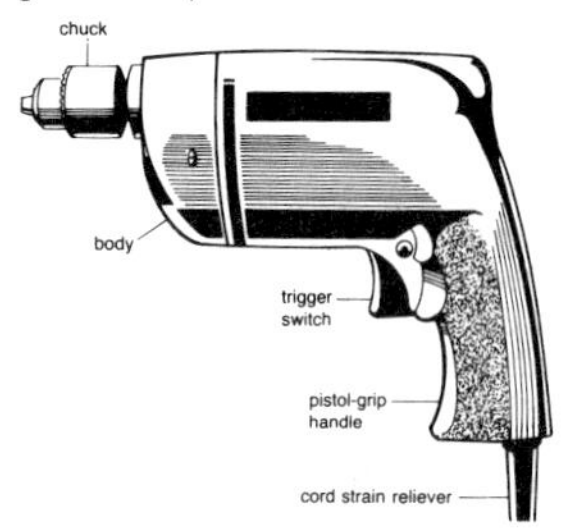

Figure 1-8. Portable electric drill.

DRILL, HAND

The hand drill is a crank-operated tool. A large wheel gear turns a smaller pinion that then turns the drill chuck holding the bit. Twist bits of various sizes are available.

The hand drill is most useful when light and delicate drilling work is needed—as in cabinetmaking situations in which a slow and steady action is required. It has largely been replaced by electric portable drills, especially battery-operated drills that can be used where power is not available.

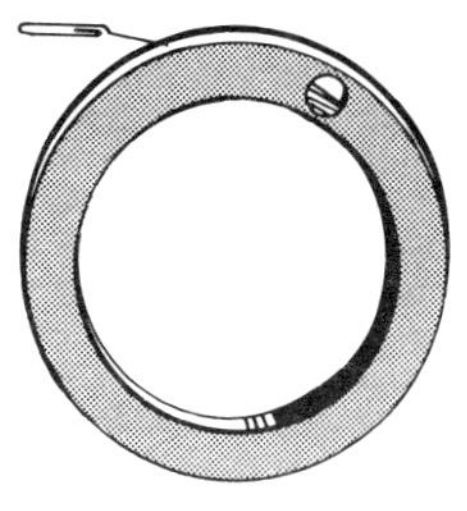
Figure 1-9. Fish tape reel.

FISH TAPE

Fish tape is thin tape used to pull wire through conduit, especially conduit that bends. Most fish tape is thin steel, about 1/8 inch wide, but some is wider, about 1/4 inch, and plastic tape is also available. The tape is usually enclosed in a plastic reel and comes equipped with a replaceable pull-eye or a loop at the end to allow easy pulling. (If the loop breaks off, a new one can be formed by heating the end of the tape with a propane torch and then forming a new loop with pliers. Do not attempt to make a new loop without heating the tape: the tape will break.) (See Figure 1-9.)

Using Fish Tape

1. Remove the insulation from about 4 inches of each wire that is to be pulled into the conduit.
2. Thread the uninsulated wire through the eye or loop of the fish tape and fold the wires back on themselves.
3. Wrap plastic tape over the wires to hold them in place. (Just a few turns of the plastic tape are necessary.)
4. Push the tape—with the wires attached—into the conduit.
5. Pull the wires through the conduit. This usually requires one person pulling at one end and another feeding the wires into the opening at the other end to make sure there are no snarls, to ease the wire in, and to prevent damage to the wire.
6. If necessary, use a lubricant to make pulling the wire through the conduit easier. However, do not use soap to lubricate since it will damage the insulation on wires and cause problems later. A wax-based lubricant works best.

IMPACT WRENCH

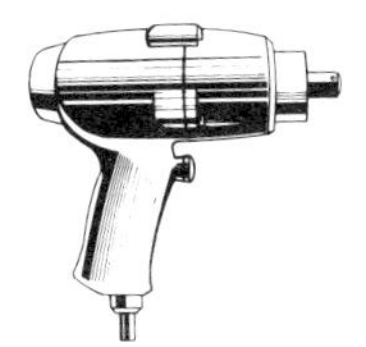

Figure 1-10. Impact wrench.

An impact wrench is a tool used to hammer a drill bit as it is turned. The hammering action keeps the bit eating away at the material, and the spiral on the bit is design to bring the removed material out of the hole being drilled. By keeping pressure on the bit, the impact wrench aids and speeds up the drilling process. This type of wrench, with its associated bit, is used to drill extremely hard materials such as masonry and very hard or very thick wood. (See Figure 1-10.)

electrician's knife

lineman's knife

skinning knife

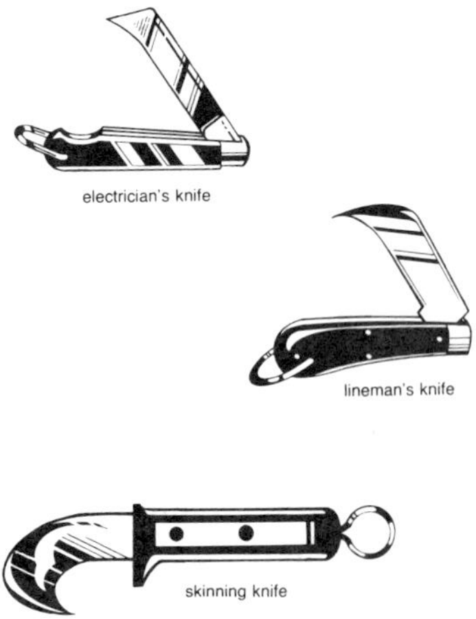

Figure 1-11. Knives: electrician's lineman's, and skinning.

KNIVES

Electricians use several types of knives—to cut wire and cable and to shape materials. (See Figure 1-11.)

Electrician's Knife. Similar to a utility pocketknife, an electrician's knife has a retractable blade and is a versatile tool. It is used to cut insulation and to remove plastic coatings on wires.

Lineman's Knife. The lineman's knife is similar to the electrician's knife, but it has a larger hook at the end of the blade. This larger hook is useful for stripping cables.

Skinning Knife. The skinning knife is used to strip large cables. Unlike the other knives discussed, the skinning knife does not have a retractable blade. It must be kept in a protective case when it is not in use, including when it is in the toolbox.

MEASURING DEVICES

Electricians need to make accurate measurements when they are laying out and installing various boxes and other devices. The two most common measuring devices used by an electrician doing residential or light commercial wiring are the extension, or folding, rule and the push-pull tape. (See Figure 1-12.)

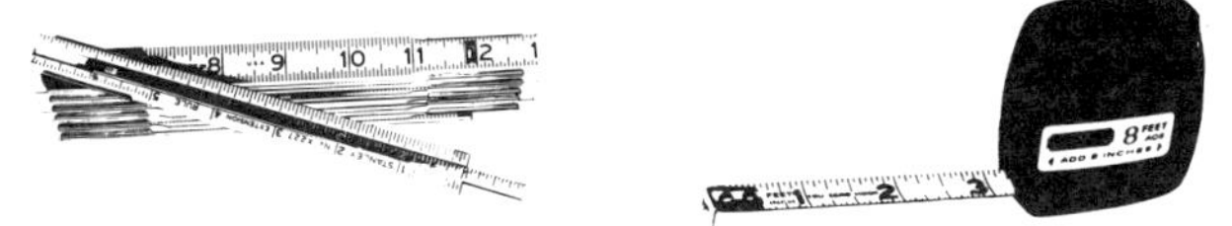

Figure 1-12. Measuring devices: zigzag extension rule and push-pull tape.

Folding Rule. A folding, or zigzag, rule is used to measure all framing lumber and to make and check other measurements up to 6 feet. A special type of folding rule is the *extension rule.* This type of rule is used to measure in tight places, as in windows or door frames. The rule is opened to within 6 inches or less of the opening being measured, and then a sliding measure is extended the remaining distance.

Push-Pull Tape. Push-pull tape is a metal tape with markings on the upper and lower edges. The tape itself is housed in a metal case. Push-pull tape, also sometimes known as *steel tape,* comes in many lengths, up to 100 feet. Some have springs that pull the tape back into the housing after it is released; others, especially the longer tapes, have a crank on the handle that is used

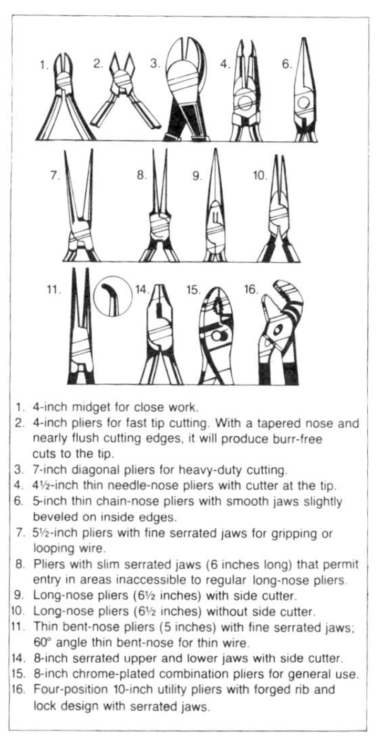

Figure 1-13. 13 types of pliers, each suited for a particular use.

to reel the tape into its housing. Push-pull tapes are often used by electricians, much more so than folding rules.

PLIERS

Pliers are used primarily for gripping. They come in a variety of sizes and shapes; the choice of which to use depends on the particular job at hand. We will mention a few of the more common types of pliers. (See Figure 1-13 for notes on many different types of pliers.)

Combination Pliers (Slip-Joint Pliers). The most common type of pliers, combination pliers have a two-position pivot construction that permits both normal and wide jaw openings. They are used for gripping and bending wire and for removing wire and nails from material. They come in lengths ranging from 8 to 10 inches.

Cutting Pliers. There are several types of cutting pliers—diagonal cutting, side cutting, and end cutting—that are not used primarily for gripping. Rather, they have a cutting surface at the tip and are used to cut wire.

Lineman's Pliers. Lineman's pliers have a wide, flat gripping surface and are used in heavy-duty wire cutting and splicing. They also cut nails and brads.

Long-Nose Pliers. Long-nose pliers are often used on electrical equipment. The long narrow nose allows one to reach into tight places and makes it easier to loop and bend wire of all sizes.

ROMEX STRIPPER

The Romex stripper is designed specifically to strip Romex cable. Romex cable consists of two or three conductors with white, red, or black plastic coatings. (Often an uninsulated conductor is also included in the bundle.) The wires are insulated by their own coatings but also have a thicker plastic jacket over them to make them easier to handle and pull through holes for branch circuits. Part of the plastic jacket has to be removed to get at the single wires. The Romex stripper is used to do this. It has a tip sticking up near the narrow end. The cable is slipped through the folded end, and the stripper is pulled while being held tightly against the cable. When the small tip has penetrated the jacket of the cable, it rips the covering as the cable is pulled through. The excess cable insulation ripped by the Romex stripper is removed by side cutters, diagonal cutters, or the electrician's knife. Once the outer cable jacket has been removed, it is possible to use wire strippers to take the insulation off the exposed individual red, black, and white plastic-coated wires. (See Figure 1-14.)

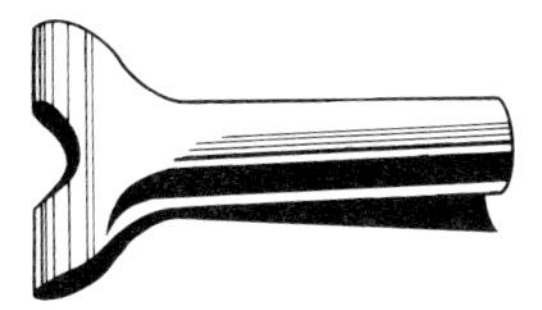

Figure 1-14. Romex stripper.

ROTO-HAMMER

A roto-hammer resembles a portable electric hand drill. It has a cam that rotates and produces a hammering action on the drill bit. The hammering action tends to loosen up the material that is being drilled.

SAWS

Electricians use a variety of saws—some powered, some nonpowered—to cut various materials to the size needed.

NONPOWER SAWS

Hacksaw. The hacksaw is primarily a metal-cutting saw, but it is often pressed into service to cut through almost anything. It is especially handy for cutting BX, conduit, and EMT.

Most hacksaws have an adjustable frame to fit several sizes of blades. The number of teeth on the blade determines the type of material that can be cut and the smoothness of the cut. Hacksaw blades come with 14, 18, 24, or 32 teeth per inch. Each blade has its own particular application in which it does the job faster and smoother than any other blade. (See Figure 1-15.)

In general, a hacksaw blade with 14 teeth per inch is used to cut soft, large sections of metal. In particular, it is used to cut material 1 inch or thicker in sections of cast iron, machine steel, brass, copper, aluminum, bronze, and slate. The blade with 18 teeth per inch is used for general cutting. It can be used to cut materials 1/4-inch to 1-inch thick in sections of annealed tool steel, high-speed steel, rail, bronze, aluminum, light structural shapes, and copper. The blade with 24 teeth per inch is used for cutting angle iron, brass, copper tubing, wrought-iron pipe, drill rod, conduit, light structural shapes, and metal trim. The blade with 32 teeth per inch is used for conduit and other thin tubing and sheet metal work.

Figure 1-15. Hacksaw.

Using a Hacksaw

1. To start an accurate cut, use the thumb as a guide and saw slowly, with short strokes.
2. Guide the blade until the cut is well established.
3. As the cut deepens, grip the front end of the frame firmly and take a full-length stroke. When sawing, stand facing the work with one foot in front of and approximately 12 inches from the other.
4. Apply pressure on the forward stroke, and release it on the return stroke, because the blade cuts only on the forward stroke.
5. Keep in mind that the teeth of the blade should always point away from the hand. Do not permit the teeth to slip over the metal, because this will dull the teeth and may break the blade.
6. Once the kerf—the slot made by the blade—is established, move the hacksaw at about 40 strokes per minute.

Handsaw. The handsaw is usually thought of as a wood-cutting tool, but it is also handy for cutting PVC conduit. There are two basic types of handsaws: crosscut, designed to cut wood across the grain; and ripsaw, designed to cut along the grain. The number of teeth per inch is usually stamped on the blade near the handle. As with other saws, the more teeth per inch, the finer the cut. Crosscut saws, which have more teeth per inch and therefore produce a finer cut, are usually better suited for the electrician's work. (See Figure 1-16.)

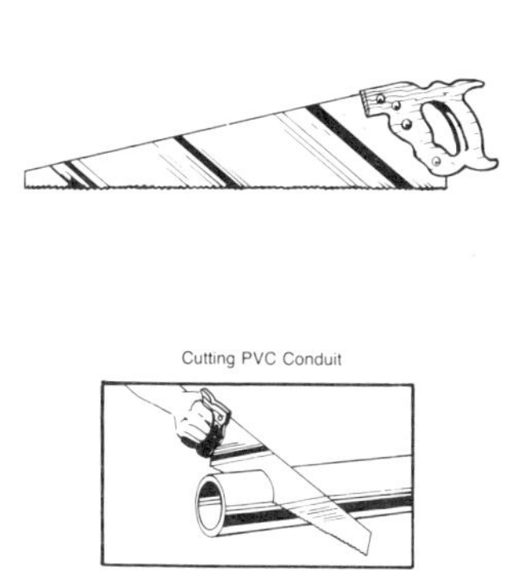

Figure 1-16. A handsaw is often used to cut PVC.

Keyhole Saw (Compass Saw). Keyhole, or compass, saws are used mainly to cut holes such as those for electrical outlets. Because these saws usually do not have a frame to hold the blade, they are versatile and can be angled to cut curves and circles as desired. The blade can be changed to fit the type of cut needed. Keyhole saws can be used to rough out round, square, or oblong holes in a floor or wall—as for electrical outlets—with the cut starting from a drilled hole. Compass saws come in lengths of 10, 12, and 14 inches and usually have 10 points (teeth) to the inch. (See Figure 1-17.)

Figure 1-17. Compass, or keyhole, saw.

POWER SAWS

Bayonet Saw (Saber Saw). The bayonet, or saber, saw, also sometimes called the *portable jigsaw*, is used for a wide variety of light work. Perhaps the best feature of this tool is its ability to cut sharp angles and curves. It is often used to cut holes after a drill has been used to start the hole. It is particularly convenient in close-work cuts or when irregular cuts are needed—as for electrical boxes.

Saber saws usually have an orbital blade with a stroke ranging from 7/16 inch to 1 inch. The top of the blade is pointed, so it can be used to start its own hole, though this is not often done. Many specialty blades are available for use with saber saws.

Hacksaw. The power hacksaw is more suited for cutting metal than is the bayonet saw. Electricians use it primarily to cut large-diameter conduit used in some commercial and industrial jobs.

Portable Electric Handsaw (Portable Circular Saw). The most commonly used portable power saw, the circular saw, has a circular blade that varies in diameter from 6 to 12 inches. The diameter of the blade determines the maximum depth of the cut. Primarily designed for cutting wood and soft material such as plastics and particleboard, this saw is used by electricians to cut boards for mounting equipment or to remove framing parts when that is necessary to install electrical devices or conduit. (See Figure 1-18.)

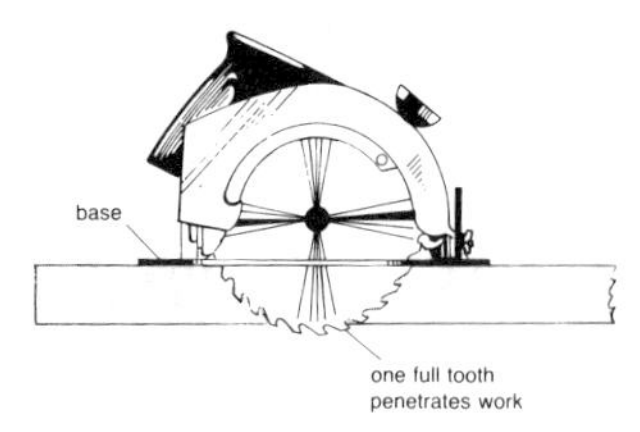

Figure 1-18. Portable circular saw.

Reciprocating Saw. A general-purpose power saw, the reciprocating saw is used for cutting wood, plastic, and metal, depending on the blade used. It is called a reciprocating saw because an up-and-down (reciprocating) cutting action is used. This saw can start its own hole, and the cutting stroke is about 1 inch. (The saber saw is a type of reciprocating saw.)

SCREWDRIVERS

Screwdrivers are designed to work with slotted fasteners such as screws. Since there are a number of different types of fastener heads, there is a different screwdriver for use on each. For heavy work, screwdriver bits are used in a ratchet brace.

Conventional Screwdriver. Conventional screwdrivers are available in many lengths, blade widths, and types of handles. (Screwdrivers with plastic handles should be used around electrical equipment.) Match the blade size to the screw slot. The tip should fit snugly and not be any wider than the diameter of the screw head so that surrounding surfaces will not be damaged. (See Figure 1-19.)

Figure 1-19. Standard-tip screwdriver and Phillips screwdriver.

Phillips Screwdriver. The Phillips screwdriver is designed to drive screws with a crosslike design. They come in at least six sizes, No. 1 through No. 6. (See Figure 1-19.)

Specialty Screwdrivers. Specialty screwdrivers are available to fit screws with head designs that cannot accept conventional or Phillips-head blades. Some of these special screw designs are found in electrical instruments and in switch covers that should be removed only by an electrician. Some circuit breakers also come with special-head screws to limit access to only qualified persons with the appropriate specialized equipment.

SOLDERING DEVICES

Electricians are increasingly called on to work with small electronic control units. Consequently, they must be able to solder and desolder. Several devices are available for on-site soldering. (See Figure 1-20.)

Soldering Iron. A soldering iron is used to melt solder (a metal or metallic alloy) and heat the joint to the temperature needed to make the solder flow around the wires to be connected. It is the preferred tool for most soldering jobs.

Soldering Gun. A soldering gun is quicker than a soldering iron. The tip heats up almost immediately inasmuch as it is the short-circuited end of a transformer. However, be careful that the heated end is touching something when the trigger is on or else the heat will build up and the tip will open, thereby putting the gun out of operation until another tip can be found. Tips are available in copper and iron-clad copper. The soldering gun is very handy for desoldering, but it takes much practice to be able to use it to make a proper soldering joint, and most electricians prefer the soldering iron.

Figure 1-20. Soldering devices.

STRIPPERS

See Romex Stripper; Wire Stripper.

TOOL HOLDER

Many electrical supply catalogs call an electrician's tool holder the *electrician's pocket.* Usually made of leather and sturdy enough to hold a number of tools, the holder has slits at the top so that it can be hung on a belt buckled around the waist. A strong leather belt is needed to support the weight of the tools and the tool holder.

TORCH, PROPANE

A propane torch provides a quick source of heat when there is no electrical power available to heat something. The torch consists of a small tank of propane and a tip that screws on. The liquid propane vaporizes and produces a gas that burns cleanly. Electricians find many uses for a propane torch, among them providing heat to solder or desolder and heating fish tape to make a loop at the end. Different-size screw-on tips are available for different jobs.

WIRE CUTTER

A wire cutter is a tool every electrician and apprentice must have on the job. A number of sizes are available, some so large that they require hydraulic units to apply enough pressure to cut the wire. Some not only cut wire but also have a dimple that can be used to crimp solderless connections. (See Figure 1-21.)

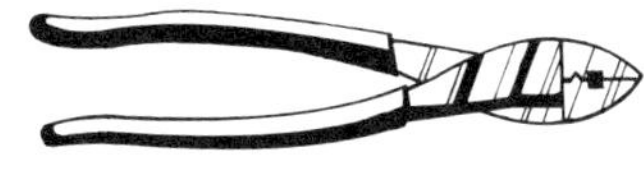

Figure 1-21. Wire cutters.

WIRE GAGE

A wire gage is an instrument used to check the size of a wire. The wire being checked is pushed from the outside to the inside of the gage slot. The size of the slit is the actual size of the wire. However, if the wire is enameled or covered with varnish, you must allow for the thickness of the covering. For example, an insulated wire than reads No. 13 on the gage is really a No. 14 wire with a one-size coating.

Most wire gages are made of steel and are sturdy. On the back, many provide a table giving the diameter of wire in decimal form. The most commonly used wire gage in the trade is the AWG, or American Wire Gage. (See Figure 1-22.)

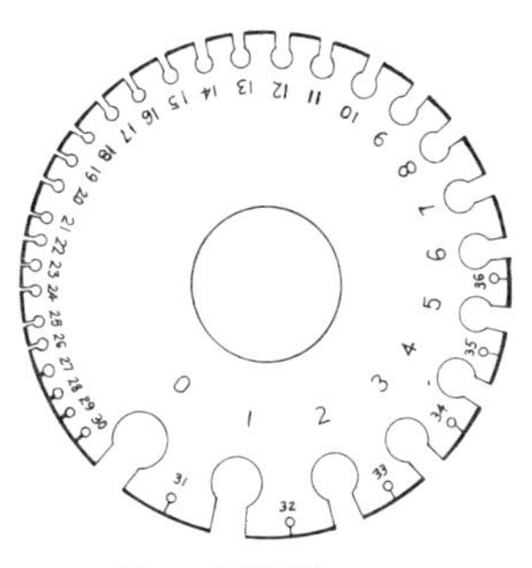

Figure 1-22. Wire gage.

WIRE STRIPPER

A wire stripper is a tool used to remove insulation from wire. Wire strippers come in many sizes and shapes for specific jobs. Diagonal cutters can be used to strip insulation from wire found in a Romex cable; small spring-loaded handles on some strippers make it possible to strip wires very quickly, and in some strippers a screw adjustment permits wires of different diameters to be stripped without fear that the cutter action that removes the plastic coating will score or damage the wire. Removing insulation without scoring the wire is very important. If copper wire is scored and then bent, it has a tendency to break at the scored spot. This can cause service problems later and create the need for expensive reworking. (See Figure 1-23.)

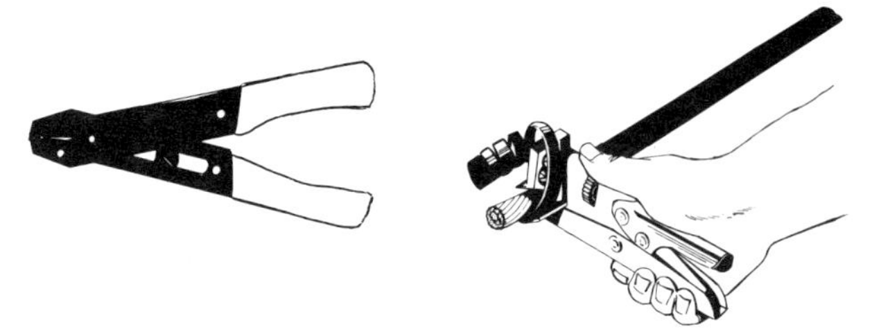

Figure 1-23. Sample wire stripper and another type of stripper, with spring-loaded handles and screw adjustment, being used to strip insulation.

2

Hand Tools for Construction Wiring

The construction electrician uses both hand and power tools on the job. Some of the wire used at the site of a new building is large enough so that special tools have to be used to bend it. Cutting, bending, crimping, and pulling wire are primary concerns of the construction electrician.

ALUMINUM CABLE CRIMPING TOOLS

The construction electrician often needs to crimp aluminum cable. Aluminum is very difficult and sometimes impossible to solder. That means that the ends of the wire have to be terminated with some type of device that will permanently hold to the aluminum wire and make a good electrical connection. Crimping tools are used to place a lot of pressure on the terminal lugs so that they will grip the aluminum and make a good connection. The crimping tools must be closed firmly to produce enough pressure—and thus high-quality terminals. To obtain the leverage to do this, crimping tools must have long handles. If crimping tools are not closed tightly enough, unsatisfactory and weak joints will result. (See Figure 2-1.)

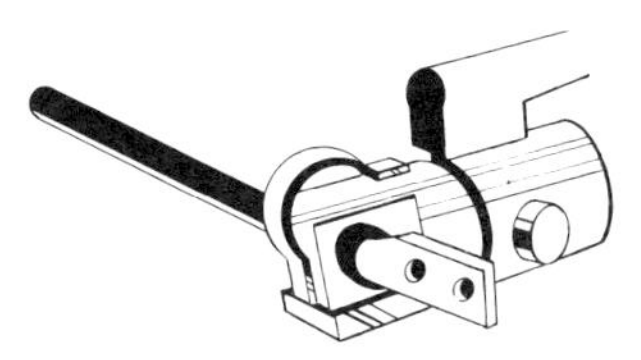

Figure 2-1. Aluminum crimping tool.

ANCHORING SYSTEMS AND TOOLS

When installing conduit, boxes, and trays, electricians use different types of anchoring systems and tools, depending on the building material and structure, the size of the device, and other factors. Some anchoring systems can be used in a wide variety of situations; others are designed for use in particular types of material; and still others are designed for specific uses, such as telephone wiring.

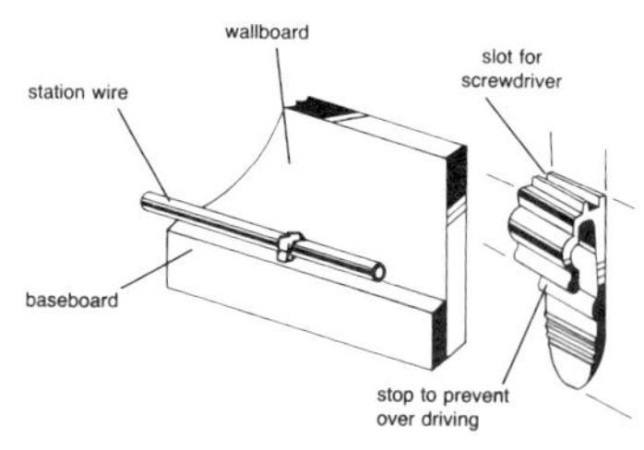

Figure 2-2. Baseboard clip.

GENERAL ANCHORING SYSTEMS

Baseboard Clips. A push-in clip for use on baseboard is available. The colorless clip slips between the wood or vinyl baseboard and the wall and can be tapped into place with a screwdriver. (See Figure 2-2.)

Cable Rings. A drive ring is used to support large-diameter cables or a number of small cables. Drive rings with 1/2-inch, 5/8-inch, 7/8-inch, and 1 1/4-inch eye capacities are available. One type of cable ring is a toggle bridle ring. It is installed where the cable has to be supported between studs or from the ceiling. A bridle ring with wood screw threads may also be used in wood where there is a need. (See Figure 2-3.)

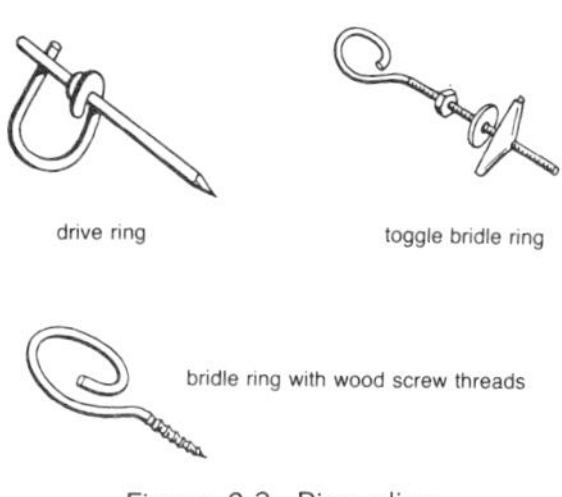

Figure 2-3. Ring clips.

Lead Anchor. A lead anchor is made of an antimony and lead-alloy sleeve with an internally threaded zinc-alloy cone that has a series of integrated ribs, or lugs. The ribs prevent the cone-shaped nut from turning in the sleeve. All bolt size anchors, except the 3/4-inch bolt size, are preassembled in a single unit, a self-contained anchor. (See Figure 2-4.)

The 3/4-inch bolt comes as three separate units—a cone-shaped nut and two sleeves. The volume of lead alloy required for the proper installation of this bolt size is broken into two units to ensure ease of caulking the nut into the concrete hole. A tool for setting the anchor is included with each box of anchors, and additional tools are available as separate items.

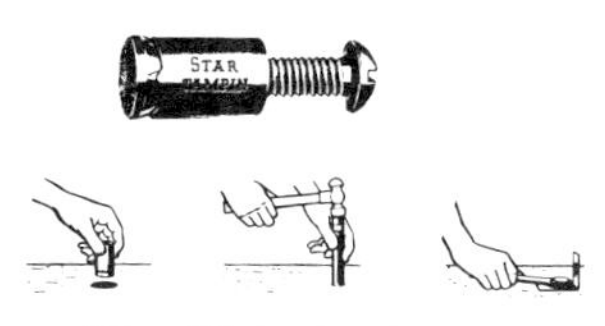

Figure 2-4. Lead anchor use.

Masonry Attachments. Wire loops and wire loop fasteners are used to anchor telephone wire to concrete, steel, mortar, cement, or cinder block. Once selected, the wire loop is passed over the cable, and the loop fastener is placed through the top of the loop. A hammer is used to drive the fastener into the material to support the cable.

ITW Linx makes the Tapin® masonry clip. (See Table 2-1.) It is used to simplify the routing and fastening of inside wires and cables to masonry and other hard surfaces. The drive pin points create a zone of compaction that generates pull-out strengths of 15 to 75 pounds, depending on the site. This securely holds the wires in place. There are four varieties available. Be sure to pull the wire or cable in the clips so that there is no sag between clips. (See Figure 2-5.)

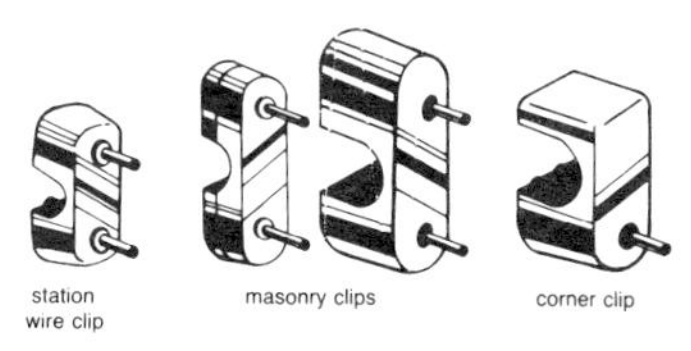

Figure 2-5. Masonry clips.

Plastic Anchor. Plastic anchors are made of nylon. They are shock- and vibration-resistant and are well suited for use in almost any material. They can be conical in shape, with a steel tip and a screw, or cylindrical, with a screw to allow expansion once it is placed in the hole. Plastic anchors are commonly used for static loads for conduit straps, electrical fixtures, and fan installations. (See Figure 2-6.)

TABLE 2–1*

ITW Linx Tapin® Masonry Clips**

Essex Wire or Cable Type	Number of Conductors	Overall Diameter Inches	Tapin Clip Inside Diameter Inches	Tapin Designation
Polyolefin or Polyvinyl Chloride Insulations				
Station Wire — Type SW (Individual Conductors)				
SW-222	2	0.11	.13	TA-1
SW-322	3	0.13	.13	TA-1
SW-422	4	0.14	.19	TA-2
Station Wire — Type SWT (Paired Conductors)				
SWT 2/22	4	0.16	.19	TA-2
SWT 3/22	6	0.20	.19	TA-2
SWT 4/22	8	0.22	.19	TA-2
Inside Cables — Type ICPI (Paired Conductors)				
2 pair 24 AWG	4	0.12	.13	TA-1
3 pair 24 AWG	6	0.14	.19	TA-2
6 pair 24 AWG	12	0.20	N/A[1]	
12 pair AWG	24	0.26	N/A[1]	
25 pair 24 AWG	50	0.37	.35	TA-5 or TA-6
50 pair 24 AWG	100	0.51	N/A[1]	

Fluoropolymer Insulations				
Station Wire — Type SWP (Individual Conductors)				
SWP-222	2	0.12	.13	TA-1
SWP-322	3	0.13	.13	TA-1
SWP-422	4	0.15	.19	TA-2
Station Wire — Type SWTP (Paired Conductors)				
SWTP 2/22	4	0.15	.19	TA-2
SWTP 3/22	6	0.17	.19	TA-2
SWTP 4/22	8	0.20	.19	TA-2
Inside Cables — Type ICPIP (Paired Conductors)				
2 pair 24 AWG	4	0.15	.19	TA-2
3 pair 24 AWG	6	0.17	.19	TA-2
6 pair 24 AWG	12	0.21	N/A[1]	
12 pair 24 AWG	24	0.27	N/A[1]	
25 pair 24 AWG	50	0.38	.35	TA-5 or TA-6
50 pair 24 AWG	100	0.50	N/A[1]	

*From "Customer Provided Telephone Wiring." Courtesy of Essex Telecommunications Products Division, Decatur, IL.
**Tapin is understood to be a proprietary trademark to ITW Linx, Division of Illinois Tool Works, Inc.
[1]N/A — not applicable or none available.

Stud Bolt Anchor. This type of anchor is designed for medium and heavyweight fastening in solid materials. It is made of steel and finished with a rust-resistant zinc plate. It works on a true expansion principle: tightening the nut forces the outer sleeve over the tapered portion of the bolt. (See Figure 2-7.)

Toggle Bolt Anchor. The two-part toggle bolt consists of a screw and toggle head. The steel machine screw has a head, and the screw length is specified. The toggle head is a permanently preassembled unit. The toggle bolt is used for fastening in vertical hollow walls of plaster over wood or metal lathe, wallboard, gypsum, steel, plywood, hollow tile, or cinder block. (See Figure 2-8.)

Wall-grip Anchor. This type of anchor is used in residential and commercial situations. It is well suited for hanging mirrors, pictures, cabinets, and draperies. However, it can also be used by the electrician to hang small boxes to hollow walls of plaster over wood or metal lathe, wallboard, gypsum, or hollow tile. It is available in three sizes: 6-32, 10-24, and 1/4-20; each is available in three lengths.

One version of the wall-grip anchor has a pointed end and can be driven through the wall with a hammer for quick installation.

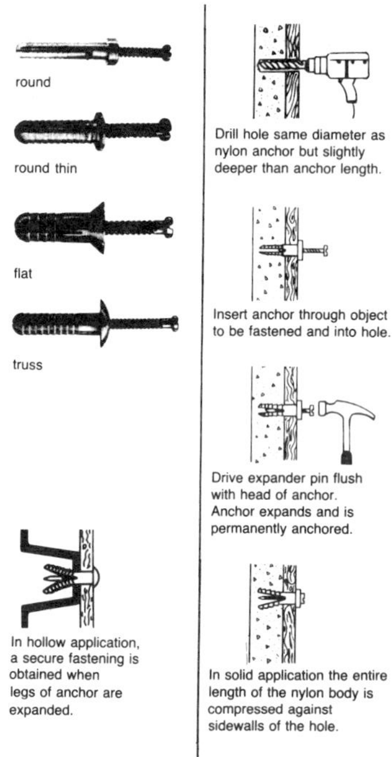

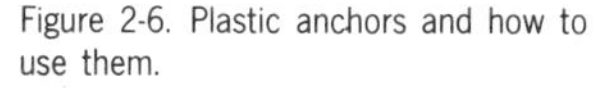

Figure 2-6. Plastic anchors and how to use them.

If the anchor slips in soft wall material, a small wrench, which is packed in each box of anchors, can be used to hold the anchor steady while the screw head is tightened. (See Figure 2-9.)

TELEPHONE WIRING ANCHORS

Adhesive Clips. In telephone wiring, adhesive clips are used when it is necessary to place station wire and cable that must not be crimped or marred. (A staple gun should not be used because it can crimp or mar the wiring and cause an open wire, and in some cases, a short between wires.) Adhesive clips are used to carry cables with 1/8-inch to 3/8-inch diameters. The clips come in various sizes and types, and some have a pressure-sensitive adhesive backing that allows for quick mounting to most dry, clean surfaces. (See Table 2-2 and Figure 2-10.)

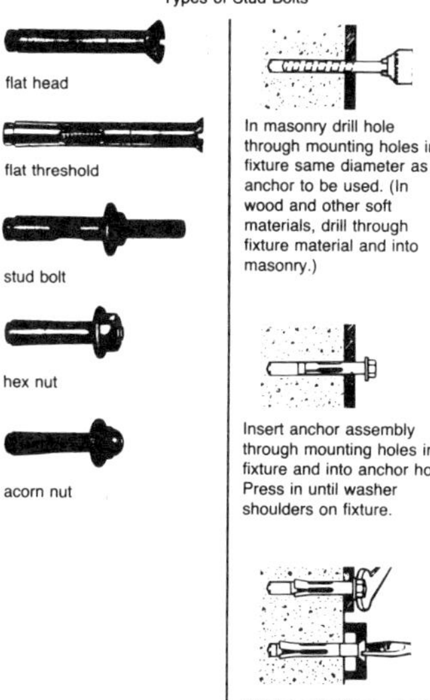

Figure 2-7. Stud bolts and how to use them.

Offset Wiring Clamps. Offset wiring clamps are used on either solid or hollow walls. They are usually installed at stud locations along the walls. Wood screws are used to anchor them and to clamp the wire. In between studs, they are mounted using plastic anchors. Offset clamps are available in a number of sizes for different conductor cable pairs. (See Figure 2-11 and Table 2-3.)

Station Wire Nails. This type of nail is used to attach station wiring to wood surfaces. The curved nailhead fits over the cable,

TABLE 2–2*

Adhesive Clip Sizes for Varying Sizes and Type of Station Wiring

Essex Wire or Cable Type	Number of Conductors	Overall Diameter Inches	Adhesive Clip Size (Inches)			
			3M[1]		ITW Linx[2]	
			Inches	No.	Inches	No.
Polyolefin or Polyvinyl Chloride Insulations						
Station Wire — Type SW (Individual Conductors)						
SW-222	2	0.11	3/16	708	.155	C1
SW-322	3	0.13	3/16	708	.155	C1
SW-422	4	0.14	3/16	708	.155	C1
Station Wire — Type SWT (Paired Conductors)						
SWT 2/22	4	0.16	3/16	708	.310	C2
SWT 3/22	6	0.20	1/4	710	.310	C2
SWT 4/22	8	0.22	1/4	710	.310	C2
Inside Cables — Type ICPI (Paired Conductors)						
2 pair 24 AWG	4	0.12	3/16	708	.155	C1
3 pair 24 AWG	6	0.14	3/16	708	.155	C1
6 pair 24 AWG	12	0.20	1/4	710	.310	C2
12 pair 24 AWG	24	0.26	5/16	712	.310	C2
25 pair 24 AWG	50	0.37	3/8	713	.525	C3A
50 pair 24 AWG	100	0.51	—	—	.525	C3A

Fluoropolymer Insulations						
Station Wire — Type SWP (Individual Conductors)						
SWP-222	2	0.12	3/16	708	.155	C1
SWP-322	3	0.13	3/16	708	.155	C1
SWP-422	4	0.15	3/16	708	.155	C1
Station Wire — Type SWTP (Paired Conductors)						
SWTP 2/22	4	0.15	3/16	708	.155	C1
SWTP 3/22	6	0.17	3/16	708	.310	C2
SWTP 4/22	8	0.20	1/4	710	.310	C2
Inside Cables — Type ICPIP (Paired Conductors)						
2 pair 24 AWG	4	0.15	3/16	708	.155	C1
3 pair 24 AWG	6	0.17	3/16	708	.310	C2
6 pair 24 AWG	12	0.21	1/4	710	.310	C2
12 pair 24 AWG	24	0.27	5/16	712	.310	C2
25 pair 24 AWG	50	0.38	—	—	.525	C3A
50 pair 24 AWG	100	0.50	—	—	.525	C3A

*From "Customer Provided Telephone Wiring." Courtesy of Essex Telecommunications Products Division, Decatur, IL.

[1] 3M's Scotchflex Brand Cable Clips, TelCom Products Division. Scotchflex is understood to be a proprietary trademark to 3M.

[2] ITW Linx Cord Clips

TABLE 2–3*

Offset Wiring Clamps and Attaching Hardware

Essex Wire or Cable Type	Number of Conductors	Overall Diameter Inches	Offset Wiring Inches	Clamp** No.	Screw Size	Mounting Hole Dia. Inches
Polyolefin or Polyvinyl Chloride Insulations						
Station Wire — Type SW (Individual Conductors)						
SW-222	2	0.11	5/32	13G	8	5/16
SW-322	3	0.13	5/32	13G	8	5/16
SW-422	4	0.14	5/32	13G	8	5/16
Station Wire — Type SWT (Paired Conductors)						
SWT 2/22	4	0.16	1/4	15G	8	5/16
SWT 3/22	6	0.20	1/4	15G	8	5/16
SWT 4/22	8	0.22	1/4	15G	8	5/16
Inside Cables — Type ICPI (Paired Conductors)						
2 pair 24 AWG	4	0.12	5/32	13G	8	5/16
3 pair 24 AWG	6	0.14	1/4	15G	8	5/16
6 pair 24 AWG	12	0.20	1/4	15G	8	5/16
12 pair 24 AWG	24	0.26	5/16	16G	12	7/16
25 pair 24 AWG	50	0.37	3/8	16	12	7/16
50 pair 24 AWG	100	0.51	9/16	18	12	7/16

Fluoropolymer Insulations						
Station Wire — Type SWP (Individual Conductors)						
SWP-222	2	0.12	5/32	13G	8	5/16
SWP-322	3	0.13	5/32	13G	8	5/16
SWP-422	4	0.15	5/32	13G	8	5/16
Station Wire — Type SWTP (Paired Conductors)						
SWTP 2/22	4	0.15	5/32	13G	8	5/16
SWTP 3/22	6	0.17	1/4	15G	8	5/16
SWTP 4/22	8	0.20	1/4	15G	8	5/16
Inside Cables — Type ICPIP (Paired Conductors)						
2 pair 24 AWG	4	0.15	5/32	13G	8	5/16
3 pair 24 AWG	6	0.17	1/4	15G	8	5/16
6 pair 24 AWG	12	0.21	1/4	15G	8	5/16
12 pair 24 AWG	24	0.27	5/16	16G	12	7/16
25 pair 24 AWG	50	0.38	1/2	17	12	7/16
50 pair 24 AWG	100	0.50	9/16	18	12	7/16

*From "Customer Provided Telephone Wiring." Courtesy of Essex Telecommunications Products Division, Decatur, IL.

**M. M. Rhodes and Sons.

and the wire is pulled snugly before the nail is tapped into place while the cable is held securely. (See Figure 2-12.)

TOOLS FOR ANCHORING

Some anchoring can be done by using a fastener in some type of handle while a drive pin is hit with a heavy hammer. This can, for example, be done to drive in a pin for the mounting of conduit, boxes, and trays. However, the depth to which the pin is driven may vary, and concrete may chip away and cause a weak connection. For these and other reasons, anchoring is often done with a power tool.

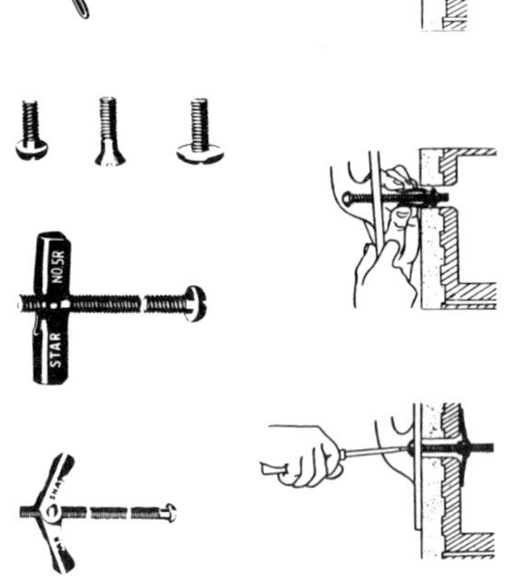

Figure 2-8. Toggle bolt and how to use it.

Anchoring Gun. A typical anchoring gun is shown in Figure 2-13. It weighs about 5 pounds and uses either .27- or .22-caliber ammunition that is made especially for the device. The tool drives fasteners up to 3 inches long with consistent power and performance. Guns that use .38-caliber ammunition are also available.

An anchoring gun is just as dangerous as a handgun. It should be operated only by a trained operator who has passed a written test and has been certified by the manufacturer. Safety goggles and spall guard are required during the gun's operation. Be sure to put the pin in before the load.

BLOWER, HOT AIR

Motor repairs are sometimes difficult at high voltages of 480 and 600 volts, and all connections must be well insulated. This can be done by using a hot air blower, similar to a hair dryer, and shrink tubing applied over the cloth tape and mastic. The heat gun shrinks the caps over the terminals, thus providing better coverage of the ends. (See Figure 2-14.)

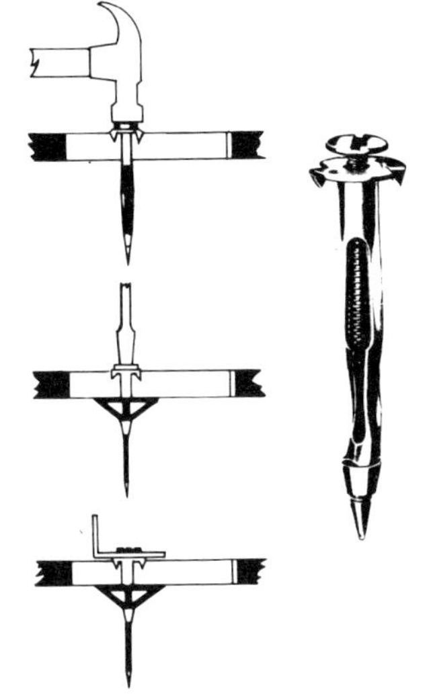

Figure 2-9. Wall-grip anchor and how to use it.

CABLE CUTTERS

Cable cutters for soft copper and aluminum have long fiberglass handles to give good leverage. They can cut cable up to 1000 MCM, or 1 million circular mils. Two models of cable cutters are available; they differ in the cable size they can cut, measured in MCMs, or thousand (M) circular mils. One model cuts cable up to 350 MCM; the other model cuts cable up to 1000 MCM. The cutting blades are replaced in the field. (See Figure 2-15.)

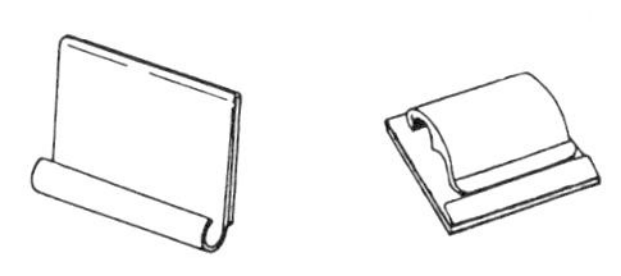

Figure 2-10. Adhesive strip and clip.

CABLE STRIPPERS

A cable stripper removes insulation from a cable. It is a hand-operated tool that cuts plastic insulating material much the same way as peeling an apple. Strippers adjust easily for cable size and insulation thickness. Some can strip cable sizes of # 1/0 through 1000 MCM, and some can strip from the end or the middle of the span.

Figure 2-11. Offset wiring clip.

Figure 2-12. Station wire nail.

CONDUIT BENDER (HICKEY)

There are a number of ways to bend conduit. Rigid (pipe) conduit is hard to bend, and usually power tools are needed to get the job done correctly. A hand bender is suited for EMT, thinwall conduit. An EMT offset bender has recently become available. It is able to bend the offset needed for flush mounting of utility boxes and panel boxes. (See Figure 2-16.)

CONDUIT THREADER

Figure 2-13. Anchoring gun.

Rigid pipe tools are valuable in the field when heavy conduit is used. Several devices are used for cutting and threading rigid pipe tools. (See Figure 2-17.)

Large Pipe Threading Equipment. Large pipe threading equipment is used in some cases and is mounted on a tripod for support in the field. This equipment is usually powered by connecting it to an extension cord.

Pipe Cutter. A pipe cutter also becomes part of the electrician's equipment when working with rigid conduit.

Portable Electric Pipe Threaders. Portable electric pipe threaders are also available for field use. Some threaders can be used to drive hoists, operate large valves, and provide power for many other applications. The threading capacity of most portable electric models is 1/8-inch through 2-inch pipe conduit.

Reamers. Reamers are used to clean the inside of pipe or conduit after cutting and threading. Some are furnished with ratchets to make them easier to handle.,

Three-way Threader. A three-way threader for small pipe and conduit is convenient and compact. It has a handy reversal of chasers for close-to-the-wall threads, and an alloy tool-steel chaser is easily removed for grinding.

Tubing Cutter. A tubing cutter can be used to cut thinwall and EMT conduit.

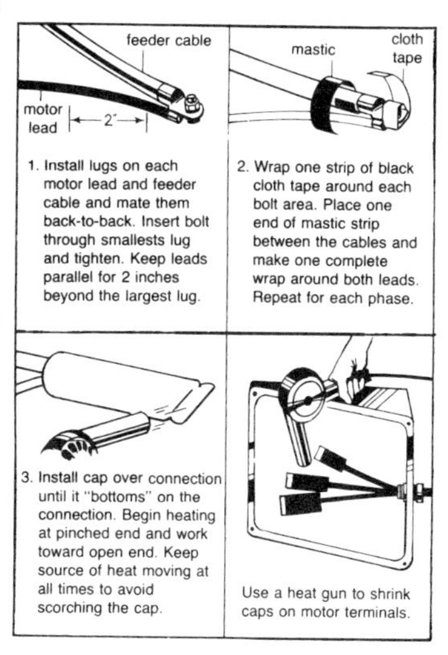

Figure 2-14. Hot air blower.

FUSE PULLER

As its name implies, a fuse puller is a tool used to pull fuses. Figure 2-18 shows a typical fuse puller designed for use with cartridge fuses that have blades at each end or rings around the ends to fit into open fuse holders. Fuse pullers are made in three sizes, and the two ends of each are of different sizes so that a wide range of fuses used in industrial and commercial installations can be accommodated. Fuse pullers are of sturdy, laminated construction with high dielectric qualities to withstand atmospheric conditions, and they usually have a fiber handle to prevent such dangerous situations as slipping and touching a "hot" terminal when pulling or replacing a fuse.

Figure 2-15. Cable cutter.

GROUND MONITOR

The ground monitor is needed in all toolboxes. (See Figure 2-19.) It provides accurate indications of ground and AC polarity in 120-volt circuits instantly, without the inconvenience of handling probes and meters. It may be left in the power outlet permanently if continuous indications are desired. It has a bulb life of 25,000 hours. Just plug it into a 3-wire, 120-volt AC outlet and the light will indicate the following:

O & K	Circuit OK
O only	Improper ground
K only	Improper neutral
O & X	Reversed polarity
K & X	Reversed ground and power
None	No power

HAMMERS

Hammers of different types are used by electricians to install equipment in residential and commercial facilities. Nail, or claw, hammers are the most commonly used. Other specialized hammers include the ball peen hammer and the mallet. (See Figure 2-20).

The handle of a hammer may be made of wood, tubular or solid steel, or fiberglass. Steel, fiberglass, and some wooden handles are usually finished with a rubber grip at the end of the handle. A hammer has to be balanced by the person using it. Learn to use it by holding it at the end, thus avoiding damage to your wrist by allowing the hammer to absorb all of the energy put into the blow.

A hammer should be discarded if the striking face or any beveling shows dents, chips, mushrooming, or excessive wear. Similarly, if the claws on claw hammers show indentations or nicks inside the nail slot or if the claws are broken, the hammer should be discarded. However, if only a handle is damaged, it can usually be replaced by an equivalent one available at a local hardware store. Many electricians use the same hammer for their entire working career.

Nail, or Claw, Hammers. Nail, or claw, hammers are designed for driving nails and nailsets. This is done by striking the nailhead at the center of the hammer face. The face on a claw hammer usually has a slight crown and beveled edges. Some heavy-duty types have checkered faces that are designed to reduce glancing blows and flying nails. However, the claw hammer should not be struck against a metal surface or used to drive metal cutting chisels.

The claws are designed for pulling nails and staples and for ripping woodwork. They come in two patterns: curved claw and straight, or ripping, claw.

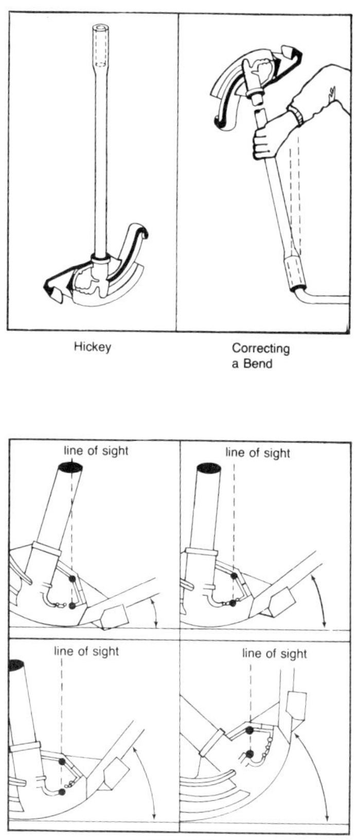

Figure 2-16. Using a hickey to bend conduit.

Other Types of Hammers. Electricians sometimes use some of the speciality hammers. The *ball peen hammer,* although used more often by machinists, is used by electricians for heavy-duty work in installing certain types of equipment. The *mallet,* a hammer with a rubber or plastic face, is useful in applying pressure and driving force where you don't want to mar the surface of the work.

KNOCKOUT PUNCHES

See Punches.

NAIL PULLER

A nail puller is designed to remove nails easily and quickly with very little effort. Its all-metal construction permits hammer blows in certain areas to seat the pincers under the nail head for easy removal. Use of a nail puller saves a lot of broken claw hammers. (See Figure 2-21.)

PUNCHES

A punch is used to make a hole—for example, in steel—to allow installation of an electrical box. The choice of punch depends on the material in which a hole is needed. Small hand punches can be used with a wrench to turn a bolt head to cause a cutter to punch a hole in thin material. For other jobs a hand-operated hydraulic pump punch is needed, and for still other types of jobs, such as punching holes in heavy-gauge metal boxes installed in factories and stores, a hydraulic punch operated by an electric motor is used. (See Figure 2-22.)

Enclosed Ratchet Drop Head Threader

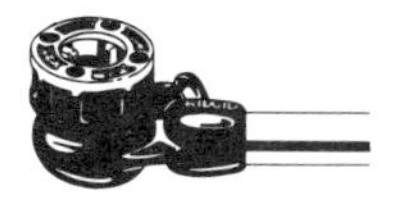

Exposed Ratchet Drop Head Threader

3-Way Threader

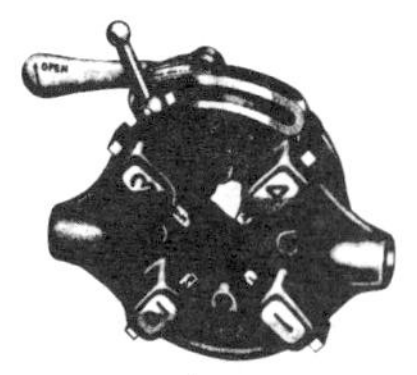

Quick Opening Threader

Figure 2-17. Conduit threaders.

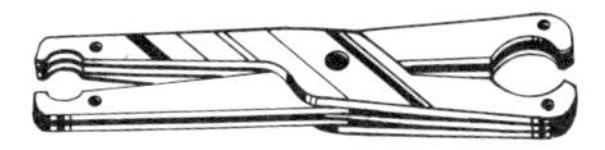
Figure 2-18. Fuse puller.

In some situations, a ratchet knockout driver is used. This type of punch will make a hole for 3-inch conduit in 10-gauge mild steel. This system, which uses a recirculating ball-screw drive system, delivers a mechanical advantage better than 220 to 1 and can operate three times as fast as using a wrench and punch method.

Figure 2-19. Ground monitor.

Some punches—for example, the slug-splitting design—punch tougher, heavier-gauge materials quickly and easily. The split slugs fall freely from the die and stud.

PVC BENDER

There are advantages and disadvantages to using PVC (polyvinylchloride) for conduit. NEC-listed PVC conduit is inherently able to withstand atmospheres containing common industrial corrosive agents as well as vapors or mists of caustic, pickling acids, plating baths, and hydrofluoric and chromic acids. Temperature is, however, an important consideration. At higher temperatures, PVC distorts and melts.

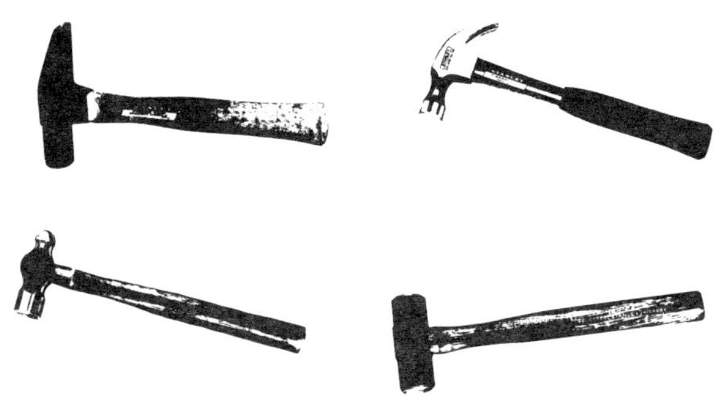
Figure 2-20. Hammers: straight claw, curved claw, ball peen, and rubber mallet.

PVC is designed for connection to couplings, fittings, and boxes by use of a suitable solvent-type cement. Follow the instructions supplied by the manufacturer for method of assembly and for precautions to be followed.

Electricians frequently need to bend PVC. To do this, it is necessary to heat it to a temperature at which it will deform easily. A heating blanket that does the job is now available. Use the blanket to heat the PVC. Then remove the blanket and bend the PVC by hand. Wear gloves because the pipe is hot to the touch. (See Figure 2-23.)

Figure 2-21. Nail puller.

RATCHET-KNOCK OUT DRIVER

See Punches.

WIREMOLD® TOOLS

Special tools are needed to do the job correctly when working with Wiremold® raceway. Wiremold is a trade name for a particular type of steel raceway. Metal raceways are channels for holding wires on exterior surfaces; they have been used in commercial and industrial wiring for many years. Various fittings are available to meet National Electrical Code requirements.

Figure 2-22. Knockout punches.

Surface raceways, along with their fittings, must be installed so that they are electrically as well as mechanically well connected. The main reason for using a raceway is the ease with which wiring can be reached. The steel also protects the wire from accidental damage.

Since Wiremold is so commonly used, we will mention some special tools to be used with it. (See Figure 2-24.) A *shear* mounted on a board can be used to cut the raceway to the length desired. A *canopy cutter* can be used for cutting by hand. (Wiremold series 200-1000 is commonly cut with a handsaw.) If the raceway needs to be bent to fit, it can be done with a *bender* designed to work with Wiremold.

In addition to Wiremold, other types of raceway are available. Made from plastic, they are used for telephone, computer, and electrical wiring. The National Electrical Code has a section dealing with this type of flat, plastic raceway.

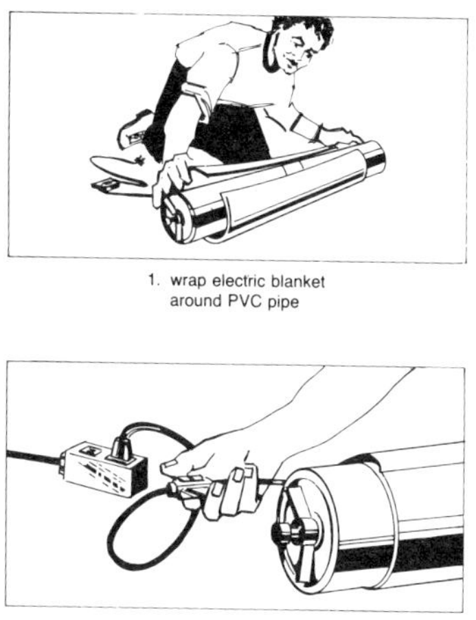

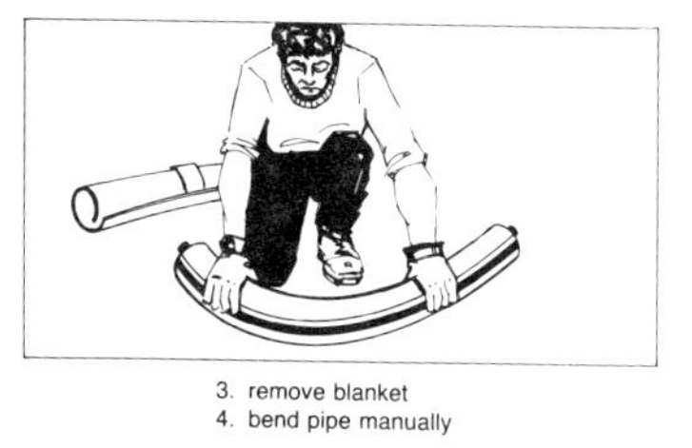

Figure 2-23. Bending PVC.

WIRE-PULLING TOOLS

The construction electrician works with heavy equipment and large conduit with many conductors. To pull the conductors through the conduit without wrecking the insulation or stretching the conductors to the breaking point, special equipment and adaptations of common equipment have been made. Some of this special wire-pulling equipment is detailed here.

Banding Tools. Bands are applied over the tail of a grip to prevent the mesh from being tripped or pulled loose and to ensure full gripping action by locking the mesh of the tail in tight contact with the cable or rope. When the tail of a grip is the leading end, the bands are particularly important to prevent accidental release caused by tripping on obstructions. A good example is a conductor-to-conductor (double-socking) pulling operation in which two grips connect two conductors to form a temporary splice. Bands should be applied to the ends of the grips. It is also common practice to tape over the banded tail areas to ensure smooth passage through the sheaves. The conductor should be installed in the grip up to the elbows of the aluminum shoulders to ensure full and complete gripping action. The banding procedure is then followed. (See Figure 2-25.)

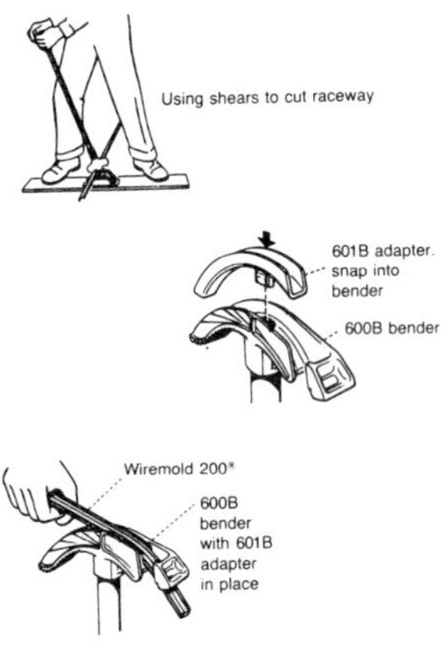

Figure 2-24. Wiremold tools.

NOTE

In all cases, two bands should be double-wrapped approximately 1 inch and 2 inches from the grip's tail. Double wrapping is required to ensure maximum reliability and to guard against accidental release.

Feeding Sheaves. A feeding sheave will guide and help protect cable and decrease friction throughout the pulling operation. Select the proper size of feeding sheave and insert it into the feeding end of the conduit run. When the pulling is completed, the sheave can be removed easily by sliding the cable through the opening in the insert tube. (See Figure 2-26.)

Grips. There are several types of grips, each designed for a specific use. (See Figure 2-27.)

Light-duty pulling grips are used in general underground electrical construction where pulling tensions are low. They are easy tools to use in wiring industrial plants and commercial buildings.

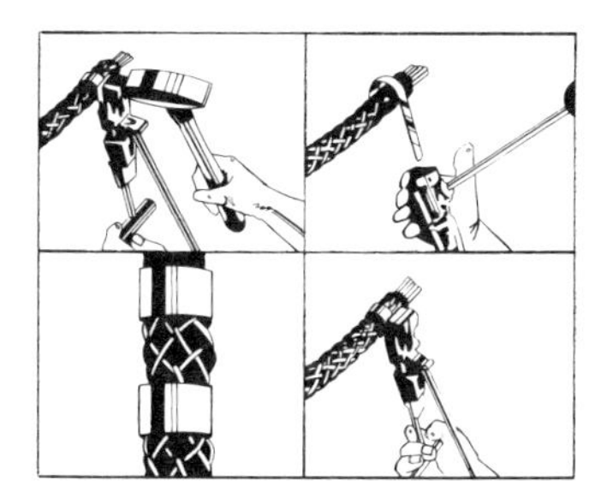

Figure 2-25. Banding tools.

T-type pulling grips are used in the installation of underground power cables and communications lines, service lines into factories and construction projects, and general underground construction. They are available in two mesh lengths—short for medium pulls and standard for general-purpose pulling.

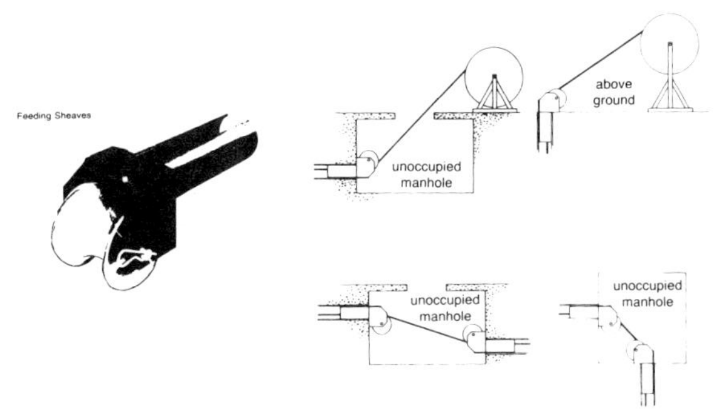

Figure 2-26. Feeding sheaves.

K-type pulling grips are specially designed for use in the installation of underground power cables and communications lines, service lines into factories, shipping centers, construction projects, and general underground construction. They are equipped with a forged-steel rotating eye that is durable, compact, and streamlined and will thread through blocks and sheaves without binding. The rotating eye is not a swivel and will not turn while under tension. However, it can turn to relieve pulling torque when the tension is relaxed. If constant swivel action is required, a swivel should be attached to the eye.

Figure 2-27. Grips.

Slack-pulling grips are widely used in pulling slack for final placement of underground cable after it has been pulled in. They are also used for removing cable. Made of galvanized steel, all slack-pulling grips have a single offset eye for easy attachment to a pulling line. Three styles are available: a closed type, used when the cable end is accessible; and split-lace and split-rod types, used when the cable end is not accessible. Standard mesh lengths are generally used in restricted space for short pulls; longer lengths are used for higher pulling loads when space is not restricted.

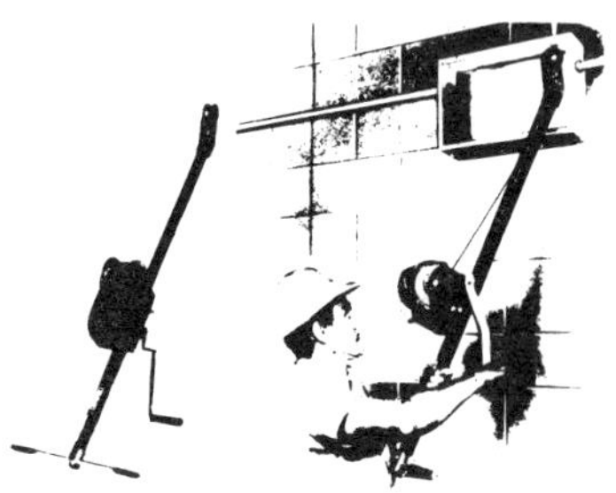

Figure 2-28. Lightweight wire puller.

Splicing grips are designed for very light light duty and small splicing jobs. Made of galvanized steel, they are available in either single- or double-weave mesh construction in various lengths and sizes to suit most applications.

Splicing grips are used as a temporary splice for rope, cable, or wire rope. They can also be used as cable reinforcement and can act as a shield to protect cables and hoses from abrasion. *Note:*

During installation, each end of the splicing grip should be taped securely to the cable to ensure smooth passage with the cable.

Lightweight Wire Puller. A hand-operated lightweight wire puller makes pulling a one-person job and does not require a power source. For small conduit and single conductors, this tool is ideal for putting tension on wire and moving it through conduit quickly and without undue force. (See Figure 2-28.)

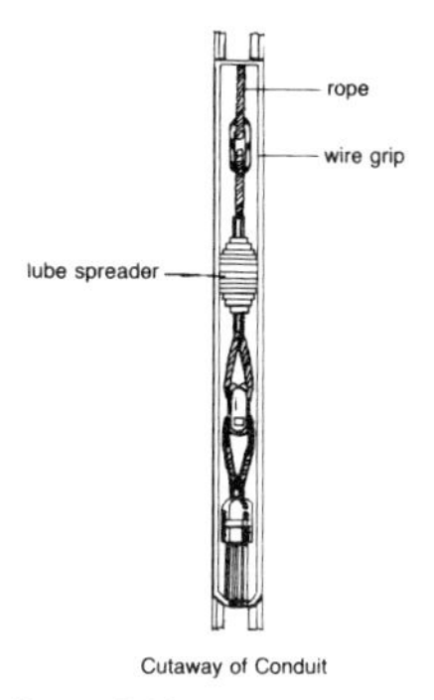

Figure 2-29. Lube spreader.

Lube Spreader. The lube spreader makes it easier for conductors to slide through conduit. Lubes were once made of liquid soap, but today's lube is made of a wax-based substance that can be applied easily because it sticks to the cable or inside the walls of the conduit. Multiple-conductor pull usually calls for a lube spreader to precede the wires. (See Figure 2-29.)

Pipe Adapter Sheave. The pipe adapter sheave is set up to feed cable or wire into conduit. The sheave guides the cable and keeps it straight for easier and faster pulling. It allows the pulling of extra cable for makeup in a single setup. (See Figure 2-30.)

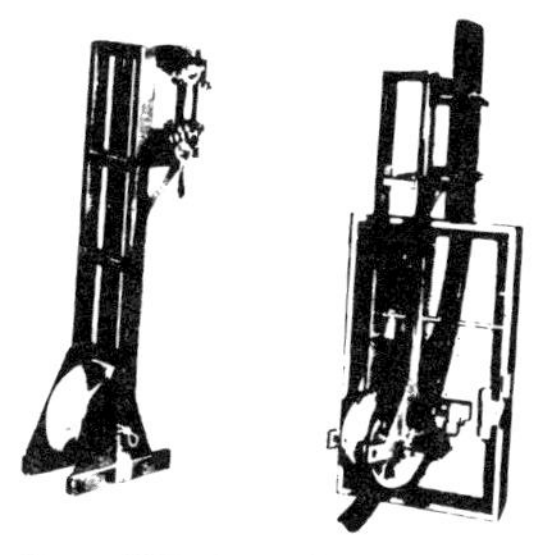

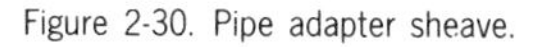

Figure 2-30. Pipe adapter sheave.

Portable Puller with a Power Package. Portable power pullers are available. They can be used anywhere and make pulling a one-person job. Some models weigh only about 40 to

45 pounds and set up in less than 5 minutes; operated by a foot switch, they can pull up to 1200 pounds. (See Figure 2-31.)

Figure 2-31. Porta-puller®.

Power Fish Tape Systems. Power fish tape systems are available; they are used for the same basic purpose as a manual fish tape system (discussed in Chapter 1) but are powered and can be used in heavy-duty operations. An industrial vacuum cleaner is often used as the source of power to blow line or tape into conduit. Some models blow line into 1/2-inch to 4-inch conduit; with accessories, into 5-inch or 6-inch conduit. The entire system is easily transported, and all accessories are held securely on the tank. (See Figure 2-32.)

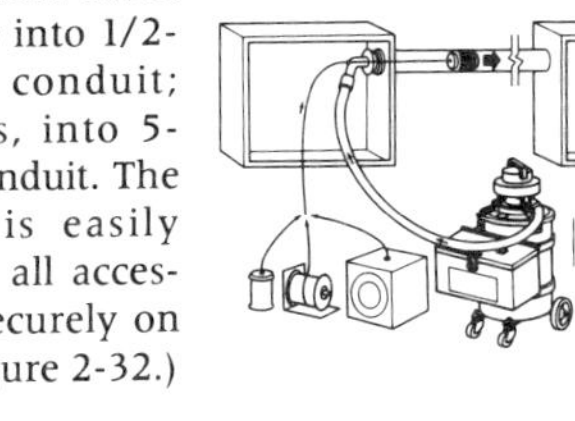

Rope Clevis. Rope clevis is a U-shape device used to pull wire. An Allen wrench that is part of the unit is used to disassemble the clevis for insertion of the rope and cable. (See Figure 2-33.)

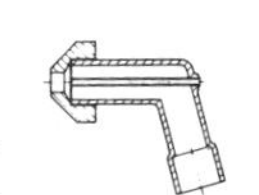

Figure 2-32. Power fish tape.

Wire-Pulling Setups. Other types of wire-pulling setups are available, some designed for specific uses. Figure 2-34 illustrates some special wire-pulling setups.

WRENCHES

Electricians frequently use wrenches and should have several types available to them at the job site. (See Figure 2-35.)

Adjustable Wrench. An adjustable wrench is used to tighten nuts in any number of situations during the installation of a wiring system. Adjustable wrenches are available in a number of sizes and should be selected to fit specific tasks.

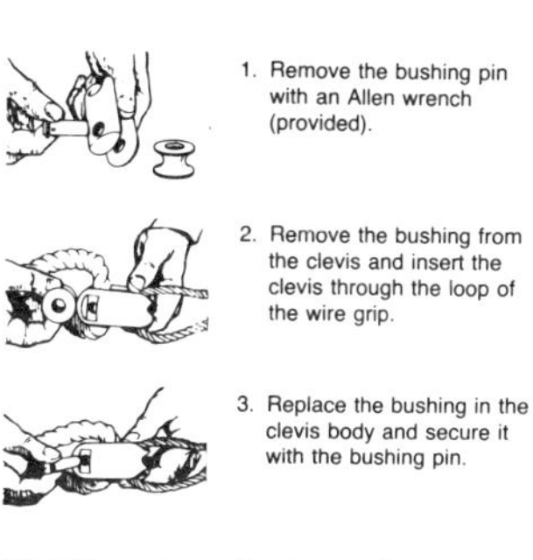

Figure 2-33. Rope clevis.

Allen Wrench. An Allen wrench is used to loosen and tighten a number of box adaptors and smaller pieces of equipment. Allen wrenches commonly come in a long-arm 10-piece set with these sizes:

5/64″	3/32″	7/64″	1/8″	9/64″
5/32″	3/16″	7/32″	1/4″	5/16″

A metric short-arm set is also commonly available.

Socket Wrench. A complete set of socket wrenches with various drives is a necessity for the construction electrician. The choice of which set to purchase or use depends on the type of nut to be tightened and loosened. Sets of many sizes are available. Drives are available in 1/4 inch, 3/8 inch, and 1/2 inch, in addition to some larger ones used for special purposes.

Sparkproof Pipe Wrench. A sparkproof pipe wrench is needed in explosive atmospheres. This type of wrench is made with a beryllium copper hook and heel jaws to guard against sparking. The beryllium copper pins hold the jaws in place and allow for fast replacement of the jaws if necessary. A sparkproof pipe wrench is very useful when working with rigid conduit.

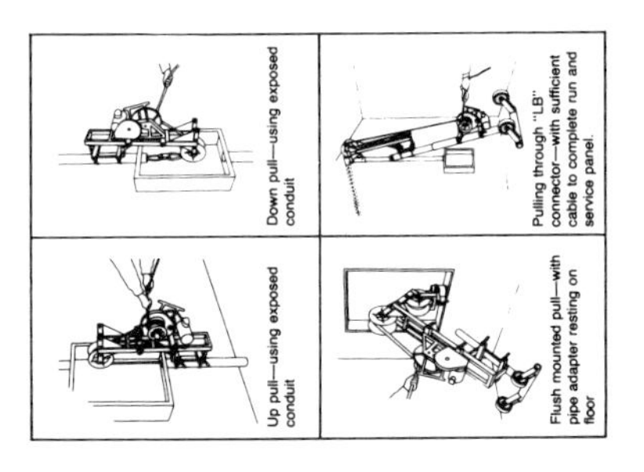

Up pull—using exposed conduit

Down pull—using exposed conduit

Flush mounted pull—with pipe adapter resting on floor

Pulling through "LB" connector—with sufficient cable to complete run and service panel.

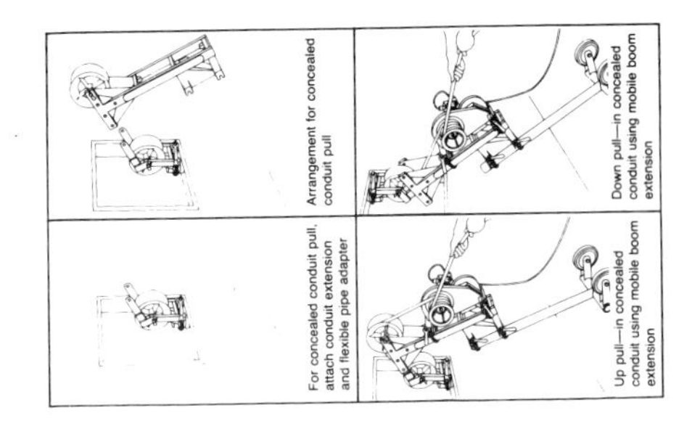

For concealed conduit pull, attach conduit extension and flexible pipe adapter

Arrangement for concealed conduit pull

Up pull—in concealed conduit using mobile boom extension

Down pull—in concealed conduit using mobile boom extension

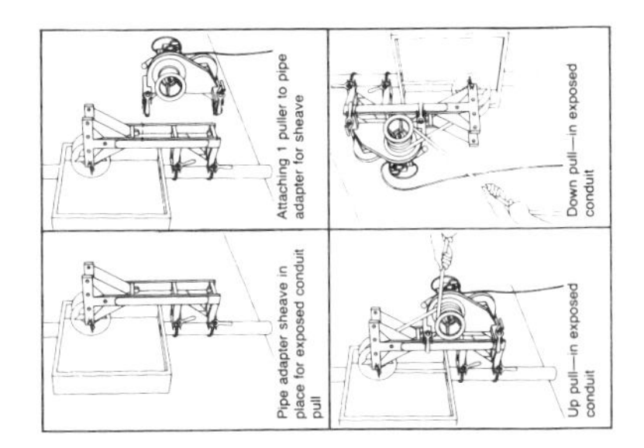

Pipe adapter sheave in place for exposed conduit pull

Attaching 1 puller to pipe adapter for sheave

Up pull—in exposed conduit

Down pull—in exposed conduit

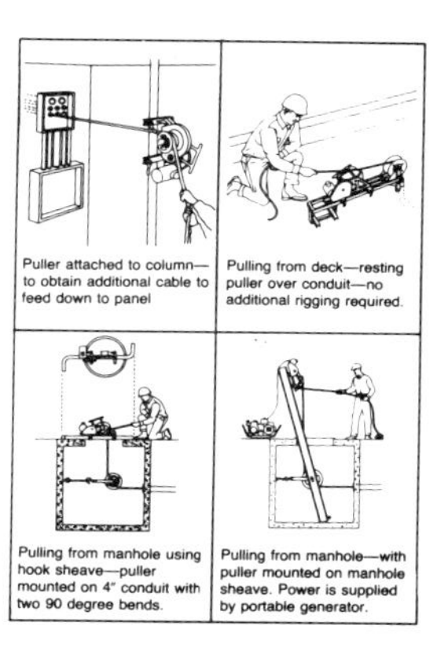

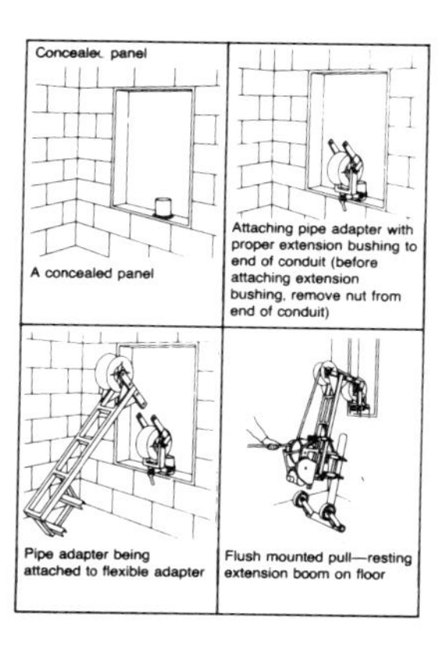

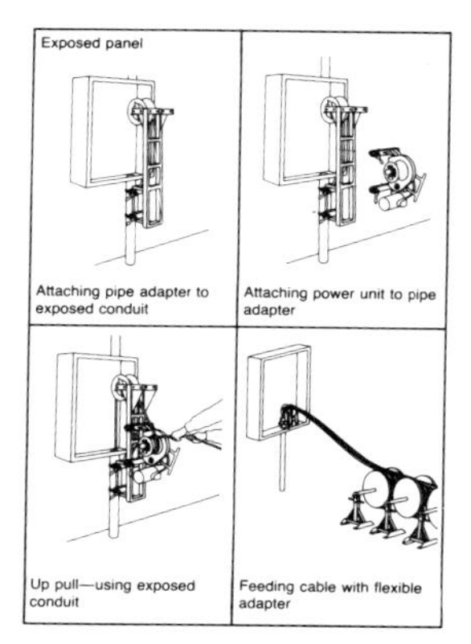

Figure 2-34. Wire-pulling setups.

Figure 2-35. Wrenches: combination, open-end, adjustable.

3

Special Equipment

Certain jobs or particular locations in which routine jobs must be done require the use of special equipment for gaining access to hard-to-reach places. Other types of jobs, especially in industrial and commercial settings, may also require the use of special or large, heavy-duty equipment. Not every electrician needs to have these tools in the toolbox. Most large, expensive, and complicated tools are provided by the electrical contractor for whom the electrician works. Directions for the use of these tools are provided by the manufacturer. However, the electrician apprentice should have familiarity with the more common special tools and what they are used for.

BENDERS

There are cable benders for large cable that cannot be bent by hand, and there are conduit benders. Since there are limitations on how big a cable or conduit can be bent by hand, some units are electrically powered. An electric motor provides power to a hydraulic pump that, in turn, causes a hydraulic fluid to become an energy multiplier. The pressure exerted by the fluid is used to bend cables, huge pipes, and conduit.

Cable Bender. A cable bender is used to bend cable in close quarters with a minimum of effort. Most cable benders have insulated handles for the user's comfort, since it takes a lot of effort to bend cable with this tool. Figure 3-1 shows a typical cable bender; this one can handle cable up to 750 MCM.

Figure 3-1. Cable bender.

Electric Bender. An electric bender can handle large rigid-steel and aluminum conduit easily. Bending shoes are sized to fit the conduit. Figure 3-2 shows a typical electric bender with different shoes and support rollers for various sizes of conduit.

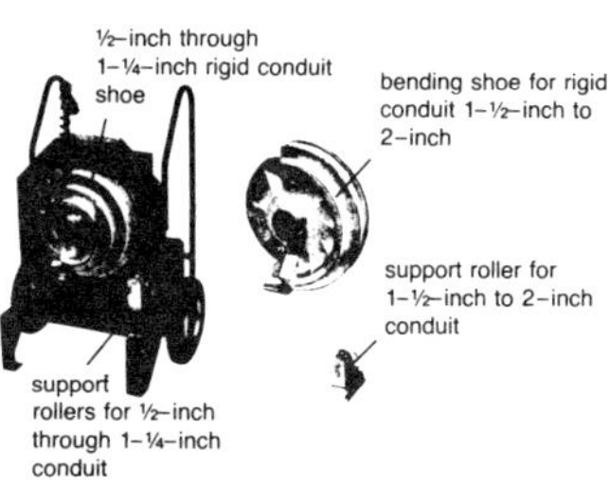

Figure 3-2. Electric bender.

Hydraulic Bender. A hydraulic bender is used to bend conduit. A steel box houses the bender unit, which can be used with two electric pumps or with a manual pump. Figure 3-3 shows a typical hydraulic bender. Become familiar with its parts so that you will be able to use it on the job site.

CABLE GUIDE

A cable guide is used to keep cable going where it is supposed to while it is being pulled through conduit, along trays, or underground through manholes in the street. Some models have an adjustable radius that permits cable of different diameters to be guided. Figure 3-4 shows a cable guide with a radius that adjusts from 19 1/2 inches to 36 inches.

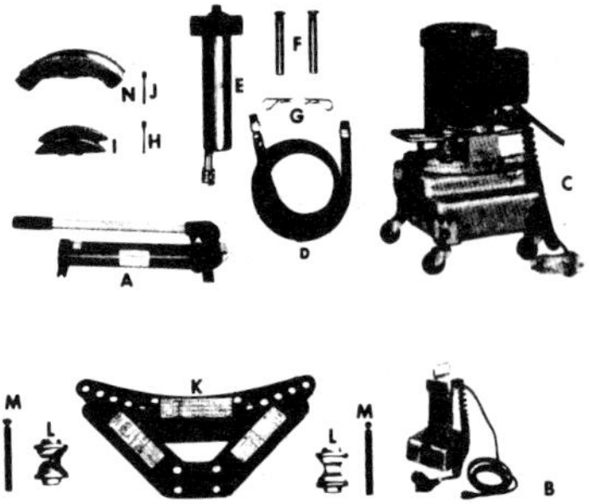

A. hydraulic hand pump
B. hydraulic power pump
C. hydraulic power pump
D. high-pressure hose with male quick coupler
E. 15-ton ram with female quick coupler
F. cylinder head pin unit with spring clip
G. spring clips
H. ram pin
I. small shoe support
J. shoe pin
K. frame unit
L. pipe support
M. pipe support pin
N. 90° aluminum bending shoes.

Figure 3-3. Hydraulic bender.

CABLE ROLLERS

Cable rollers are designed to reduce the problems associated with pulling cable over trays and ladders and to prevent snags. There are several types of cable rollers, each designed for a specific purpose. (See Figure 3-5.)

Straight Rollers. Straight cable rollers are designed to be mounted on trays of common widths or bottom-mounted in most ladder configurations. They can be used flat or vertically.

Radius Rollers. Radius rollers can be used alone or in combination with pulling offsets for turns of less than 90°. They can be top-mounted or bottom-mounted. When bottom-mounting, allow a 5 1/2-inch minimum clearance between rungs. Radius rollers have two J bolts and mounting hardware similar to that on straight rollers.

Right Angle Rollers. Right angle rollers handle even the heaviest cables. A combination of horizontal and vertical ball bearing rollers, they ensure easy pulling and added control at all points of

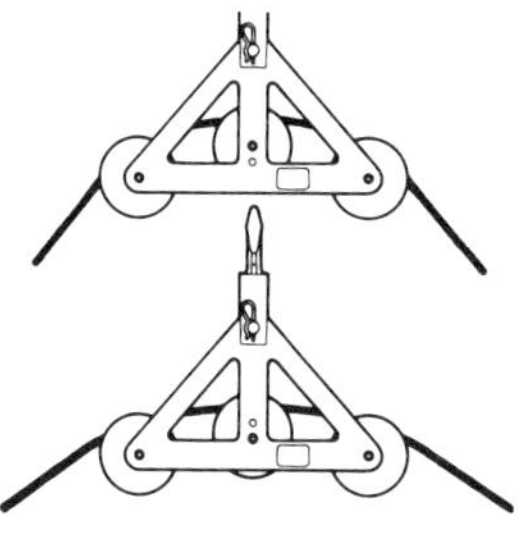

adjusts to radius from 19½ to 36 inches

Figure 3-4. Cable guide.

contact on 90° turns. They can be top-mounted or inside-mounted on any manufacturer's tray or ladder.

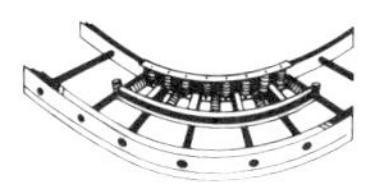

right-angle cable rollers

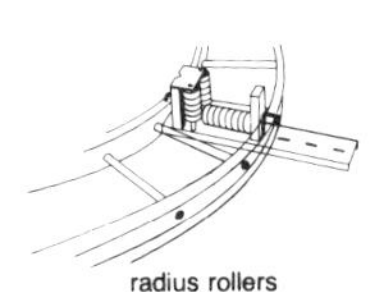

radius rollers

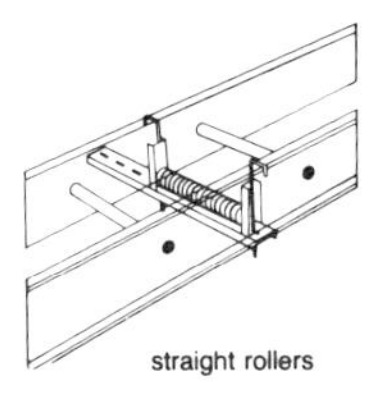

straight rollers

Figure 3-5. Types of rollers.

CONNECTOR TOOLS FOR ALUMINUM WIRE OR CABLE

Hydraulic pumps are used to ensure that all sizes of aluminum conductors can be properly terminated. They are also used in bending aluminum cable when it is of such diameter as to require the extra pressure afforded by a hydraulic device.

Basically, a hydraulic pump for aluminum connector tools exerts high pressure to crimp or squeeze lugs onto aluminum cable.

Before compression, a typical cross section of cable and connector consists of about 75% metal and 25% air. After compression, the cross section is hexagonal, and, with the force provided by a hydraulic compressor tool, 100% of the metal makes contact with the wire with no air spaces between the cable wires and the lug or terminal. (See Figure 3-6.)

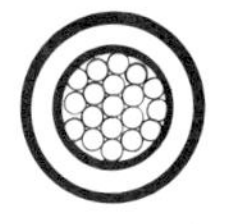

Figure 3-6. Cross section of cable.

An aluminum connector tool consists of a pump to which is attached different sizes of heads. If a source of electric power is available, an electric motor is used to drive the hydraulic pump. If no power is available, a portable generator is used to supply power, or a hand-operated pump is used.

PUMP

Hydraulic pumps for aluminum connector tools come in many sizes. Figure 3-7 shows a typical pump. It weighs only 95 pounds and operates on 120 volts. It has a 1-horsepower motor that can pump 115 cubic inches of hydraulic fluid per minute at 2,500 pounds per square inch or 57 cubic inches per minute at 10,000 pounds per square inch. Foot- and hand-operated switches are used to control the motor, which, in turn, drives the pump. The high pressure generated provides the force to crimp the aluminum cable.

HEADS

The heads are the tools that actually do the work of compressing the cables and lugs to make good, tight fits. Heads are available in different sizes to fit the requirements of particular jobs. They are identified by the amount of pressure they are able to exert on a connection. Three common types of heads are mentioned here. (See Figure 3-7.)

Twelve-ton Head. The 12-ton head is a versatile tool designed for field or bench work. The small size of the compression head and the fact that it can be operated by a switch located nearby but not on the pump allows the pump to be in one area, where there is space, and the head in a confined area, where the connection has to be made.

The head shown in Figure 3-7 is made from forged steel and weighs 7 1/2 pounds. When in use, it can apply 12 tons of pressure, or 24,000 pounds per square inch. The dies that fit into the head are color-coded to match color-coded connectors. A 12-ton head installs aluminum lugs and splices wire from # 12 AWG to 750 MCM.

Fifteen-ton Head. The 15-ton head can apply 30,000 pounds per square inch of pressure. It is used to install aluminum lugs, taps, and splices for # 12 AWG to 1,000 MCM. The head shown in Figure 3-7 weighs 15 1/2 pounds and can be operated up to 100 feet from the electric hydraulic pump. Compression dies are available in two types: one that fits directly into the head and is used for large cables and one that is used in combination with an adaptor for small conductors. All dies are color-coded to match color-coded connectors.

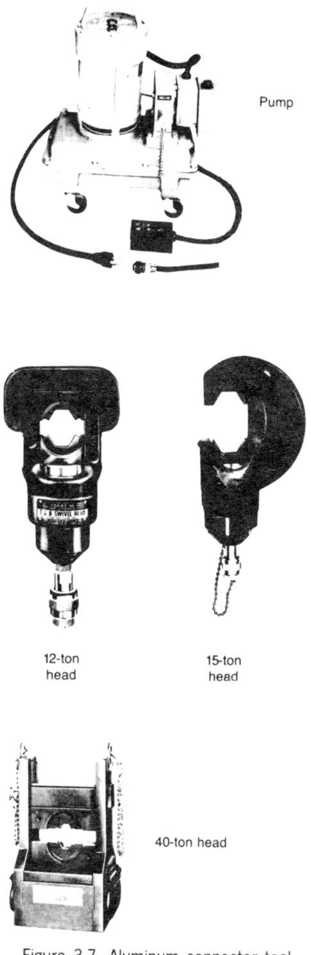

Figure 3-7. Aluminum connector tool.

Forty-ton Head. The 40-ton head (80,000 pounds per square inch of pressure) makes connections with cable up to 2,000 MCM.

PUMPS, MANUAL (HAND- OR FOOT-OPERATED)

Figure 3-8. Hydraulic pump.

A manual pump aids in getting a number of jobs done in the field. These pumps can be used to bend conduit and apply pressure on connectors for taps and lugs as well as splices. (See Figures 3-8 and 3-9.)

Manual pumps can be used in vertical or horizontal positions. The long handle on most models makes it possible to apply a great deal of pressure on the pumping apparatus and obtain the needed squeezing power to make a good electrical connection. In some cases, a hand- or foot-operated hydraulic pump can also be used to make punches for circuit breaker boxes, metal studs, or other steel boxes.

USING A MANUAL PUMP

- Before using the pump, check all hoses for leaks and cracks.
- Check that all safety valves are operating.
- Always wear goggles when using a manual pump.

PUNCH

A punch is used to make holes needed for electrical wiring in metal studs. Figure 3-10 shows one type of punch—a long-handled channel-hole punch that can punch square or round holes of 7/8-inch or 1 11/32-inch diameter or width in 24-gauge steel stud flanges.

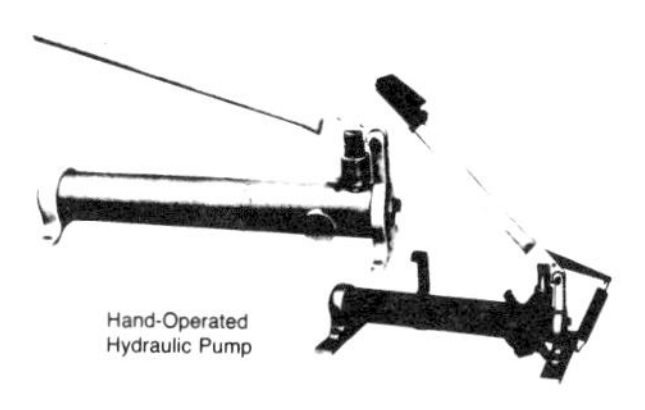

Figure 3-9. Hand-operated hydraulic pump.

TAPS

A tap is a connection brought out of a winding at some point between its ends to control voltage ratio. A tap is made of highly conductive cast aluminum and is tin-plated. Tee or parallel taps perform equally well on aluminum, copper, or combinations of aluminum and copper. All cable contact surfaces are serrated and coated with an oxide-inhibiting compound.

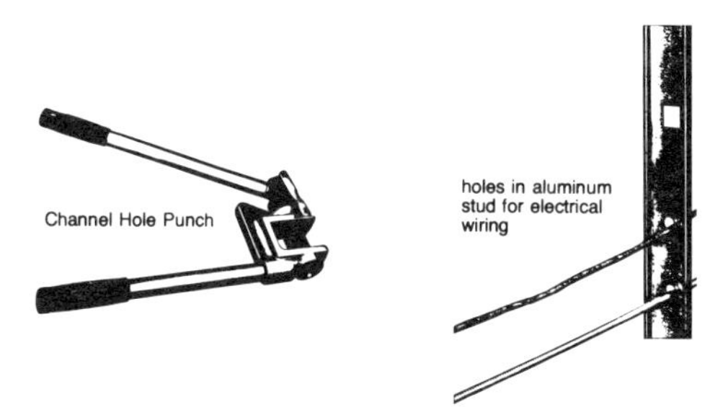

Figure 3-10. Channel hole punch.

The tap is made by first removing the insulation. Use a special tool or a knife for peeling it off. Apply compound to the conductor and lightly wire-brush or sand the outer cable strands to remove the oxide film that builds up on aluminum surfaces. Loosen the bolts; turn the cap 90° and place the tap on the main wire. Then, partially tighten the bolts and insert the tap wire. Tighten the bolts to finish the job.

New methods of making taps allow for joint making while the power is still on. This saves cutting off the power while making emergency repairs.

WIRE DISPENSER

A wire dispenser is a device that holds and dispenses wire. One common type is shown in Figure 3-11. It handles six 2,500-foot spools of wire up to 18 inches in diameter. The dispenser itself can be passed through a standard 30-inch doorway. The front wheel is swivel-mounted to allow feeding in any direction from a central location and can be locked to prevent movement. Wire can be paid out easily, and there is automatic braking. Less expensive dispensers that handle smaller spools are available. They accommodate eight 500-foot spools of wire and usually do not have automatic braking, so a wire pileup can occur.

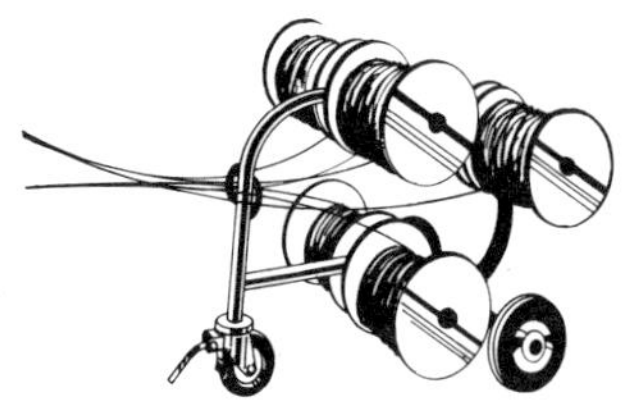

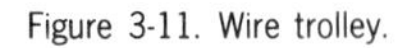
Figure 3-11. Wire trolley.

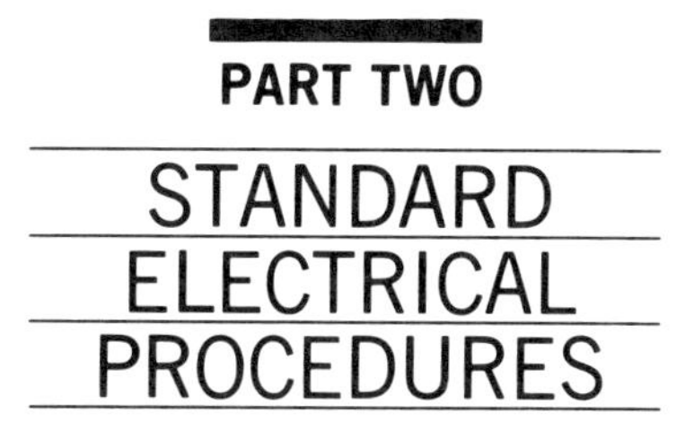

PART TWO

STANDARD ELECTRICAL PROCEDURES

4

Troubleshooting

Troubleshooting tests the electrician's ability to observe and understand how things work. Electrical problems are many: every connection and every device is a potential problem. One of the best ways to prevent trouble is to check certain items as a routine procedure—to catch trouble before it becomes serious and causes fires, other damage, or even death.

This chapter presents some general troubleshooting techniques, concentrating on common problems and causes of trouble, and then focuses on troubleshooting electric motors. Remember, though, that troubleshooting hints apply only to common problems and their probable causes and remedies. They cannot give specific solutions, since on-the-scene facts may alter a situation. It takes a trained observer to ferret out the facts in a specific situation and make a diagnosis. Once a problem is identified, it is usually easily corrected.

GENERAL TROUBLESHOOTING

There are several problems that occur frequently in electrical work. Among them are handling wet and damp areas, removing a ground and preventing accidental shock, and checking wiring installations.

HANDLING DAMP AREAS

One factor that can cause problems in any wiring system is dampness. Watertight equipment should be installed wherever there is a danger of water coming into contact with live wires.

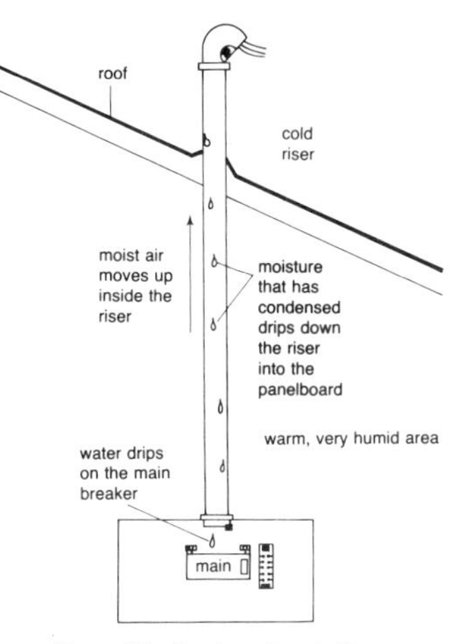

Figure 4-1. Condensation inside panelboard.

PROBLEM OF CONDENSATION

One major problem is the condensation of moisture inside panelboard. Moisture condenses when warm moist air in the basement moves up and comes into contact with cold air outside, thus making the riser cold. (See Figure 4-1.) In areas where this is a problem, an underground entrance should be made, or an outside riser should be mounted alongside the house, making its entry only when it reaches the panelboard. An entrance as low as possible is preferred so that any moisture that does condense will easily drain out at the bottom of the panelboard without contacting the hot side of the distribution panel. (See Figure 4-2.) This problem can occur in farm buildings that house livestock because the animals generate a lot of humidity.

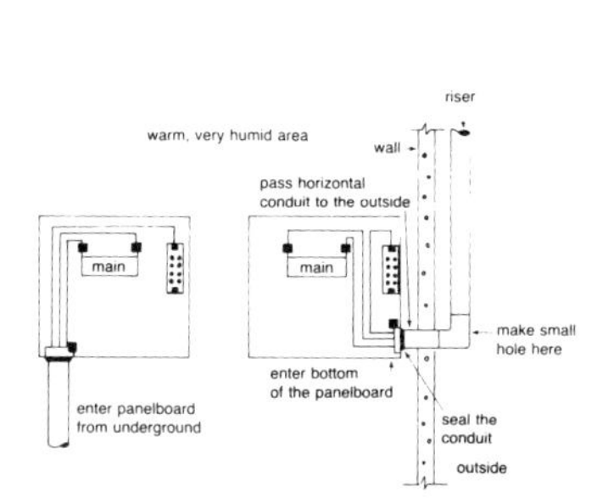

Figure 4-2. Underground entrance to try to prevent condensation from contacting hot distribution panel.

RUST AND CORROSION

Another problem is that of rust and corrosion. Anywhere there is moisture, there is the possibility of rust and corrosion. Both can cause contact problems between metals and affect resistance of a ground system.

Paint helps to prevent rust. Touch up scratched surfaces frequently. Corrosion can be prevented or at least limited by using galvanized panels or conduit, or in some cases by using plastic boxes, panels, and conduit. In some cases, metals may also be coated with an anticorrosion agent. Check NEC regulations for safe applications to prevent corrosion.

HANDLING GROUNDS AND PREVENTING ACCIDENTAL SHOCK

Removing a ground produces a potentially hazardous situation. In a properly installed 120/240-volt system, the current on the neutral line carries the difference between the current flowing on the hot lines. If rust or corrosion prevents contact with the proper grounding lugs, the ground becomes open—in effect, it is removed. (See Figure 4-3.)

One of the indications of this open-ground situation in a house is that some of the lights in the house will appear very bright and others very dim. If this occurs, turn off the main switch and locate the open or corroded ground connection before doing anything else. A situation like this is dangerous for anyone who touches any of the conductors. That person or animal then completes the ground circuit, and fatal shock may occur.

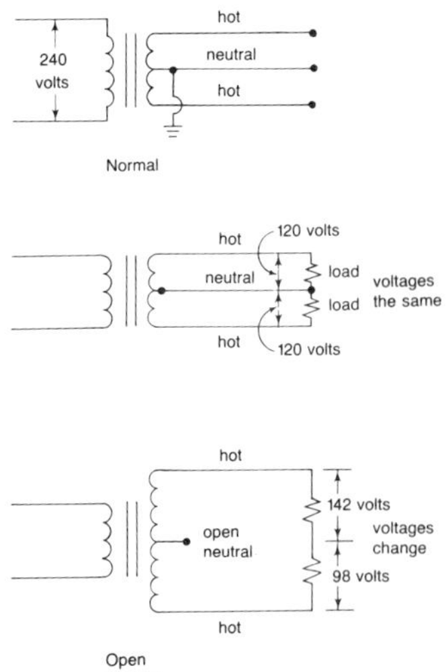

Figure 4-3. Removing a ground.

GROUND FAULT CIRCUIT INTERRUPTER

The ground fault circuit interrupter (GFCI) is one device used to prevent accidental shock. However, the use of a GFCI should never be a substitute for good grounding practices. Rather, it should test for and support a well-maintained grounding system.

Several types of GFCI devices are used to check grounding systems and thus prevent accidental shock. Figure 4-4 shows some of these devices. Some plug-in testers check polarity and grounding (A); some check the continuity of a ground path (B), which is important with many tools, especially those with metal handles. Another type is a ground loop tester (C) that measures the ground loop impedance of live circuits. Others (D) check the 500-volt and 1,000 volt DC insulation resistance of deenergized circuits and electrical equipment and the continuity in low-resistance circuits. Still others (E) provide insurance against a tool developing a fault

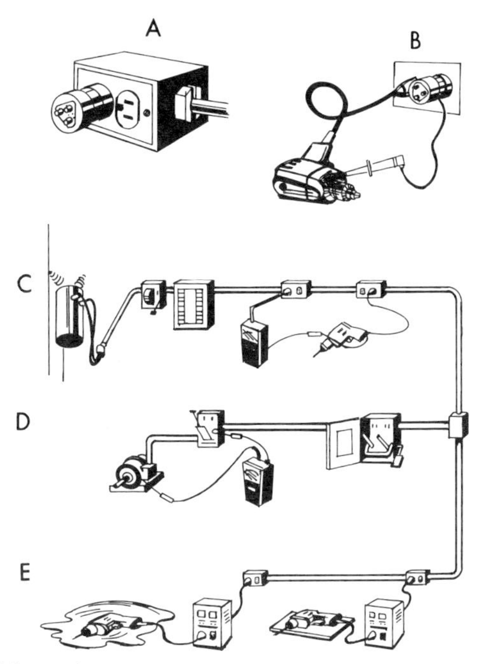

Figure 4-4. devices check the grounding system and thus prevent shock: (A) checking polarity and grounding; (B) same tester checking the continuity of the ground path of a tool; (C) ground loop tester measuring the impedance of live circuits; (D) meter checking DC insulation resistance of electrical equipment; (E) device testing tools to ensure that any current leakage is below a hazardous level.

while being used, possibly causing serious personal injury, by ensuring that any current leaking is below a hazard level.

One problem associated with ground fault circuit interrupters is nuisance tripping. Keep in mind that even a few drops of moisture or flecks of dust can trip the GFCI. One way to avoid this problem is to use watertight plugs and connectors on extension cords.

GROUND FAULT RECEPTACLES

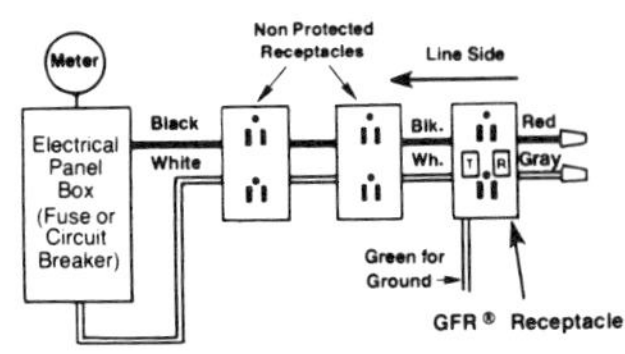

Figure 4-5. Wiring diagram for GFR that is also a receptacle and services only one outlet.

A ground fault receptacle (GFR) is a ground fault circuit interrupter with a receptacle. A GFR may be wired to protect its own outlet—terminal installation—or it may be wired to protect its outlet and other downstream receptacles. A typical wiring diagram for terminal installation is shown in Figure 4-5. One way to check for terminal installation is to check the red and gray wires. If they are capped with a wire nut, then you know that the GFR does not service any other outlets.

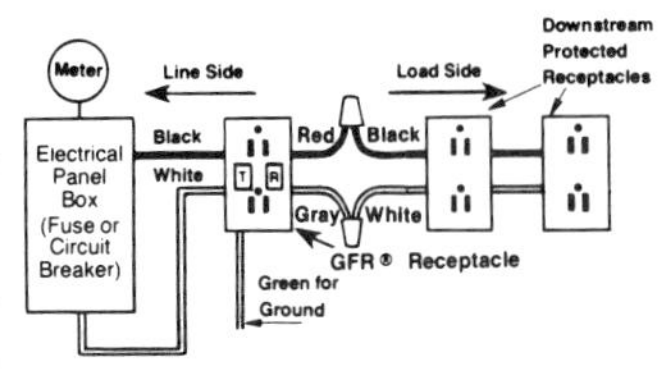

Figure 4-6. Wiring diagram for GFR that protects other downstream receptacles.

When a GFR that protects more than one outlet is tripped, it takes all of the outlets it services out of line. This can lead to some problems. For example, a GFR located in an upstairs bathroom may also protect an outside outlet and a downstairs bathroom outlet. If someone tries to use the outside outlet and it doesn't work, he or she may not relate this malfunction to the GFR in the upstairs bathroom and may call the electrician unnecessarily. All that needs to be done in this situation is to push the test (T) button on the GFR upstairs and then the reset (R) button to make sure the fault does not still exist; this puts the outlets back on line. Figure 4-6 shows a typical wiring diagram for a GFR that protects more than one outlet.

WIRING PROBLEMS

Some problems with wiring systems can be traced to the use of improper materials in the wiring devices. Therefore, the use of

TABLE 4–1

Chemical Resistance of Materials Commonly Used in Wiring Devices

Chemical	Nylon	Melamine	Phenolic	Urea	Polyvinyl Chloride	Polycar-bonate	Rubber
Acids	C	B	B	B	A	A	B
Alcohol	A	A	A	A	A	A	B
Caustic Bases	A	B	B	B	A	C	C
Gasoline	A	A	A	C	A	A	B
Grease	A	A	A	A	A	A	B
Kerosene	A	A	A	A	A	A	A
Oil	A	A	A	A	A	A	A
Solvents	A	A	A	A	C	C	C
Water	A	A	A	A	A	A	B

A—Completely Resistant. Good to excellent, general use.
B—Resistant. Fair to good, limited service.
C—Slow attack. Not recommended for use.

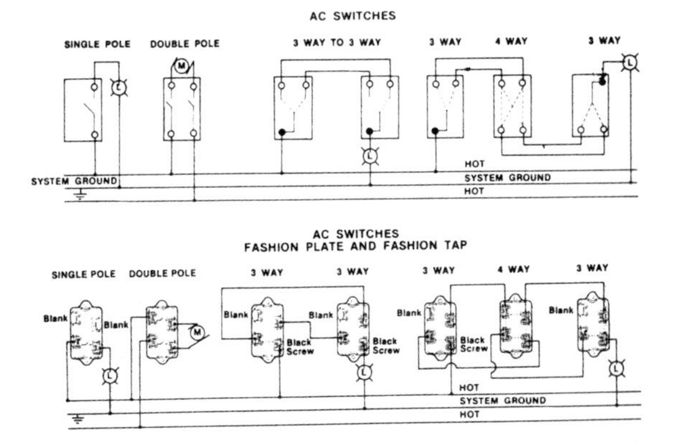

Figure 4-7. Wiring diagrams

proper materials in the installation stage is a form of preventive maintenance, decreasing the likelihood of problems later.

Shock hazards should be minimized by the dielectric strength of the material used for the molded interior walls and the indi-

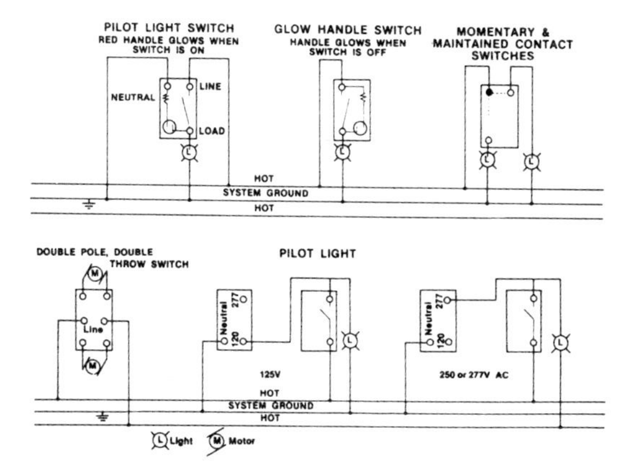

for switches in various systems.

vidual wire pocket areas. Each molded piece has to support adjacent molded pieces to result in good resiliency and strength. Nylon seems to be best for this job. Nylon devices withstand high impact in heavy-duty industrial and commercial applications. De-

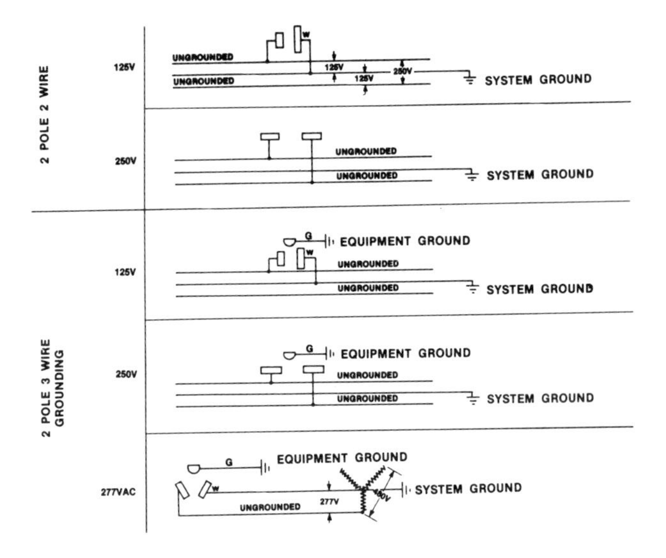

Figure 4-8. Grounding systems used

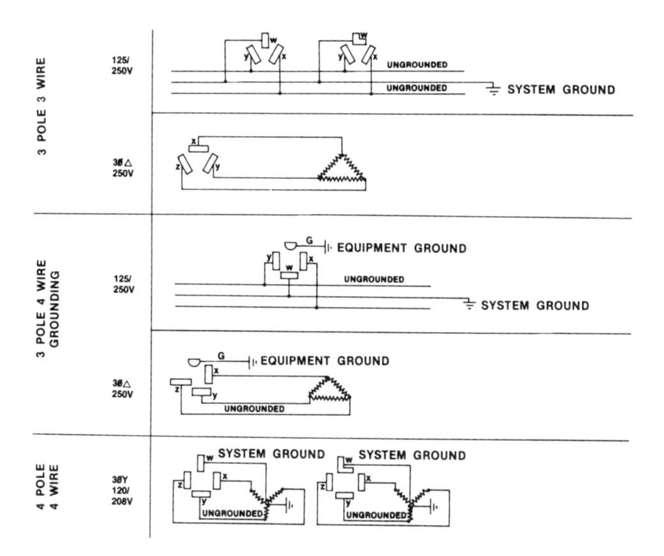

by connecting various plugs to a circuit.

TABLE 4–2

Mechanical and Electrical Properties of Materials Commonly Used in Wiring Devices

	Nylon	Melamine	Phenolic	Urea	Polyvinyl Chloride	Polycarbonate
Tensile Strength (psi)	900-12,000	7000-13,000	6500-10,000	5500-13,000	5000-9000	8000-9500
Elongation (percent)	60-300	0.6-0.9	0.4-08	0.5-1.0	2.0-4.0	60-100
Tensile Modulus (10^5 psi)	1.75-4.1	12-14	8-17	10-15	3.5-6	3.5
Compressive Strength (psi)	6700-12,500	25,000-45,000	22,000-36,000	25,000-45,000	8000-13,000	12,500
Flexural Strength (psi)	No break	10,000-16,000	8500-12,000	10,000-18,000	10,000-16,000	13,500
Impact Strength (ft-lb/in.)	1.0-2.0	0.24-0.35	0.24-0.60	0.25-0.40	0.4-20	12-16
Rockwell Hardness	R109-R118	M110-M125	M96-M120	M110-M120	70-90 shore	M70-R116
Continuous Temperature Resistance (°F)	180-200	210	350-360	170	150-175	250
Heat Distortion, 66 psi (°F)	360-365	*	*	*	179	285
Dielectric Strength (volts/mil)	385-470	300-400	200-400	300-400	425-1300	400
Arc Resistance (sec) ASTM D495	100-105	110-180	0-7	110-150	60-80	10-120
Burning Rate (in./min)	Self extinguishing	Self extinguishing	Very low	Self extinguishing	Self extinguishing	Self extinguishing

*Not applicable because these are thermosetting materials.

vices made of vinyl, neoprene, urea, or phenolic materials can crack or be damaged under pressure. Damage can be invisible and cause direct shorts and other hazards. Nylon also has the ability to withstand high voltages without breaking down. Tables 4-1 and 4-2 detail the properties of materials commonly used in wiring devices.

In troubleshooting wiring problems, it is important to check that switches have been wired properly. Figure 4-7 shows wiring diagrams for switches in various systems. Plugs and connectors must also be wired and grounded correctly. Figure 4-8 shows the grounding systems used in connecting various plugs to a circuit.

In some troubleshooting situations, the electrician will be called upon to trace wires. Most multiwire cables have color-coded conductors. Follow the color code to identify the same conductor at each end of the installation. When installing a wiring system, make a list of the color-coded conductors and where each terminates to make troubleshooting at a later date easier.

Test for wire pairs using an ohmmeter. Short two wires together on one location and use an ohmmeter at the other end to test for continuity. Then mark these wires with colored tape or a numbered label.

TROUBLESHOOTING ELECTRIC MOTORS

Electric motors should be installed and maintained by electricians. There are millions of motors in use in homes, offices, stores, and industrial plants. One major concern is choosing the right motor for a specific job. The choice depends on the intended use of the motor, the availability of power, the location, and other factors. Once installed, the motor must be properly maintained to prevent problems. The electrician and apprentice should ensure proper motor maintenance, and if problems do occur, they should be able to pinpoint the source and cause of the problem and correct it. This section discusses preventive maintenance of electric motors, danger signals to recognize in motors, and hints for correcting common problems.

PREVENTIVE MAINTENANCE

Small motors usually operate with so little trouble that they are apt to be neglected. However, they should be thoroughly inspected twice a year to detect wear and to remove any conditions that might lead to further wear or cause trouble. Special care should be taken to inspect motor bearings, cutouts, and other parts subject to wear. Also, be sure that dirt and dust are not interfering with ventilation or clogging moving parts.

CHECKING AND MAINTAINING SMALL MOTORS

- *Be sure there is adequate wiring.* After a motor has been installed or transferred from one location to another, check the wiring. Be sure that adequate wire sizes are used to feed electric power to the motor; in many cases, replacing wires at this early stage will prevent future breakdowns. Adequate wiring also helps to prevent overheating and reduces electric power costs.
- *Check internal switches.* Although switches usually give little trouble, regular attention will make them last longer. Use fine sandpaper to clean contacts. Be sure sliding members on a shaft move freely, and check for loose screws.
- *Check load conditions.* Check the driven load regularly. Sometimes additional friction develops gradually within a machine and after a period of time imposes an overload on the motor. Watch motor temperatures and protect motors with properly rated fuses or overload cutouts.
- *Provide extra care in lubrication.* Motors should be lubricated according to manufacturer's recommendations. A motor running three times more than usual needs three times as much attention to lubrication. Provide enough oil, but don't overdo it.

- *Keep commutators clean.* Do not allow a commutator to become covered with dust or oil. Wipe it occasionally with a clean, dry cloth or one moistened with a solvent that does not leave a film. If it is necessary to use sandpaper to clean the commutator, use No. 0000 or finer.
- *Replace worn brushes.* Inspect brushes at regular intervals. Whenever a brush is removed for inspection, be sure to replace it in the same axial position—that is, it should not be turned around in the brush holder when it is put back in the motor. If the contact surface that has been worn in to fit the commutator is not replaced in the same position, excessive sparking and loss of power will result. Brushes naturally wear down and should be replaced before they are less than 1/4 inch long.
- *Be sure that the motor has a proper service rating.* Whenever a motor is operated under different conditions or used in a different way, be sure it is rated properly. For example, a motor is rated for intermittent duty because the temperature rise within the motor will not be excessive when it is operated for short periods of time. Putting such a motor on continuous duty will result in excessive temperature increases that will lead to burnout or at least to deterioration of the insulation.

LUBRICATION

As stated in the list above, extra care in lubrication is essential in maintaining all motors. The exact function of the lubricant and the choice of lubricant depends somewhat on the type of motor.

Lubricating Ball-Bearing Motors. Ball-bearing motors are motors that have rollers or ball bearings supporting the rotor at each end. The lubricants in ball-bearing motors serve to

- dissipate heat caused by friction of bearing members under load,
- protect bearing members from rust and corrosion, and
- provide protection against the entrance of foreign matter into the bearings.

Lubricating Sleeve-Bearing Motors. Sleeve-bearing motors are quieter than ball-bearing motors. They operate by having a small tubelike (brass, bronze, or alloy) sleeve support the rotor shaft at each end. The lubricant used with sleeve bearings must actually provide a film that completely separates the bearing surfaces from the rotating shaft member and ideally eliminates all metal-to-metal contact.

Oil, because of its adhesive properties and viscosity (resistance to flow), is dragged along by the rotating shaft of the motor and forms a wedge-shaped film between the shaft and the bearing. The oil film forms automatically when the shaft begins to turn and is maintained by the motion. The forward motion sets up pressure in the oil film, which, in turn, supports the load. This wedge-shaped film of oil is an absolutely essential feature of effective hydrodynamic sleeve-bearing lubrication. Without it, no great load can be carried, except with high friction loss and resultant destruction of the bearing. When lubrication is effective and an adequate film is maintained, the sleeve bearing serves chiefly as a guide to ensure alignment. In the event of failure of the oil film, the bearing functions as a safeguard to prevent actual damage to the motor shaft.

Choice of Lubricant. Good lubricants are essential to low maintenance costs. An oil that provides the most effective bearing lubrication and does not require frequent renewal should be selected. In general, low-viscosity oils are recommended for fractional horsepower motors because they offer low internal friction, permit fuller realization of the motor's efficiency, and minimize the operating temperature of the bearing. Top-grade oils are recommended, since they are refined from pure petroleum; are substantially noncorrosive to metal surfaces; are free from sediment, dirt, or other foreign materials; and are stable in the presence of heat and moisture encountered in the motor. In terms of performance, the higher-priced oils often prove to be cheaper in the long run.

If extremes in bearing temperature are expected, special care should be exercised in selecting the proper lubricant. Special oils are available for motor applications at high temperatures and at low temperatures. Standard-temperature-range oils should not be used in situations with high ambient temperatures and high motor-operating temperatures because there will be a decrease in viscosity, an increase in corrosive oxidation products, and usually a reduction in the quantity of lubricant in contact with the bearing. This, in turn, will cause an increase in bearing temperature and deleterious effect on bearing life and motor performance.

PREVENTION OF WEAR

Sleeve-bearing motors are less susceptible to wear than are ball-bearing motors because the relatively soft surface of the sleeve bearing can absorb hard particles of foreign materials. Therefore, there is less abrasion and wear. However, good maintenance procedures for both types of motors require that the oil and bearings be kept clean. How often the oil should be changed depends on local conditions, such as the severity and continuity of service and operating temperatures. A conservative lubrication maintenance program should call for inspection of the oil level and cleaning and refilling with new oil every 6 months.

WARNING

Over-lubrication should be avoided at all times. Excess motor lubricant is a common cause of the failure of motor winding insulation in both sleeve- and ball-bearing motors.

DANGER SIGNALS

There are danger signals that usually appear before a motor overheats or burns out. At the first appearance of these signs, the electrician or apprentice should make every effort to correct and/or improve maintenance procedures and remedy any other problems to try to avoid permanent damage to the bearings or motor.

Major Danger Signals in Ball-Bearing Motors

1. A sudden increase in the temperature differential between the motor and bearing temperatures—usually indicating malfunction of the bearing lubricant.
2. A temperature higher than that recommended for the lubricant—warning of a reduction in bearing life. (A good rule of thumb is that grease life is halved for each 25° increase in operating temperature.)
3. An increase in bearing noise, accompanied by an increase in bearing temperature—indicating a serious bearing malfunction.

Major Causes of Ball-Bearing Failure

1. Foreign matter in the bearing from dirty grease or ineffective seals.
2. Deterioration of grease because of high temperatures or contamination.
3. Overheating, caused by too much grease.

Major Danger Signals in Sleeve-Bearing Motors

1. Rumbling noise.
2. High-pitched whistling sound.

COMMON MOTOR TROUBLES AND THEIR CAUSES

When troubleshooting an electric motor, it is advisable to keep in mind the most common problems associated with the particular type of motor.

With a fractional-horsepower motor, easy-to-detect symptoms indicate exactly what is wrong in many cases. However, at times, several types of problems have similar symptoms, and it becomes necessary to check each symptom separately to diagnose the overall problem. Table 4-3 lists some of the more common problems of small motors, along with suggestions of possible causes.

Most common motor troubles can be checked by some tests or by visual inspection. The order in which to make these test rests with the troubleshooter, but it is natural to make the simplest tests first. For instance, when a motor fails to start, you first inspect the motor connections, since this is an easy and simple thing to do. In some cases, a combination of symptoms will provide a clue to the source of trouble and eliminate some possibilities. For example, in the case cited above—a motor will not start—if heating occurs, that fact suggests that a short or ground exists in one of the windings and eliminates the likelihood of an open circuit, poor line connection, or defective starter switch.

Centrifugal starting switches, found in many types of fractional-horsepower motors, occasionally are a source of trouble. If the mechanism sticks in the run position, the motor will not start once it is turned off. On the other hand, if the switch sticks in the closed position, the motor will not attain speed and the start winding quickly heats up. The motor may also fail to start if the contact points of the switch are out of adjustment or coated with oxide. It is important to remember, however, that any adjustment of the switch or contacts should be made only at the factory or at an authorized service station.

Knowing the arrangement of coils aids in checking for possible shorts, grounds, and openings in any type of AC or DC motor. Figure 4-9 shows these motor connections.

Knowing the current expected to be drawn in the normal operation of a motor also aids in troubleshooting. Then it is possible to use a clamp-on ammeter, check the current drawn by the motor, and compare it with the standards to see if it is excessive or otherwise incorrect. Table 4-4 shows the ampere ratings of AC motors that operate on three-phase power. Remember, however, that these are average ratings and that the rating on a specific motor could be higher or lower. Therefore, the selection of heater coils on this basis alone involves some risk. For fully reliable motor protection, check the motor's nameplate and select heater coils on the basis of the full-load current rating given there.

CONNECTION DIAGRAMS

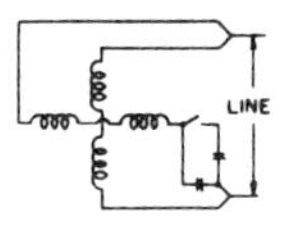

RELUCTANCE SYNCHRONOUS TWO-VALUE CAPACITOR—Reversible only from rest (by transposing leads).

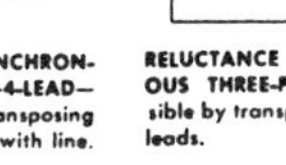

RELUCTANCE SYNCHRONOUS TWO-PHASE—4-LEAD—Reversible by transposing either phase leads with line.

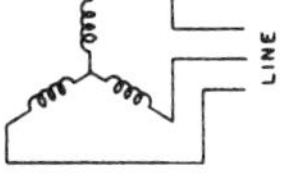

RELUCTANCE SYNCHRONOUS THREE-PHASE—Reversible by transposing any two leads.

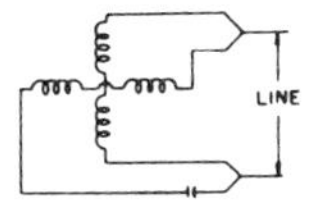

HYSTERESIS SYNCHRONOUS PERMANENT-SPLIT CAPACITOR—Reversible by transposing leads.

SHADED POLE—Non-reversible.

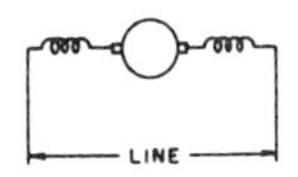

SERIES-WOUND—2-LEAD—Non-reversible.

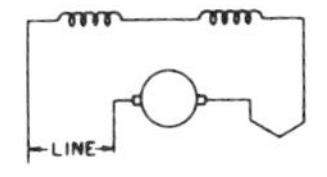

SERIES-WOUND—4-LEAD Reversible by transposing armature leads.

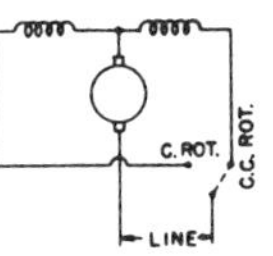

SERIES-WOUND SPLIT FIELD—Reversible by connecting either field lead to line.

ELECTRIC GOVERNOR CONTROLLED SERIES-WOUND—Non-reversible.

(A)
SHUNT MOTOR
4-WIRE REVERSIBLE

(B)
COMPOUND MOTOR
5-WIRE REVERSIBLE

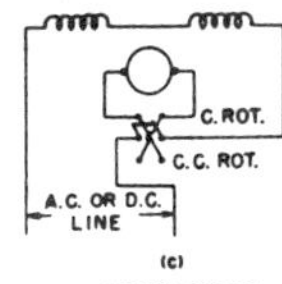

(C)
SERIES MOTOR
4-WIRE REVERSIBLE

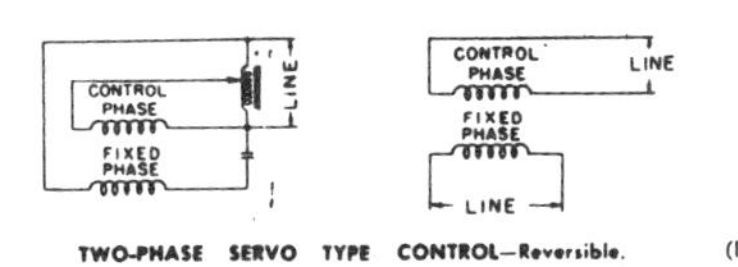

TWO-PHASE SERVO TYPE CONTROL—Reversible. (Bodine Electric Co.)

Figure 4-9. Connection diagrams showing the arrangement of coils in various

CONNECTION DIAGRAMS

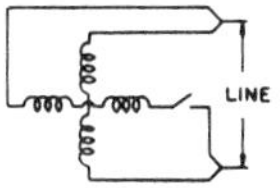

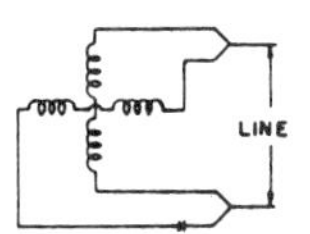

SPLIT-PHASE — Reversible only from rest (by transposing leads).

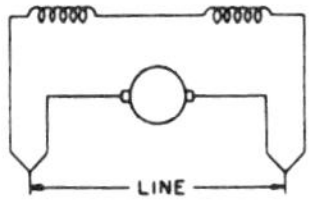

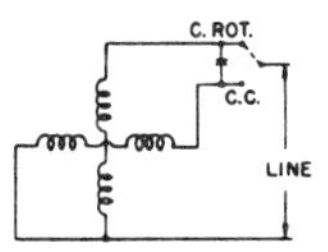

SHUNT-WOUND — Reversible by transposing leads.

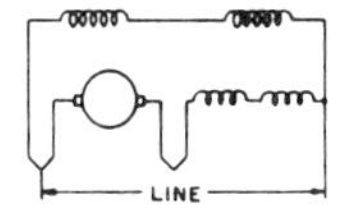

COMPOUND-WOUND — Reversible by transposing armature leads.

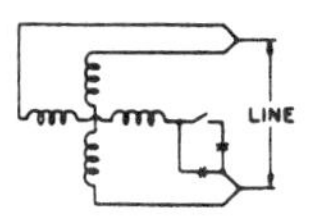

PERMANENT-SPLIT CAPACITOR 4-LEAD—Reversible by transposing leads.

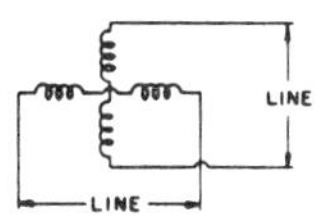

PERMANENT-SPLIT CAPACITOR 3-LEAD—Reversible by connecting either side of capacitor to line.

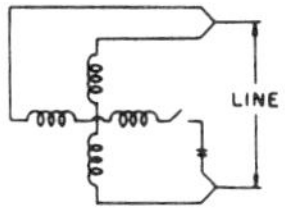

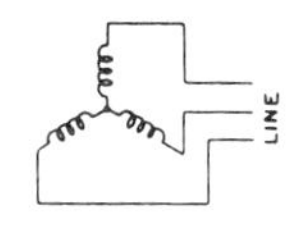

CAPACITOR-START—Reversible only from rest (by transposing leads).

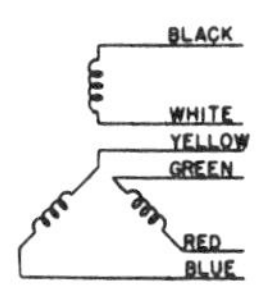

TWO-VALUE CAPACITOR—Reversible only from rest (by transposing leads).

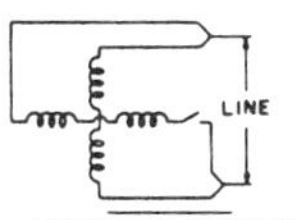

TWO-PHASE 4-LEAD—Reversible by transposing either phase leads with line.

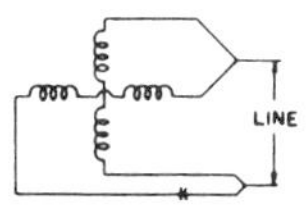

THREE-PHASE — SINGLE VOLTAGE — Reversible by transposing any two leads.

BLACK
WHITE
YELLOW
GREEN
RED
BLUE

3-PHASE, STAR-DELTA. 6-LEAD REVERSIBLE—
For 440 volts connect together white, yellow and green: Connect to line black, red and blue. To reverse rotation, transpose any two line leads. For 220 volts connect white to blue, black to green, and yellow to red: Then connect each junction point to line.

To reverse rotation, transpose any two junction points with line.

RELUCTANCE SYNCHRONOUS SPLIT-PHASE — Reversible only from rest (by transposing leads).

RELUCTANCE SYNCHRONOUS PERMANENT — SPLIT-CAPACITOR 4-LEAD—Reversible by transposing lead

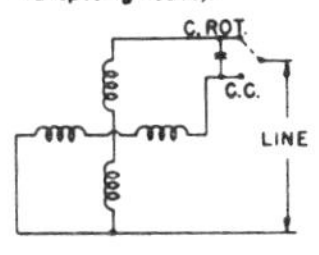

RELUCTANCE SYNCHRONOUS PERMANENT-SPLIT CAPACITOR 3-LEAD—Reversible by connecting either side of capacitor to line.

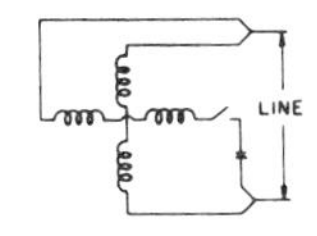

RELUCTANCE SYNCHRONOUS CAPACITOR-START—Reversible only from rest (by transposing leads).

types of motors.

TABLE 4–3[1]

Guide to Probable Causes of Motor Troubles

Motor Type	A. C. Single Phase				A.C. Polyphase (2 or 3 phase)	Brush Type (Universal, Series, Shunt, or Compound)
	Split Phase	Capacitor Start	Permanent-Split Capacitor	Shaded Pole		
Trouble	*Probable Causes					
Will not start.	1, 2, 3, 5	1, 2, 3, 4, 5	1, 2, 4, 7, 17	1, 2, 7, 16, 17	1, 2, 9	1, 2, 12, 13
Will not always start, even with no load, but will run in either direction when started manually.	3, 5	3, 4, 5	4, 9		9	
Starts but heats rapidly.	6, 8	6, 8	4, 8	8	8	8
Starts but runs too hot.	8	8	4, 8	8	8	8
Will not start but will run in either direction when started manually—overheats.	3, 5, 8	3, 4, 5, 8	4, 8, 9		8, 9	
Sluggish—sparks severely at the brushes.						10, 11, 12, 13, 14
Abnormally high speed—sparks severely at the brushes.						15
Reduction in power—motor gets too hot.	8, 16, 17	8, 16, 17	8, 16, 17	8, 16, 17	8, 16, 17	13, 16, 17
Motor blows fuse, or will not stop when switch is turned to off position.	8, 18	8, 18	8, 18	8, 18	8, 18	18, 19
Jerky operation—severe vibration.						10, 11, 12, 13, 19

***Probable Causes**

1. Open in connection to line.
2. Open circuit in motor winding.
3. Contacts of centrifugal switch not closed.
4. Defective capacitor.
5. Starting winding open.
6. Centrifugal starting switch not opening.
7. Motor overloaded.
8. Winding short-circuited or grounded.
9. One or more windings open.
10. High mica between commutator bars.
11. Dirty commutator or commutator is out of round.
12. Worn brushes and/or annealed brush springs.
13. Open circuit or short circuit in the armature winding.
14. Oil-soaked brushes.
15. Open circuit in the shunt winding.
16. Sticky or tight bearings.
17. Interference between stationary and rotating members.
18. Grounded near switch end of winding.
19. Shorted or grounded armature winding.

[1]Courtesy of Bodine Electric Company, Chicago.

TABLE 4–4[1]

Ampere Rating of Three-Phase, 60 Hertz, AC Induction Motors[2]

Hp	Syn. Speed RPM	Current in Amperes 115 Volts	230 Volts	380 Volts	460 Volts	575 Volts	2200 Volts
1/4	1800	1.90	.95	.55	.48	.38	....
	1200	2.80	1.40	.81	.70	.56	....
	900	3.20	1.60	.93	.80	.64	....
1/3	1800	2.38	1.19	.69	.60	.48	....
	1200	3.18	1.59	.92	.80	.64	....
	900	3.60	1.80	1.04	.90	.72	....
1/2	1800	3.44	1.72	.99	.86	.69	....
	1200	4.30	2.15	1.24	1.08	.86	....
	900	4.76	2.38	1.38	1.19	.95	....
3/4	1800	4.92	2.46	1.42	1.23	.98	....
	1200	5.84	2.92	1.69	1.46	1.17	....
	900	6.52	3.26	1.88	1.63	1.30	....
1	3600	5.60	2.80	1.70	1.40	1.12	....
	1800	7.12	3.56	2.06	1.78	1.42	....
	1200	7.52	3.76	2.28	1.88	1.50	....
	900	8.60	4.30	2.60	2.15	1.72	....
1 1/2	3600	8.72	4.36	2.64	2.18	1.74	....
	1800	9.71	4.86	2.94	2.43	1.94	....
	1200	10.5	5.28	3.20	2.64	2.11	....
	900	11.2	5.60	3.39	2.80	2.24	....
2	3600	11.2	5.60	3.39	2.80	2.24	....
	1800	12.8	6.40	3.87	3.20	2.56	....
	1200	13.7	6.84	4.14	3.42	2.74	....
	900	15.8	7.90	4.77	3.95	3.16	....
3	3600	16.7	8.34	5.02	4.17	3.34	....
	1800	18.8	9.40	5.70	4.70	3.76	....
	1200	20.5	10.2	6.20	5.12	4.10	....
	900	22.8	11.4	6.90	5.70	4.55	....

Table 4–4[1] (*continued*)

Hp	Syn. Speed RPM	Current in Amperes 115 Volts	230 Volts	380 Volts	460 Volts	575 Volts	2200 Volts
5	3600	27.1	13.5	8.20	6.76	5.41	....
	1800	28.9	14.4	8.74	7.21	5.78	....
	1200	31.7	15.8	9.59	7.91	6.32	....
	900	31.0	15.5	9.38	7.75	6.20	....
7 1/2	3600	39.1	19.5	11.8	9.79	7.81	....
	1800	43.0	21.5	13.0	10.7	8.55	....
	1200	43.7	21.8	13.2	10.9	8.70	....
	900	46.0	23.0	13.9	11.5	9.19	....
10	3600	50.8	25.4	15.4	12.7	10.1	....
	1800	53.8	26.8	16.3	13.4	10.7	....
	1200	56.0	28.0	16.9	14.0	11.2	....
	900	61.0	30.5	18.5	15.2	12.2	....
15	3600	72.7	36.4	22.0	18.2	14.5	....
	1800	78.4	39.2	23.7	19.6	15.7	....
	1200	82.7	41.4	25.0	20.7	16.5	....
	900	89.0	44.5	26.9	22.2	17.8	....
20	3600	101.1	50.4	30.5	25.2	20.1	....
	1800	102.2	51.2	31.0	25.6	20.5	....
	1200	105.7	52.8	31.9	26.4	21.1	....
	900	109.5	54.9	33.2	27.4	21.9	....
25	3600	121.5	60.8	36.8	30.4	24.3	....
	1800	129.8	64.8	39.2	32.4	25.9	....
	1200	131.2	65.6	39.6	32.8	26.2	....
	900	134.5	67.3	40.7	33.7	27.0	....
30	3600	147.	73.7	44.4	36.8	29.4	....
	1800	151.	75.6	45.7	37.8	30.2	....
	1200	158.	78.8	47.6	39.4	31.5	....
	900	164.	81.8	49.5	40.9	32.7	....
40	3600	193.	96.4	58.2	48.2	38.5	....
	1800	202.	101.	61.0	50.4	40.3	....
	1200	203.	102.	61.2	50.6	40.4	....
	900	209.	105.	63.2	52.2	41.7	....

Table 4–4[1] (*continued*)

Hp	Syn. Speed RPM	Current in Amperes 115 Volts	230 Volts	380 Volts	460 Volts	575 Volts	2200 Volts
50	3600	241.	120.	72.9	60.1	48.2	....
	1800	249.	124.	75.2	62.2	49.7	....
	1200	252.	126.	76.2	63.0	50.4	....
	900	260.	130.	78.5	65.0	52.0	....
60	3600	287.	143.	86.8	71.7	57.3	....
	1800	298.	149.	90.0	74.5	59.4	....
	1200	300.	150.	91.0	75.0	60.0	....
	900	308.	154.	93.1	77.0	61.5	....
75	3600	359.	179.	108.	89.6	71.7	....
	1800	365.	183.	111.	91.6	73.2	....
	1200	368.	184.	112.	92.0	73.5	....
	900	386.	193.	117.	96.5	77.5	....
100	3600	461.	231.	140.	115.	92.2	....
	1800	474.	236.	144.	118.	94.8	23.6
	1200	478.	239.	145.	120.	95.6	24.2
	900	504.	252.	153.	126.	101.	24.8
125	3600	583.	292.	176.	146.	116.	....
	1800	584.	293.	177.	147.	117.	29.2
	1200	596.	298.	180.	149.	119.	29.9
	900	610.	305.	186.	153.	122.	30.9
150	3600	687.	343.	208.	171.	137.	....
	1800	693.	348.	210.	174.	139.	34.8
	1200	700.	350.	210.	174.	139.	35.5
	900	730.	365.	211.	183.	146.	37.0
200	3600	904.	452.	274.	226.	181.	....
	1800	915.	458.	277.	229.	184.	46.7
	1200	920.	460.	266.	230.	184.	47.0
	900	964.	482.	279.	241.	193.	49.4
250	3600	1118.	559.	338.	279.	223.	....
	1800	1136.	568.	343.	284.	227.	57.5
	1200	1146.	573.	345.	287.	229.	58.5
	900	1200.	600.	347.	300.	240.	60.5

Table 4–4[1] (*continued*)

Hp	Syn. Speed RPM	Current in Amperes 115 Volts	230 Volts	380 Volts	460 Volts	575 Volts	2200 Volts
300	1800	1356.	678.	392.	339.	274.	69.0
	1200	1368.	684.	395.	342.	274.	70.0
400	1800	1792.	896.	518.	448.	358.	91.8
500	1800	2220.	1110.	642.	555.	444.	116.

[1]Courtesy Bodine Electric Company, Chicago.
[2]Ampere ratings of motors vary somewhat. The values given here are for drip-proof, Class B insulated (T frame) where available, 1.15 service factor, NEMA Design B motors. The values represent an average full-load motor current that was calculated from the motor performance data published by several motor manufacturers. In the case of high-torque squirrel cage motors, the ampere ratings will be at least 10% greater than the values shown.

ANALYZING SYMPTOMS

Table 4-5 presents a comprehensive guide to troubleshooting three-phase motors. Find the symptoms you are dealing with in a particular motor, study the possible causes listed, eliminate any that you can because of the presence or absence of other identifying signs, and then try the recommended corrections.

Be sure that in searching for symptoms and analyzing possible causes, you have correct and complete information. That means that all measurements—for example, of volts or amperes—must be made with accurate meters. All other information from visual inspection or any other test should also be done firsthand and carefully.

SAFETY PRECAUTIONS

When working with motors, it is important that certain safety procedures be followed at all times.

- Be sure that all motor installation, maintenance, and repair work is performed only by qualified people.
- Disconnect and lock all input power before doing any work on motors (or any electrical equipment).
- Whenever equipment is to be lifted, follow the specific procedures for using the lift bail.
- Follow National Electrical Code rules and all local codes when installing any electrical equipment.
- Be sure that all equipment is properly grounded in accordance with the NEC.
- Keep hands, hair, clothing, and tools away from all moving parts when operating or repairing equipment.
- Provide proper safeguards to prevent contact with rotating parts.
- Be familiar with and adhere to recommendations outlined in "Safety Standards for Construction and Guide for Selection, Installation, and Use of Electric Motors and Generators" (NEMA MG-2).

WARNING: WHEN USING LIFT BAIL

Do not use the lift bail on the motor to lift the motor along with additional equipment, such as pumps, compressors, or other driven machinery. In the case of assemblies on a common base, do not lift with the motor lift bail but rather use a sling around the base or the lifting means provided on the base. In all cases, take care to ensure lifting only in the direction intended in the design of the lifting means. Also, be careful to avoid hazardous overloads due to deceleration, acceleration, or shock forces.

TABLE 4–5[1]

Motor Troubleshooting Guide[1]

Symptom	Possible Causes	Correction
High Input Current (all three phases)	Accuracy of ammeter readings.	First check accuracy of ammeter readings on all three phases.
Running Idle (Disconnected from load)	High line voltage 5 to 10% over nameplate.	Consult power company—possibly decrease by using lower transformer tap.
Running Loaded	Motor overloaded.	Reduce load or use larger motor.
	Motor voltage rating does not match power system voltage.	Replace motor with one of correct voltage rating.
		Consult power company—possibly correct by using a different transformer tap.
Unbalanced Input Current (5% or more deviation from the average input current)	Unbalanced line voltage due to: a. Power supply. b. Unbalanced system loading. c. High resistance connection. d. Undersized supply lines.	Carefully check voltage across each phase *at the motor terminals* with good, properly calibrated voltmeter.
NOTE: A small voltage unbalance will produce a large current unbalance. Depending on the magnitude of unbalance and the size	Defective motor.	If there is doubt as to whether the trouble lies with the power supply or the motor, check per the following:

of the load the input current in one or more of the motor input lines may greatly exceed the current rating of the motor.		Rotate *all three input power lines* to the motor by one position—i.e., move line #1 to #2 motor lead, line #2 to #3 motor lead, and line #3 to #1 motor lead. a. If the unbalanced current pattern follows the *input power lines*, the problem is in the power supply. b. If the unbalanced current pattern follows the *motor leads*, the problem is in the motor. Correct the voltage balance of the power supply or replace the motor, depending on answer to a. & b. above.
Excessive Voltage Drop (more than 2 or 3% of nominal supply voltage)	Excessive starting or running load.	Reduce load.
	Inadequate power supply.	Consult power company.
	Undersized supply lines.	Increase line sizes.
	High resistance connections.	Check motor leads and eliminate poor connections.
	Each phase lead run in separate conduits.	All 3-phase leads shall be in a single conduit, per National Electrical Code. (This applies only to metal conduit with magnetic properites.)

Table 4–5[1] (*continued*)

Symptom	Possible Causes	Correction
Overload Relays Tripping Upon Starting (Also see "Slow Starting")	Slow starting (10–15 seconds or more) due to high inertia Load.	Reduce starting load. Increase motor size if necessary.
	Low voltage at motor terminals.	Improve power supply and/or increase line size.
Running Loaded	Overload.	Reduce load or increase motor size.
	Unbalanced input current.	Balance supply voltage.
	Single phasing.	Eliminate.
	Excessive voltage drop.	Eliminate (see above).
	Too frequent starting or intermittent overloading.	Reduce frequency of starts and overloading or increase motor size.
	High ambient starter temperatures.	Reduce ambient temperature or provide outside source of cooler air.
	Wrong size relays.	Correct size per nameplate current of motor. Relays have built in allowances for service factor current. Refer to National Electrical Code.
Motor Runs Excessively Hot	Overloaded.	Reduce load or load peaks and number of starts in cycle or increase motor size.

	Blocked ventilation: a. TEFC's.	Clean external ventilation system — check fan.
	b. O.D.P.'s.	Blow out internal ventilation passages.
		Eliminate external interference to motor ventilation.
	High ambient temperature over 40°C (104°F).	Reduce ambient temperature or provide outside source of cooler air.
	Unbalanced input current.	Balance supply voltage. Check motor leads for tightness.
	Single phased.	Eliminate single phase condition.
Won't Start (just hums and heats up)	Single phased.	Shut power off. Eliminate single phasing. Check motor leads for tightness.
	Rotor or bearings locked.	Shut power off. Check shaft for freeness of rotation.
		Be sure proper sized overload relays are in *each of the 3 phases* of starter. Refer to National Electrical Code.
Runs Noisy Under Load (excessive electrical noise or chatter under load)	Single phased.	Shut power off. If motor cannot be restarted, it is single phased. Eliminate single phasing.

Table 4–5[1] (*continued*)

Symptom	Possible Causes	Correction
		Be sure proper sized overload relays are in *each of the 3 phases* of the starter. Refer to National Electrical Code.
Slow Starting (10 or more seconds on small motors—15 or more seconds on large motors)		
Across the Line Start	Excessive voltage drop (5–10% voltage drop causes 10–20% or more drop in starting torque).	Consult power company. Check system. Eliminate voltage drop.
	High inertia load.	Reduce starting load or increase motor size.
Reduced Voltage Start	Excessive voltage drop. Loss of starting torque.	Check and eliminate.
Y-Delta	Starting torque reduced to 33%.	Reduce starting load or increase motor size.
PWS	Starting torque reduced to 50%.	Choose type of starter with higher starting torque.
Auto. Transformer	Starting torque reduced—25–64%.	Reduce time delay between 1st and 2nd step on starter—get motor across the line sooner.

Load Speed Appreciably Below Nameplate Speed	Overload.	Reduce load or increase voltage.
	Excessively low voltage.	**NOTE:** A reasonable overload or voltage drop of 10–15% will reduce speed only 1–2%. A report of any greater drop would be questionable.
	Inaccurate method of measuring RPM.	Check meter using another device or method.
Excessive Vibration (mechanical)	Out of balance: a. Motor mounting.	Be sure motor mounting is tight and solid.
	b. Load.	Disconnect belt or coupling—restart motor—if vibration stops, the unbalance was in load.
	c. Sheaves or coupling.	Remove sheave or coupling—securely tape ½ key in shaft keyway and restart motor—if vibration stops, the unbalance was in the sheave or coupling.
	d. Motor.	If the vibration does not stop after checking a, b, and c above, the unbalance is in the motor—replace the motor.
	e. Misalignment on close coupled application	Check and realign motor to the driven machine.

Table 4–5[1] (*continued*)

Symptom	Possible Causes	Correction
Noisy Bearings (listen to bearings)		
Smooth Mid-Range Hum	Normal fit.	Bearing OK.
High Whine	Internal fit of bearing too tight.	Replace bearing — check fit.
Low Rumble	Internal fit of bearing too loose.	Replace bearing — check fit.
Rough Clatter	Bearing destroyed.	Replace bearing — Avoid: a. Mechanical Damage. b. Excessive Greasing. c. Wrong Grease. d. Solid Contaminants. e. Water running into motor. f. Misalignment or close coupled application. g. Excessive belt tension.
Mechanical Noise	Driven machine or motor noise.	Isolate motor from driven machine — check difference in noise level.
	Motor noise amplified by resonant mounting.	Cushion motor mounting or dampen source of resonance.
	Driven machine noise transmitted to motor through drive.	Reduce noise of driven machine or dampen transmission to motor.
	Misalignment on close coupled application.	Improve alignment.

5

ABCs Of The Trade

There are many specific functions and jobs that an electrician and apprentice should be familiar with. It is impossible in a book such as this to explain and discuss all of them. We will, however, discuss six major topics that include most of the tasks an electrician may be called on to do. These are the installation and repair (if necessary) of alarm systems, lighting systems, motors, power systems, programmable controllers, and general wiring systems.

ALARM SYSTEMS

There are a number of different types of alarm systems. Some are hard-wired, with wires running from switches to a central location where a relay is energized or deenergized to cause an alarm to be sounded. However, some types of alarm systems require less wiring. They rely on electronics for sensing and sending signals to central units. They may be digitized, with programmed codes that disable the system temporarily to allow entry into the premises or to allow zone changes from time to time.

A TYPICAL HARD-WIRED SYSTEM

In a typical hard-wired system, fire and security systems are wired together. Figure 5-1 shows the wiring diagram of a house with such a system. A detailed diagram of the wiring for the security/fire system is given in Figure 5-2. Notice the closed-loop intruder-detection circuit and how the switches work. Then examine the fire-detection circuit and how it is wired into the

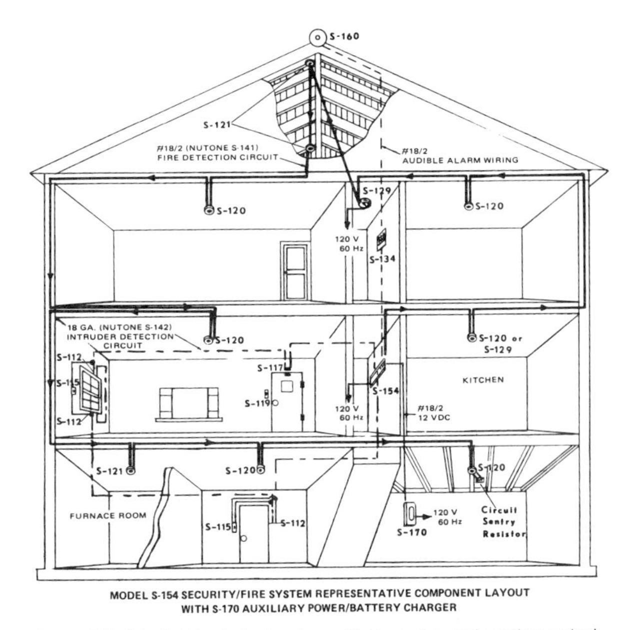

Figure 5-1. A typical hard-wired system with fire and security systems wired together.

system. Except for the 120-volt line to the power supply, the rest of the wiring is low voltage.

The control unit should be mounted between studs in a location with easy access from all rooms in the house; this is usually in a hallway. The control box should be mounted out of the reach of small children and not readily accessible to casual visitors. (See Figure 5-3.)

When possible, the control unit, the smoke/heat detectors, and the auxiliary power/battery charger (if used) should be connected to an individually fused 120-volt, 60-hertz power source, according to local codes.

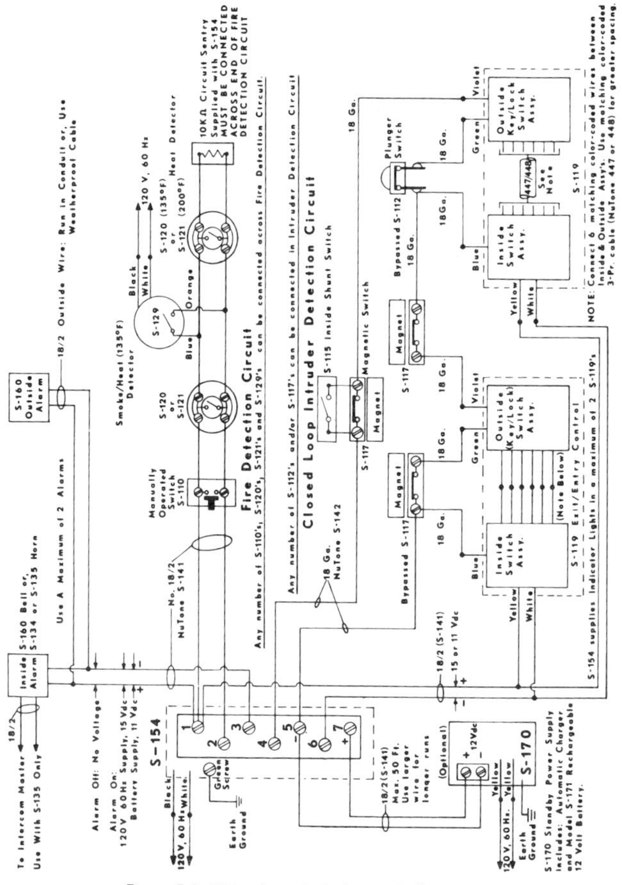

Figure 5-2. Wiring for a typical security/fire system.

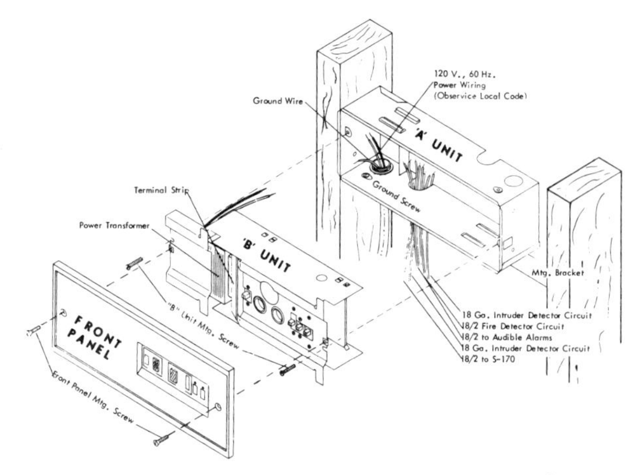

Figure 5-3. The control unit should be mounted between studs.

PARTS OF THE SYSTEM

Different hard-wired systems are available with various components in addition to the usual smoke/heat and security parts. The choice of which parts and system to use depends on the purpose of the system, what type of building it is in, location, and other factors. A discussion of common parts of a typical hard-wired system follows.

Motion Detectors. A motion detector operates on ultrasonic frequencies. An ultrasonic frequency is above the human hearing range. The receiver transducer and transmitter are approximately the size of a quarter. Any motion within the range of the unit causes a slight change in frequency reflected back to the receiving unit. This difference in frequency causes a relay or a transistor circuit to energize and complete the circuit to the alarm control center.

Ultrasonic motion detectors are quite sensitive. They can be activated by the movements of a dog or cat or even by a hot air furnace causing curtains to move. These units need little or no maintenance if wired into a household circuit.

Heat/Smoke Detectors. Heat detectors sense heat above a certain temperature—usually 136°F—and activate an alarm. Most heat detectors contain bimetallic switches that remain open below a temperature of 135°F but close above that temperature, triggering an alarm. Some are activated by higher temperatures—200°F—and these switches should be used in areas with higher than normal temperatures, such as boiler rooms and attics.

With most smoke detectors, a concentration of less than 4% smoke in the air will sound the alarm. What is 4%? Examine a cigarette smoking in an ashtray. The plume of smoke rises and coils and spreads. It diffuses into the air. Somewhere above the cigarette is a region where the smoke is drawn out and wispy, just barely visible. That concentration is close to 4%. If that amount of smoke were concentrated in the chamber of a smoke detector, it would set off the alarm. However, under normal conditions, even a roomful of people smoking won't trigger an alarm because the smoke spreads out so that the concentration is very low. Figure 5-4 shows how the detector is wired to the control unit.

There are two types of smoke detectors. In the *photoelectric smoke detector*, the smoke is detected as it interrupts a small light from a bulb inside the unit. This type is best at detecting thick smoke because that kind of smoke causes the light beam to be interrupted. This type of detector is used primarily in the home, since house fires are usually of the smoldering type with thick smoke.

Less dense smoke triggers the *radioactive detector*. This type of detector contains a dime-sized capsule of the radioactive element americium. Smoke interrupts the beam of ions produced by the low-level radiation and thus triggers the alarm. This type is best for detecting the kind of smoke generated in business or commercial fires. It is, however, more susceptible to false alarms than is the photoelectric type.

For maximum protection, heat/smoke detectors should be placed in every enclosed area, including bathrooms and closets. If any single dimension of a room exceeds 20 feet, additional detec-

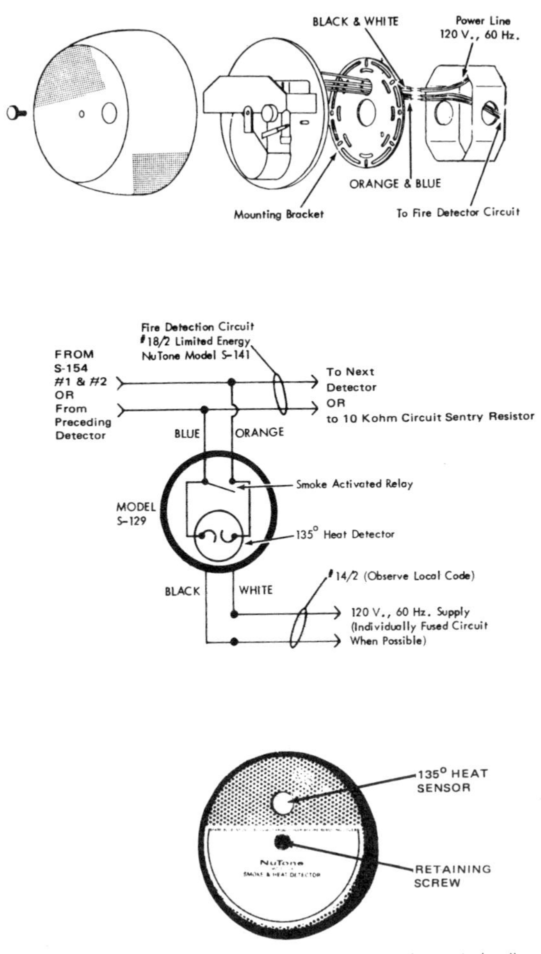

Figure 5-4. How a smoke/heat detector is wired to the control unit.

tors should be installed. For unfinished attic spaces with gable roofs, detectors should be installed 10 feet from the end and 20 feet apart on the bottom of the ridgepole. Figure 5-5 shows how to locate heat sensors.

Heat detectors should be installed on the ceiling but, when necessary, may be mounted on a side wall. When installed on a smooth ceiling, each detector will protect a 20-square-foot area. When installed on a wall, detectors should be at least 6 inches and no more than 12 inches below the ceiling and at least 12 inches from any corner.

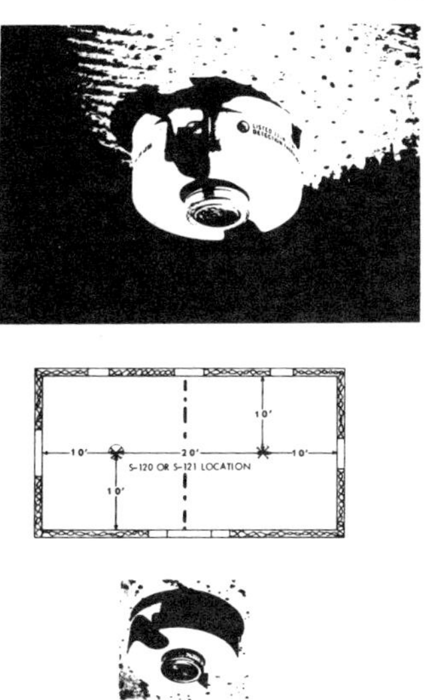

Figure 5-5. How to locate heat sensors.

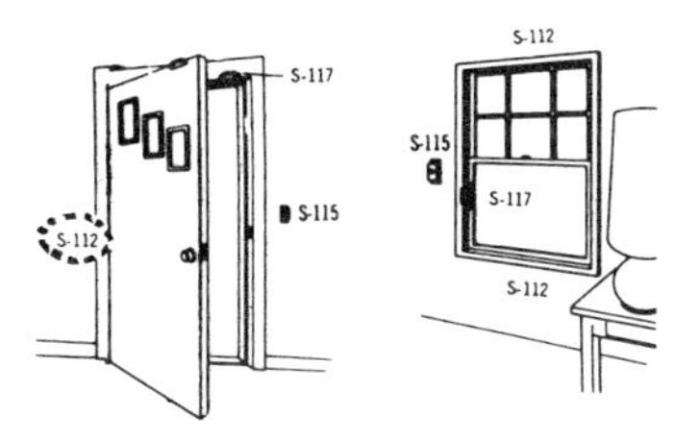

Figure 5-6. Plunger-type switches can be used on wooden door frames and window sashes.

Intruder Detectors. A motion detector functions as an intruder detector, but other intruder-detecting systems are also available. A typical intruder-detection circuit is a closed-loop system, meaning that the entry-detection switches are connected in a series. When a protected door or window is opened, the circuit between the two terminals of the control unit is broken and the intruder alarm is activated. Several types of switches are available.

Plunger-type switches can be used in wooden door frames or window sashes. The exact positioning of the switch depends on the type and design of the door or window. If possible, the switch should be installed in the frame on the hinge side of a door, and on double-hung windows, both top and bottom sections should be protected. Remember, the mounting area must be large enough to accommodate the flange of the switch. (See Figures 5-6 and 5-7.)

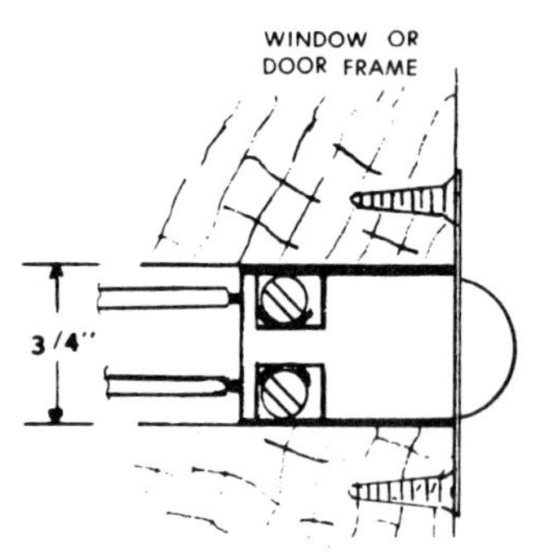

Figure 5-7. When mounting plunger-type switches, be sure the mounting area is large enough to accommodate the flange of the switch.

Magnetic-operated reed switch assemblies are available for horizontal or vertical surface-mounting on wood or metal door frames or window sashes. Install the magnetic section on the door or window and the switch on the frame or sash. Remember, when installing reed switches, that they are easily damaged if dropped. Use shims if necessary to install them, and be sure they are parallel and flush. (See Figure 5-8.)

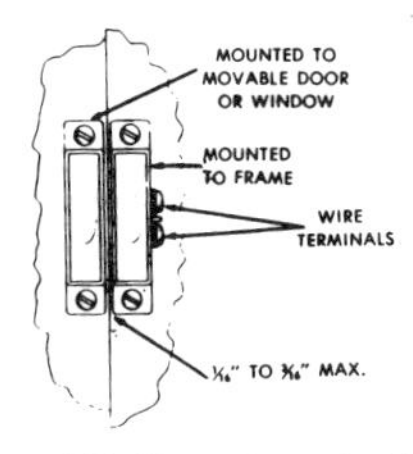

Figure 5-8. Magnetic reed switch mounted on a door or window.

An inside shunt switch is used to bypass switches used in doors and windows. (See Figure 5-9.) This type of switch is surface-mounted and connected to bypass the entry detector switch, as previously shown in Figure 5-2.

Exit/entry control units include inside and outside switch plate assemblies to be installed on 2 1/2-inch-deep single-gang wall boxes, above the reach of small children. The indicator lights will

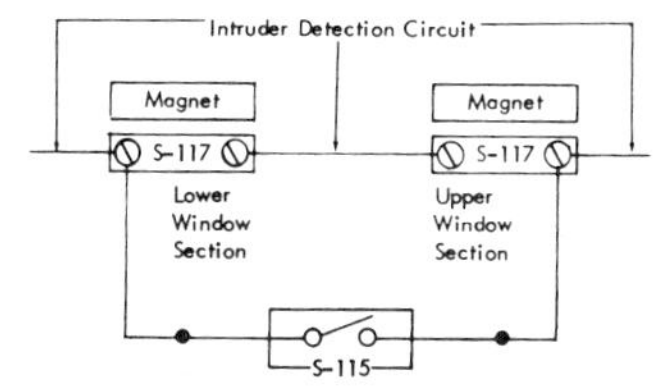

S-115 BYPASSING 2 ENTRY DETECTOR SWITCHES (DOUBLE HUNG WINDOW INSTALLATION)

Figure 5-9. A shunt switch used to bypass entry detector switches on a window.

burn at a somewhat lower level when operating on the 12-volt DC power supply activated by failure of the 120-volt supply. New models rely on a light-emitting diode (LED) for low power consumption. Figure 5-10 shows the location of inside-outside switches.

The wiring diagram in Figure 5-11 shows that changing the position of either S_1 or S_2 bypasses the entry detector switch and turns the indictor lights off so that the door can be opened without activating the alarm. The system can then be rearmed by changing the position of S_1 or S_2.

OUTSIDE INSIDE

Figure 5-10. The location of inside-outside switches.

Pushbuttons for Emergencies. A manually operated pushbutton for emergency operation of the fire detection circuit should be installed near the outside (including basement) doors on the knob side at a height convenient for adults but out of the reach of small children. Such pushbuttons may also be installed in bedrooms or other areas where emergency operation of the alarm is desired.

Warning Devices. An alarm horn (see Figure 5-12) is a warning device. Surface-mounted units should be installed in a central area for greatest coverage throughout the house. The intercom can be connected so that the alarm is broadcast throughout the house when the speakers are on. Wiring to an outside bell should be hidden if possible. Wiring that is exposed should be run in a conduit, or a waterproof cable should be used. In the system illus-

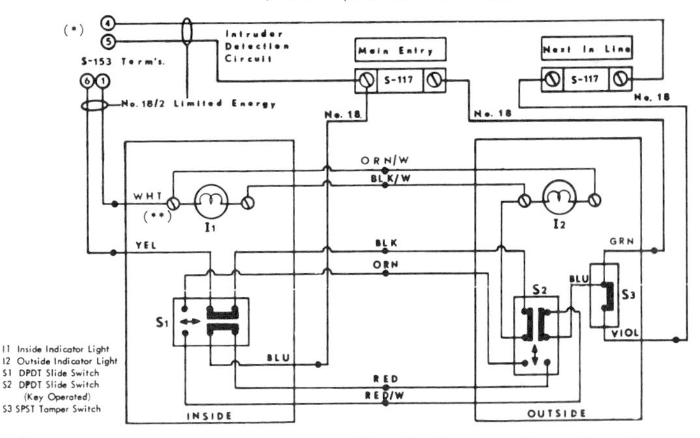

Figure 5-11. Wiring diagram showing how S_1 or S_2 changes can bypass the entry detector switch.

trated, the alarm circuit to energize the bell is approximately 15 volts DC, with normal 120-volts AC supplied to the control unit. When there is no 120-volt supply, the emergency unit (e.g., an auxiliary power/battery charger) is in operation. This voltage will be about 11 volts from the battery. Therefore, any warning device, such as a bell, has to be able to operate on a 10- to 16-volt range. (See Figure 5-2.)

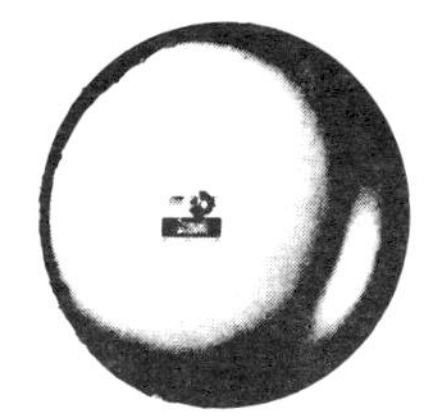

Figure 5-12. Typical alarm bell.

Auxiliary Power/Battery Charger. A complete alarm system should have an auxiliary power/battery charger so that the system can operate in situations where the normal 120-volt power supply is cut off. (See Figure 5-2.)

A TYPICAL ELECTRONIC SECURITY SYSTEM

A number of electronic security systems are available. Most have standard parts that can be programmed to operate in a number of different configurations. The installation instructions that come with the system indicate how the system is to be wired.

PARTS OF THE SYSTEM

A typical microcomputer-based system has about eight fully programmable zones, plus keypad and fire, police, and medical emergency functions. Each installation may be custom-designed to meet specific requirements. These systems are nonvolatile electrically erasable read-only memory (EEROM) chips to store all data. This memory is retained even during complete power failures.

All programming is accomplished by entering data through remote digital keypads. These keypads are available in different styles, providing flexibility in system design. Some include a speaker that may be used as an indoor siren and/or an intercom station. (See Figure 5-13.)

The control panel houses the electronics for controlling the burglary/fire system and the digital communicator. The communicator itself is also keypad-programmable and may be programmed for most transmission formats. The control panel also contains a 12-volt DC, 1.5-ampere power supply that will provide 900 milliamperes of power for auxiliary devices.

Entry Delay Lines. Each zone in the system may be programmed for two entry times (from 000 to 225 seconds duration). After a time is programmed into each delay time's memory location, each separate zone may then be programmed for entry delay time # 1 or # 2. For example, a long delay time (# 1) might be used for a garage door to allow sufficient time for car entry or exit, whereas a shorter delay time (# 2) would be programmed for front door entry.

Figure 5-13. Remote digital keypad.

Loop Response Time. The system may also be programmed for fast or slow loop response times, or how long it takes for the switch on or off condition to be acted upon by the control unit. These range from 40 milliseconds to 10 seconds in 40-millisecond increments. After the fast and slow loop response times are programmed, each separate zone may then be programmed for slow or fast loop response time.

Keyswitch Zones. If the application calls for arm/disarm control from a keyswitch, the system may be programmed for a keyswitch zone (a zone operated by using a key to turn a switch). Through keypad programming, the installer assigns one zone as a keyswitch zone. This then can be controlled by a momentary normally open or a momentary normally closed keyswitch. In some cases a key-operated switch is desired for the garage door or a gate.

Bypassing Zones. Zones may be individually bypassed (shunted) with a two-keystroke operation at the remote control keypad. All manually bypassed zones are cleared when the system is disarmed. They must then be bypassed again before the system is armed. Zones cannot be bypassed once the system is armed. For special applications, fire zones may be programmed to be shuntable. The digital communicator can also be programmed to send a bypassed-zone-arming report to the central station.

Fire Zones and Day Zones. Both fire zones and day zones sound an audible prealarm for a trouble condition. The user can easily silence a fire prealarm at the keypad by pressing the CLEAR key, while an LED will remain blinking to remind the user that a trouble conditions exists.

Alarm Outputs. A number of alarm outputs may be wired to operate two on-board auxiliary relays. The burglar alarm output can be programmed to be steady or pulsing. The auxiliary relay outputs can be connected to a self-contained siren, a two-tone siren driver, or an eight-tone siren driver.

Alarm outputs can be programmed to provide a one-second siren/bell test upon arming the system. To extend the system's capability, the alarm outputs can be wired directly to a siren driver with "trigger inputs" (such as the eight-tone driver), freeing the on-board auxiliary relays for other tasks.

A typical electronic system can also be programmed for silent or audible alarms. There are separate outputs for burglary, fire, police, and medical. Each can be individually programmed for a cutoff time of 001 to 255 minutes or for continuous operation.

A lamp output terminal is activated for 2 minutes each time a key is pressed during entry and exit delay and during any alarm condition. It can be used as light control for a hall light or other applications.

Arm/Disarm Codes. Most systems have arm/disarm codes and a master programming code that can be changed by the user. The codes can be one to five digits long and digits may be repeated. Each of the four arm/disarm codes in the system illustrated can be programmed to be:

1. An arm/disarm code only.

2. An access code only. (The system includes an access control feature that can be used for an electric door release or other device. The access control output is timed. It is programmable for 1 to 254 seconds in duration.)
3. An arm/disarm code if "1" is entered first; an access code if "0" is entered first.
4. Arm/disarm and access simultaneously.
5. Same as 4, plus just "0" and code allows access.
6. A duress code.

The fourth code (#4 above) can be programmed to function for a predetermined number of times (1 to 254). This code may be used to allow entry by housekeeping, service personnel, and the like. After the code is used the allowed number of times, it will no longer work.

Most systems also include a programmable short-arming feature that allows one- or two-digit arming. If the first digit of an arm/disarm code is "0," then one-button arming is possible—just enter "1." This feature also allows all other command keys to function simply by pressing the desired command key. However, the full code must be entered to disarm the system.

Burglary Alarms. A burglary causes an audible prealarm. An alarm causes the ARMED LED to flash. Entering the arm/disarm code silences the audible prealarm, but the flashing LED remains on. Pressing the CLEAR button resets the flashing LED. During an alarm, entry of an arm/disarm code will abort the digital communicator unless the code is a distress code. A fire or medical alarm can be silenced by pressing the CLEAR key.

Fire Alarms. Fire alarms can be silenced by entering an arm/disarm code by pressing the CLEAR key. The fire zone LED remains on for an alarm or remains blinking for a trouble condition. If a trouble condition clears itself, the FIRE LED automatically resets. The FIRE alarm LED can be cleared by entering the arm/disarm code.

When a fire alarm is silenced by pressing the CLEAR key, this does not reset the smoke detectors. After the smoke detector that caused the alarm has been identified (by viewing the alarm memory LEDs on the smoke detectors), pressing the SMOKE RESET key and the arm/disarm code interrupts power to the smoke detectors for 5 seconds, resetting them.

SMOKE RESET also performs a battery test. It removes AC power from the control panel and dynamically tests the battery. If the battery fails the test, the POWER LED will flash. A low-battery condition will be reported to the central station. The flashing POWER LED can be reset only by a subsequent test that recognizes a good battery.

INSTALLING THE SYSTEM

The panel box of a typical electronic security system is easily mounted and contains the terminal strips required to complete all of the prewiring and terminal wiring connections. (See Figure 5-14.) The terminal strips themselves are marked for identification. All input connections are made on one terminal strip. All output connections are made on the other. This arrangement allows the installer to complete the prewiring, make the terminal connections, and then safely lock the box.

The main circuit board is a separate module. It is mounted with four hex screws. The board is mounted to the terminal strips and is simply plugged in with one connector. This arrangement allows for prewiring and protects the electronics from damage during wiring.

Once the installation and wiring are complete, power is supplied by a flip of the power switch. This eliminates the need to disconnect the power wiring during installation or service.

When the system is powered up, the installer may program the system directly from digital remote control keypads. Programming worksheets (complete with zone identification, programming memory locations, and factory program values) are provided to aid the installer in programming. A detailed programming manual is also provided.

After the programming is complete, the installation can be checked out from the keypad. The installer simply uses the system status LEDs, the zone status command key, and the detector check key to verify the system installation. Then all user operations can be performed to verify the correctness of the program.

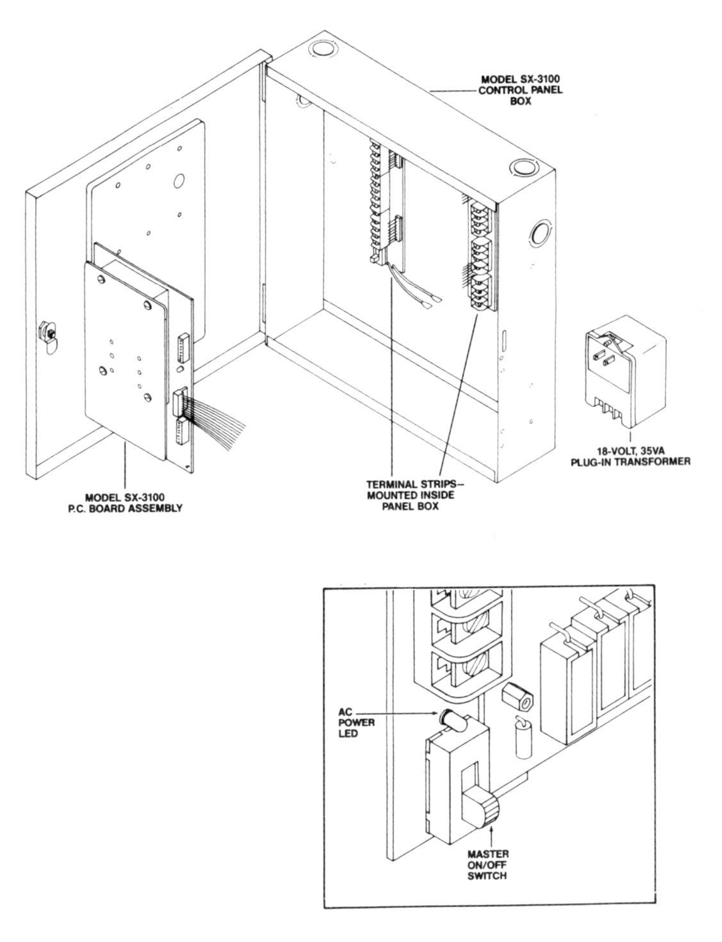

Figure 5-14. Exploded view of the panel box and master power switch location.

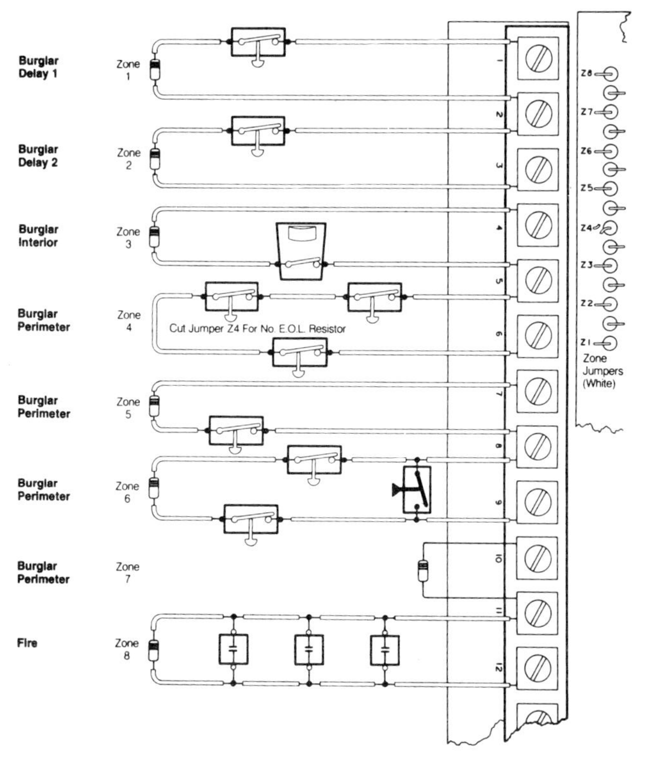

Figure 5-15. Typical eight-zone wiring.

TROUBLESHOOTING THE SYSTEM

In a typical eight-zone system, the alarm memory function automatically reduces troubleshooting by one-eighth. Simply pressing the alarm memory command key reveals which zones were last alarmed. Then, within that zone, the detector check command key may be used to troubleshoot specific detectors. For example, a typical system may contain forty-five contacts, and the keypad can be used to troubleshoot the entire installation. With a conventional one-zone system, it would be necessary to troubleshoot most of the forty-five contacts by trial and error. Figure 5-15 shows typical wiring for an eight-zone system.

LAMPS AND LIGHTING SYSTEMS

The installation, troubleshooting, and repair of lighting systems are some of the most common tasks of an electrician. Familiarity with the basic types of lamps and lighting fixtures and with systems for controlling lighting is essential for any electrician or apprentice.

The National Electric Code (NEC) classifies lamps as incandescent or electric discharge. Incandescent lamps have a filament that glows white hot. Lamps that produce light without a filament are classified as electric discharge lamps. Fluorescent lamps, mercury vapor lamps, metal halide lamps, and many other types of lamps are classified as discharge types. Electric discharge lamps operate by passing current through a gas-filled envelope. (In some cases they may have a filament to get started.)

INCANDESCENT LAMPS

An incandescent lamp has a filament made of tungsten. When the tungsten becomes white hot, or incandescent, it gives off light and

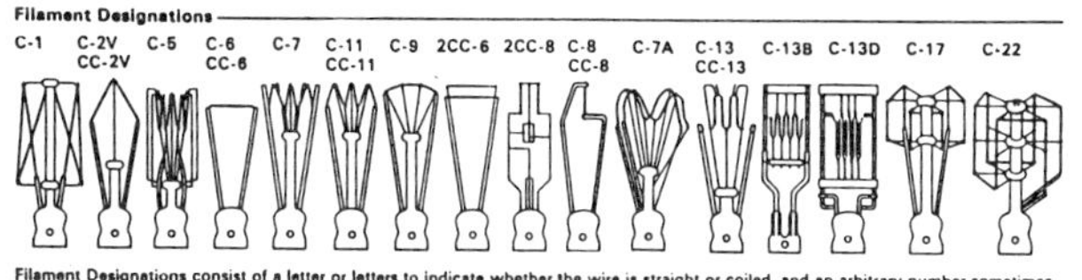

Filament Designations consist of a letter or letters to indicate whether the wire is straight or coiled, and an arbitrary number sometimes followed by a letter to indicate the arrangement of the filament on the supports. Prefix letters include: S (straight)—wire is straight or slightly corrugated; C (coil)—wire is wound into a helical coil or it may be deeply fluted; CC (coiled coil)—wire is wound into a helical coil and this coiled wire again wound into a helical coil. Some of the more commonly used types of filament arrangements are illustrated.

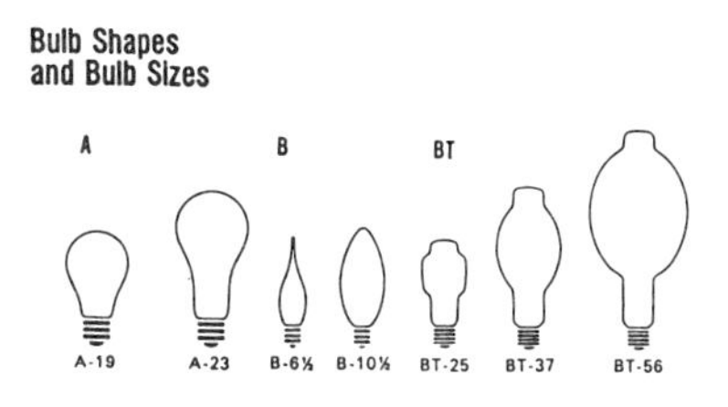

Figure 5-16. Incandescent lamp filaments.

heat. Incandescent lamp filaments are available in a number of sizes and shapes. (See Figure 5-16.)

Incandescent lamps differ in their output, bases, wattage, and life expectancy.

OUTPUT

Most incandescent lamp output is in the infrared portion of the electromagnetic spectrum. (See Figure 5-17.) This means that larger amounts of light are available with incandescent lamps than with other light sources. (See Figure 5-18.)

One type of incandescent lamp that has a very high output is the quartz bulb. This bulb has a filament that glows inside a quartz tube. It operates at extremely high temperatures. The main advantage of the quartz bulb is its fairly constant light output over its entire operating life. Quartz bulbs are used in automobile headlights and for dome lights inside a car.

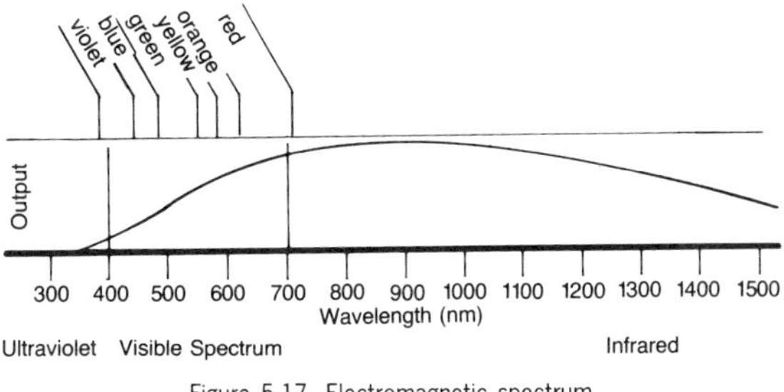

Figure 5-17. Electromagnetic spectrum.

A quartz bulb cannot be touched when hot or cold because the quartz bulb will absorb any oil on the hand, and its surface will be damaged. So remember to use gloves when handling quartz bulbs.

BASES

Most incandescent lamps for home use have a threaded base that is referred to as the *Edison base.* The threads on the base screw into matching threads on a socket. A base of the same shape but much larger in diameter is called a *mogul base.* It is usually found on bulbs that have high wattage capacity, draw high currents, and operate on 220 volts or higher. Mogul bases are also used on some mercury vapor lamps.

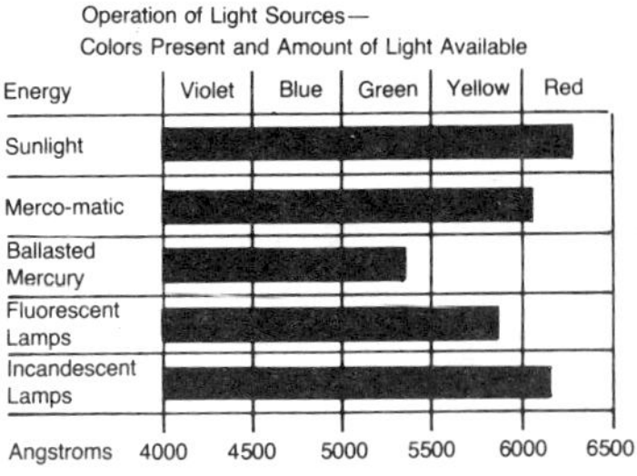

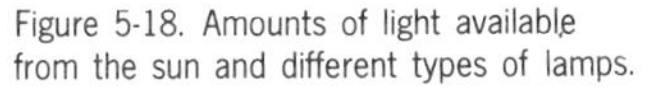
Figure 5-18. Amounts of light available from the sun and different types of lamps.

Some incandescent lamps have a bayonet base: two pins protruding on opposite sides of a smooth metal base. The socket into which bayonet-base lamps fit has a spring-metal contact at the bottom and two L-shaped channels on opposite sides of the socket wall that correlate to the pins on the lamp. The lamp is inserted by sliding the pins into the channels as far as they will go, which includes pushing against the spring-metal contact, and rotating one quarter turn. With this action, the pins then slide into the horizontal leg of the L-shaped channel, holding the lamp firmly against the spring-metal contact.

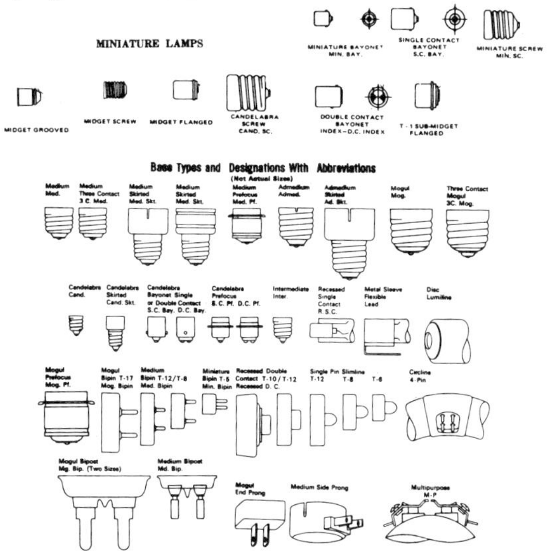

Figure 5-19. Popular incandescent lamp bases.

Figure 5-19 shows various bases, including miniature and other types of lamps. Each is designed to make good electrical contact and provide support for the bulb.

WATTAGE AND LIFE EXPECTANCY

The wattage rating of a bulb is directly related to its light output: the higher the wattage, the greater the light output. Wattage, in turn, is directly related to the voltage applied to the filament and changes with any increase or decrease in voltage. (See Table 5-1.)

TABLE 5–1
An Ordinary 120-V, 100-W Incandescent Lamp Operating at Overvoltage and Undervoltage

Voltage	Watts	Life (hours)	Lumens	Lumens/Watt
110	88	2325	1250	14.2
115	94	1312	1450	15.2
120	100	750	1690	16.9
125	106	525	1960	18.5
130	112	375	2210	19.7

An increase in voltage over the rated wattage of the bulb decreases the operating life of the bulb. Lower voltage gives longer life—but less light.

In an effort to increase the life expectancy of a bulb, some manufacturers have placed a diode in series with the filament. Some of these are washer-type diodes, designed to be inserted in the base of a lamp to provide direct current (DC) to the lamp. This causes a change in the spectrum light and a loss in brilliance. In addition, engineers report that the diodes can inject a DC component into the power supply inside the house, and this can damage computers, videocassette recorders, and other sensitive electronic equipment. Considering the loss in brilliance and change in spectrum light, the addition of a diode to try to extend the life of a bulb is not worthwhile; the diode does not provide sufficient benefits to warrant its use.

INCANDESCENT FIXTURES

Incandescent fixtures are usually mounted directly over a junction box. Be sure that there is proper support so that the weight is held in place without movement. Fixtures that weigh over 50 pounds have to be mounted by an independent box.

Incandescent fixtures can be recess-mounted, but these lamps generate a lot of heat and can cause a fire. For this reason insulation is not permitted within 3 inches of a fixture. This is to prevent the insulation from trapping the heat and increasing the operating temperature of the fixture.

The wiring used in the recess-mounted incandescent lamp fixtures must be able to withstand higher-than-normal temperatures. Branch circuit wires cannot be mounted in the recessed fixture unless their temperature rating is sufficient to take the heat.

FLUORESCENT LAMPS

A fluorescent lamp is an arc-discharging device that has no inherent resistance. Therefore, unless it is controlled, current flow will rapidly increase until the lamp burns out. This problem is solved by the use of a device called a ballast that is connected between the lamp and the power supply to limit the current to the correct value for proper lamp operation. (To "ballast" something is to stabilize or steady it; the ballast in a fluorescent lamp stabilizes, or limits, the amount of current flow.)

BALLAST

One of the most practical ways to limit current to a fluorescent lamp in an AC circuit is to use a coil or inductor. Simple inductive ballasts are basically coils of copper wire wound around iron cores. Alternating current passes through the turns of the copper wire, creating a strong magnetic field. The magnetic field reverses its polarity 120 times per second when operating on 60 hertz AC. The resulting reactance opposes a change in current flow and limits the current to the lamp. Only a few ballasts are this simple, but the basic principle of operation is the same in all ballasts. (See Figure 5-20.)

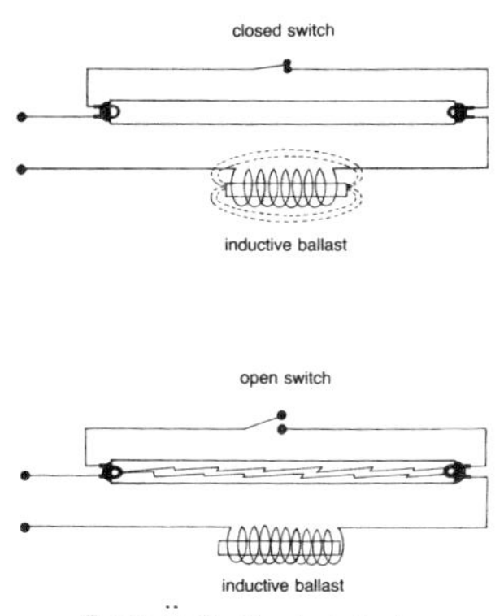

Figure 5-20. Simple ballast.

Ballast Life. The medium life expectancy of fluorescent ballasts is approximately 12 years. This means that 50% of the units installed will have failed after a period of 12 years. There are, however, several factors that affect ballast life, among them temperature, voltage, hours of use, and lamp condition.

The best way to ensure full ballast life is to apply the ballast/fixture combination properly so that ballast case temperatures do not exceed 90°C. Ballasts, like other electromagnetic devices such as motors and transformers, generate heat during operation. Ballast heat is transferred from the internal components to the ballast case through the filling medium. Once the heat reaches the ballast case, it is dissipated by conduction, convection, and radiation to the surrounding air or mounting surfaces. If these normal means of heat dissipation are not adequate, the ballast may overheat. If this occurs, coil insulation may break down, resulting in internal short circuits and ballast or capacitor failure.

Applied voltage also affects the life of a ballast. A 1% increase in voltage over the nominal design voltage will cause the ballast's operating temperature to rise 1°C. If voltage increases continue, the ballast may overheat. Some ballasts (in Class P) now have an automatic resetting thermostat or ballast protector to prevent overheating. (See Figure 5-21.)

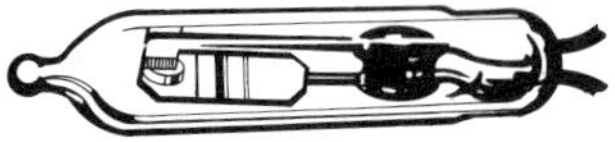

Figure 5-21. Automatic resetting thermostat ballast.

Duty cycle is also a consideration in the expected 12-year life expectancy. The 12-year expectancy is based on a duty cycle of 16 hours a day, 6 days a week, 50 weeks a year. This cycle, totaling 60,000 hours of operation, would cause half of the ballasts to wear out in 12 years. However, about 4 hours of each day's cycle are required to bring the ballast up to temperature, so the duty cycle described really totals only 12 hours a day, or about 45,000 hours, when the ballast is operating at full 90°C case temperature, during 12 years. Significant variations from the typical cycle will affect life expectancy, either increasing or decreasing it.

Failed lamps in most instant-start and preheat (switch-operated) fluorescent systems should not be allowed to remain in the fixture for extended periods of time because this too will cause ballast overheating and result in significant reduction in ballast life. Ballasts are designed to withstand abnormal conditions resulting from a failed lamp for up to about four weeks, but not for extended periods. There is, however, one type of ballast—rapid start ballast—that is not affected by failed lamps in a fixture.

Types of Ballasts. The overheating of fluorescent ballasts has caused a number of fires, and electrical codes now specify the use of thermally protected ballasts—Class P ballasts—in some situations. See NEC Section 410-73(e) for detailed information. All ballasts installed indoors are now required to have thermal protection integral to the ballast itself. Previously installed, old ballasts without thermal protection must be replaced with ballasts that have thermal protection.

Most standard indoor-type ballasts are designed to start and operate reliably in ambient temperatures of 50°F or above. For lamp-starting in outdoor waterproof fixtures or plastic signs, special low-temperature ballasts are available. Many of these will reliably start lamps in ambient temperatures as low as −20°F.

There are also special ballasts designed for flashing rapid-start lamps without reducing the lamp life, which would normally occur if the lamp were turned on and off frequently. These special ballasts maintain a constant voltage on the cathode and interrupt the arc current to produce the flashing. This helps ensure satisfactory lamp life for the flashing lamp.

Ballast Capacitor to Reduce Loss in the Circuit. Ballasts can be equipped with a capacitor to reduce loss of current in the line and make the operation of the fluorescent lamp circuit more efficient. An example will help explain this. Most standard fixtures with two 40-watt lamps draw about 800mA, or 0.8A. If you multiply 120 volts by 0.8 amperes, you get 96 VA. (This is a reactive load since the inductors represent the power-consuming part of the circuit.) The two lamps total 80 watts (40×2). Subtracting this from the 96 VA means that 16 VA is lost in the circuit. A capacitor in the ballast can help bring the phase angle back in phase and reduce line current.

Ballast Role in Maintaining High Power. High power factor means that the line current is almost in phase with the voltage. This means near unity, or 100%. A resistive load presents a 100% (or unity) power factor, whereas the inductive load presented by the fluorescent lamp ballast is normally about 0.40 to 0.60 (53°

to 66°). This can be corrected with the addition of a capacitor to within 5% (0.95 or about 18°). It is advisable to correct the power factor when a large number of fluorescent lamps are used in building. Buying high-power-factor ballasts also improves the efficiency and lowers operating costs.

Sound. The hum produced by ballasts varies with the way the ballast is mounted on or in the fixture. The sound is rated from A to F, A being the quietest. Factories can often use ballasts with an F rating, but libraries need an A-rated ballast. The problem of ballast hum has been investigated extensively. By installing ballasts in accordance with sound-control recommendations provided by the manufacturer, it is possible to predict the likelihood of audible hum and try to reduce or eliminate sound problems.

TYPES OF FLUORESCENT LAMPS

There are several types of fluorescent lamps available, among them rapid-start and instant-start cold-cathode lamps that do not require a starter and preheat lamps that do require a starter.

Preheat Fluorescent Lamp. The preheat, or switch start, circuit is the oldest one in use today. It is used with preheat or general-line lamps requiring starters and is particularly well suited for low wattage, low cost applications. Pushing the switch button causes the starting switch to close. This allows the current to pass through the ballast, the starting switch, and the cathodes at each end of the lamp. The current heats the cathodes until electrons "boil off" and form clouds around each cathode. When the push button is released, the starting switch is opened. Then a higher-voltage inductive kick, caused by the sudden collapse of the ballast's magnetic field, strikes the arc and lights the lamp. The ballast then limits, through its inductive reactance, the amount of current that flows through the circuit.

Instant-Start Cold-Cathode Fluorescent Lamp. Cold-cathode lamps do not have to be heated before starting. One type, the instant-start type, was developed in the 1940s. It requires no starter. The lamp is started by "brute force," by applying a high starting voltage between the lamp cathodes—with no preheating.

There are two types of two-lamp, instant-start circuits. One, called the series circuit, has a lower initial cost but does not provide independent lamp operation. The other type, the more expensive lead-lag circuit, provides not only independent lamp operation but also reliable starting, even in low temperatures. (See Figure 5-22.)

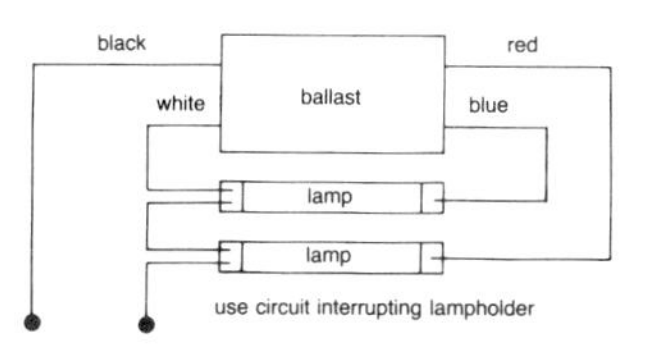

Figure 5-22. The lead-lag circuit is one type of instant-start cold cathode fluorescent lamp.

Rapid-Start Fluorescent Lamp. The rapid-start ballast circuit is the most popular in use today. It is used with rapid-start (430mA), high-output (800mA), and extra-high-output (1500mA) lamps that use cathode heating continuously during lamp operation. No starter is required. The cathodes are designed to heat quickly once the switch is turned on. Current begins to flow through the tube almost instantly. Rapid-start fluorescent lamps require a special ballast so that the lamp cathodes are quickly heated and are kept heated while they are turned on. (See Figure 5-23.)

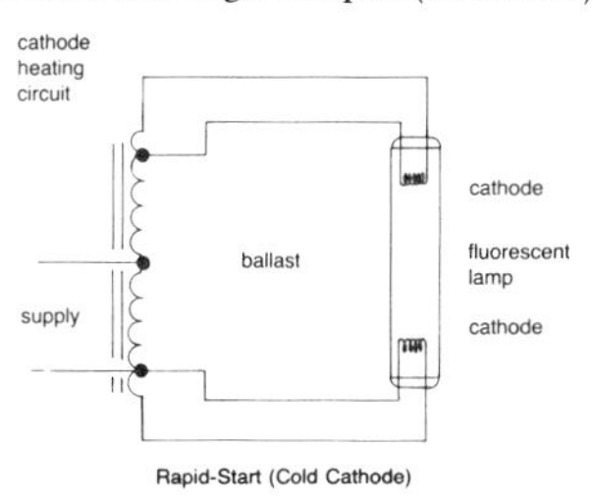

Figure 5-23. A rapid-start fluorescent circuit, the most popular type of ballast circuit in use today.

SIZE OF FLUORESCENT LAMPS

Fluorescent lamps come in a number of different diameters. Code letters and numbers stamped on the end of each lamp indicate a number of details. Let's use an example of F24T12/CW/HO to explain the code. The letter "F" indicates that the lamp is fluorescent. The number immediately following the initial letter gives the length of the lamp in inches—in this case, 24 inches. The letter "T" tells you that the lamp is tubular. The numbers following this letter give the diameter of the lamp in eighths of an inch. The tubular fluorescent lamp in our example has a diameter of 12/8, or 1 1/2 inches. The letters "CW" following the slash indicate that the light it gives off is a cool white, and the letters "HO" following the next slash indicate that it is a high-output lamp.

STARTERS

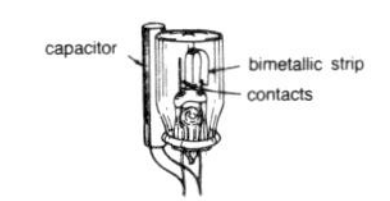

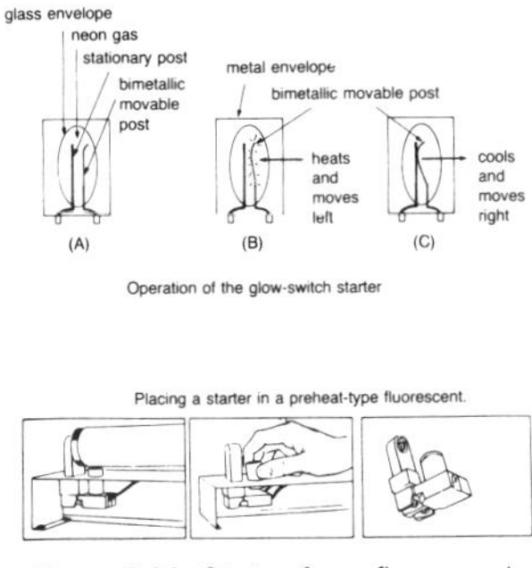

Figure 5-24. Starter for a fluorescent lamp.

A starter is needed to automatically start some preheat types of fluorescent lamps. They can be manually started if you use a switch or automatic if you use a sealed glow switch. (See Figure 5-24.) The glow switch consists of a can-covered glass envelope. Inside the envelope are a stationary post and a movable bimetallic post. Neon or argon gas under pressure is injected into the envelope and it is sealed. (If it has an orange glow it is neon.) When the lamp is turned on, the bimetallic strip begins to glow. When it glows, it heats up. As it heats up, the moveable end begins to move toward the stationary contact pole. Once the bimetallic strip makes contact with the stationary pole, current travels through the cathode at each end of the tube, causing the cathodes to heat up.

Once the bimetallic strip makes contact with the stationary pole, the bimetallic strip begins to cool. As it cools, the movable end moves away from the stationary post and the circuit is broken. By the time the circuit is broken, the gas in the lamp has ionized and started to conduct and produce light, and there is no longer a need for the starter in the circuit. It is now ready to be used again once the circuit has been turned off and then on again.

FLUORESCENT FIXTURES AND BASES

Fluorescent fixtures are usually sold by the manufacturer already wired, and they need only be connected to a source of AC power, usually 120 volts. The rating of the wire used to serve these fixtures has to be such that it can withstand 90°C (194°F), or the wire cannot be placed within 3 inches of the ballast. (Type THW wire can be used for this purpose.)

Very little heat is generated by a fluorescent fixture, but it has to be taken into consideration whenever the fixture is mounted in hazardous environments. Special types of fixtures are available for use in spray-painting booths and similar environments, and fixtures to be used in explosive atmospheres or agricultural environments must meet special requirements set by the NEC (see Section 410-4).

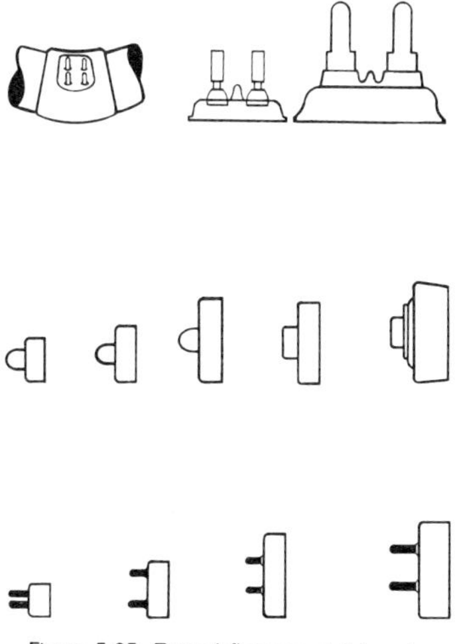

Figure 5-25. Typical fluorescent tube pin base.

Fluorescent fixtures may be flush-mounted or recessed. Fluorescent lamp fixtures that are flush-mounted to a box on the ceiling have to be arranged so that they cover the box completely. A hole in the fixture is needed to provide access to the junction box. If the hole is not located in the proper place, use a hole punch of some type to make a hole for access to the wiring. (See NEC Sections 410-14(b) and 410-12 for futher details.)

Special fixtures are made for a recessed installation. T-bars are used to fasten the fixture to the ceiling joists. In a commercial location, conduit usually feeds a junction box and flexible conduit or BX is used from the junction box to the fixture.

Fluorescent lamps that are installed in an industrial plant may be mounted to a raceway or to overhead pipes. In these cases the lamps must be cord-connected to an overhead junction box directly in the fixture.

Fluorescent tubes are available with single-pin and bi-pin bases in different sizes. The type of base must be matched to the fixture installed. (See Figure 5-25.)

TROUBLESHOOTING FLUORESCENT LAMPS

Hints for troubleshooting fluorescent lamps are given in Table 5-2 shown below.

TABLE 5–2
Troubleshooting Chart

Condition	Possible Causes to Investigate
I **Lamps won't start**	1. Lamp failure 2. Poor lamp-to-lampholder contact 3. Incorrect wiring 4. Low voltage supply 5. Dirty lamps or lamp pins 6. Defective starters* 7. Low or high lamp bulb-wall temperature 8. High humidity 9. Fixture not grounded 10. Improper ballast application 11. Ballast failure

Table 5–2 (*continued*)

Condition	Possible Causes to Investigate
II **Short lamp life**	1. Improper voltage 2. Improper wiring 3. Poor lamp-to lampholder contact 4. Extremely short duty cycles (greater than average number of lamp starts per day; check with lamp manufacturer) 5. Defective starters* 6. Defective lamps 7. Improper ballast application 8. Defective ballast
III **Lamp flicker (spiraling or swirling effect)**	1. New lamps (should be operated 100 hours for proper seasoning) 2. Defective starters* 3. Drafts on lamp bulb from air-conditioning system (lamp too cold) 4. Defective lamps 5. Improper voltage 6. Improper ballast application 7. Defective ballast
IV **Audible ballast "hum"**	1. Loose fixture louvers, panels, or other parts 2. Insecure ballast mounting 3. Improper ballast selection 4. Defective ballast
V **Very slow starting**	1. Improper voltage—too low 2. Inadequate lamp-starting-aid strip** (refer to fixture manufacturer) 3. Poor lamp-to-lampholder contact 4. Defective starter* 5. Defective lamp 6. Improper circuit wiring 7. Improper ballast application 8. High humidity

TABLE 5–2 (*continued*)

Condition	Possible Causes to Investigate
	9. Bulb-wall temperature too low or too high
VI **Excessive ballast heating (over 90° C ballast case temperature)**	1. Improper fixture design or ballast application (refer to fixture manufacturer) 2. High voltage 3. Improper wiring or installation 4. Defective ballast 5. Poor lamp maintenance (instant-start and preheat systems) 6. Wrong type of lamps 7. Wrong number of lamps
VII **Blinking**	1. Improper fixture design or ballast application (refer to fixture manufacturer) 2. High voltage 3. Improper wiring or installation 4. Defective ballast 5. Poor lamp maintenance (instant-start and preheat systems) 6. Wrong type of lamps 7. Wrong number of lamps 8. High ambient temperature

*Applies only to preheat (switch-start) circuits
**Applies only to rapid-start and trigger-start circuits

OTHER TYPES OF LAMPS

In addition to incandescent and fluorescent lamps, there are other sources of light, among them mercury vapor lamps, metal halide lamps, and sodium lamps.

MERCURY VAPOR LAMPS

Mercury vapor lamps are commonly used in shopping centers, in parking lots, in large farmyards, and along highways. A mercury vapor lamp is an arc-discharging, nonfilament device. Once the arc is struck and the unit is emitting light, some type of ballast is usually required to limit the current flow to the device. There are two general types of mercury vapor lamps: those that require the addition of a ballast some place on the fixture and those that are self-ballasted. The two types are similar in appearance, and you must be careful to screw the bulb only into a socket designed for its specific type. Figure 5-26 shows a typical screw-in mercury lamp that is not self-ballasted.

Many mercury vapor lamps mounted on poles in parking lots, in farmyards, or along the highways have a photoelectric eye that turns the fixture on at dusk and off at dawn. Often the lamp needs time to warm up to full brilliance.

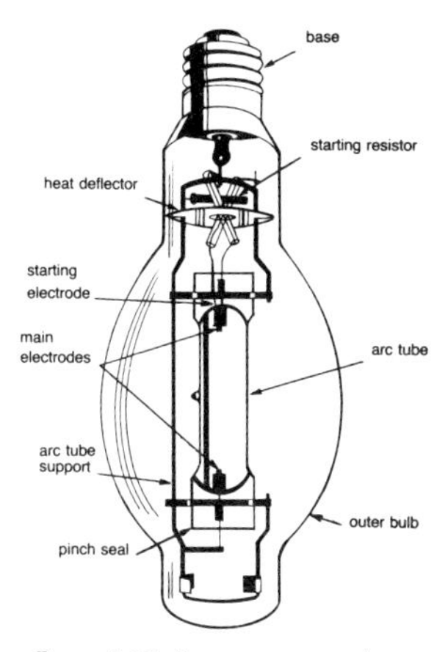

Figure 5-26. A mercury vapor lamp.

Possible problems with mercury vapor lamps include a burned-out lamp, a loose or broken wire, a bad photoelectric switch, and, if the lamp is not self-ballasted, a burned-out ballast. The more often the lamp is started, the shorter its life expectancy. Most mercury vapor lamps used on farms and in commercial parking lots last about 24,000 hours, or 7 years, when used with an automatic on-off switch for operation every night.

The best way to troubleshoot a mercury vapor lamp is with a meter. Check the bulb first to see if it is operational. Then check the on-off switch or photoelectric switch. To do this, short the two wires to the switch together. If the light comes on, the switch is not defective. Remember, however, that it may take a few minutes for the lamp to begin to glow. If the bulb is not self-ballasted, disconnect the ballast and test it with an ohmmeter for shorts and opens.

A cracked mercury vapor bulb can give off harmful radiation

and damage the eyes and skin. Most, however, are safety bulbs that will burn out almost immediately if the bulb is cracked.

METAL HALIDE LAMPS

The halide family of elements consists of mercury, sodium, thallium, indium, and iodine. The metal halide lamp contains many of these metals and puts out a better-quality light than a lamp containing only mercury. Colored objects appear their normal color under this type of lamp—another advantage. Metal halide lamps are used for sports lighting, outdoor produce market lighting, and wherever a bright light with true colors is needed.

Metal halide lamps operate with the base up or down. However, you must know which; otherwise you can ruin the lamp because the starting switch inside the lamp will not operate properly unless it is mounted correctly. Do not place a metal halide lamp in a fixture designed for a mercury vapor lamp, and be sure the metal halide lamp will operate in the fixture that you have for it. (See Figure 5-27.)

Although the metal halide lamp provides bright, true-color light there are some disadvantages associated with it. It is more expensive than the mercury vapor lamp, lasts a shorter time, and produces a steadily decreasing amount of light with age.

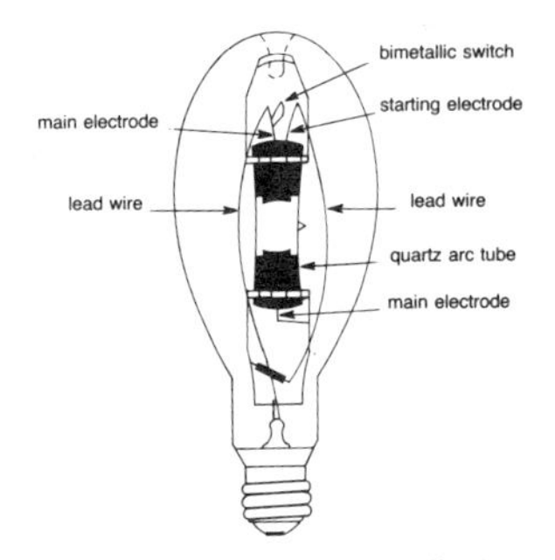

Figure 5-27. Metal halide lamp (for base-down mounting).

SODIUM LAMPS, HIGH PRESSURE

A high-pressure sodium lamp contains xenon gas, along with mercury and sodium. When the gases are vaporized by an electric arc, the glow gives a very intense light. The color of the light is somewhat yellowish-orange, but objects under it appear almost their natural color. (See Figure 5-28.)

High-pressure sodium lamps are used primarily as farm lights. They are more efficient than mercury lamps, give off about twice as much light for the same wattage, and can produce light much faster (within 2 minutes) than mercury vapor lamps. They do, however, cost more and need a special ballast.

SODIUM LAMPS, LOW PRESSURE

A low-pressure sodium lamp produces an orange color and gives objects under it an off-color cast. It can be used where high illumination is needed with little need for true color.

A low-pressure sodium lamp makes a good security lamp. Lamp life is about 18,000 hours with special ballast and fixture.

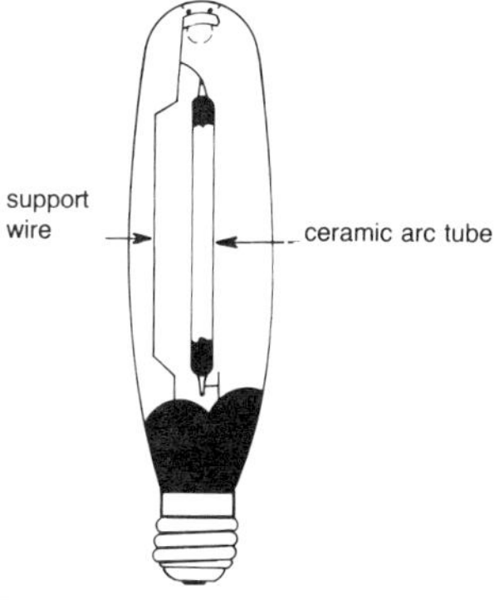

Figure 5-28. High-pressure sodium lamp.

OTHER TYPES OF LAMPS

There are many other types of lamps, including ballasted mercury vapor lamps; compact-source iodide lamps; high-pressure mercury vapor lamps; repro-lamps; black light lamps; water-cooled, super-high-pressure mercury lamps; air-cooled, super-high-pressure mercury lamps; quartz tubes; pulsed xenon lamps; super actinic lamps; spectral lamps; and neon lamps. (See Figure 5-29.)

ANALYZING A LIGHTING SYSTEM

The type of lighting system needed depends on the facility—residential, commercial, recreational, or industrial—and on the activities of that facility. Use Table 5-3 as a guideline for determining the amount of light needed for various activities. Table 5-4 provides information on the color of light produced, the length of the

TABLE 5–3
Amount of Light Needed for Various Activities

10fc	20fc	30fc	50fc	70fc	100fc	150fc	200fc*	500fc*
Concourses Freight Cars	Corridors Stairways Elevators	Rough Assembly-easy seeing Reading Good printed matter Rest Rooms Ship Fabricating	Rough Assembly-difficult seeing Rough Bench and machine work Bank Lobbies Ordinary Inspection	Reading Hand-writing-pen or soft pencil Intermittent Filing Folding Textiles	Reading Hand-writing-Hard pencil Active Filing Mail Sorting Medium Bench and Machine Work Indoor Tennis Car Repairing Fine Painting Store Merchandise Service Areas	Bookkeeping Accounting Bank Teller Stations Posting and Keypunching Reading-poor Reproductions	Detailed Drafting Store Merchandise-Self-Serve Areas Sheet Metal Scribing	Fine Bench and Machine Work Fine Foundry Inspection Store Feature Displays Jewelry and Watch Manu-facturing Fine Automatic Machines
Service Station-Entrance Drive Football Field (recreational) Baseball Outfield (recreational)	Swimming Exhibitions Loading & Freight Docks Active Storage Yard	Buildings-med. dark surfaces, bright surroundings Tennis-Tournament	Buildings-dark surfaces, bright surroundings		Football-major stadium	Major League Baseball-infield		

fc-footcandle, a unit of illuminance.
*Often obtained with a combination of general and supplementary lighting.

TABLE 5–4

Lumens, Life, and Efficacy for the Lamps.

Watts	Color	Length (feet)†	Initial Lumens	Life (hours)	Efficacy (lumens/watt)
Incandescent, standard inside frosted					
25			235	1000	9
40			480	1500	12
60			840	1000	14
75			1210	850	16
100			1670	750	17
150			2850	750	19
200			3900	750	19
300			6300	1000	21
500			10750	1000	21
1000			23100	1000	23
Incandescent PAR-38					
150 Spot			1100	2000	7
150 Flood			1350	2000	9
Quartz incandescent					
500			10550	2000	21
1000			21400	2000	21
1500			35800	2000	24
Fluorescent, rapid start					
40	CW & WW	4	3150	20000	78
40	CWX	4	2200	20000	55
60	CW & WW	4	4300	12000	41
60	CWX	4	3050	12000	50
85	CW & WW	4	2850	20000	81
85	CWX	4	2000	20000	57
Fluorescent, U-line					
40	CW & WW	2	2850	12000	71
40	CWX	2	2020	12000	50
Fluorescent, instant start					
60	CW & WW	8	5600	12000	93
60	CWX	8	4000	12000	66
75	CW & WW	8	6300	12000	84
75	CWX	8	4500	12000	60
95	CW & WW	8	8500	12000	89
95	CWX	8	6100	12000	64
110	CW & WW	8	9200	12000	83
110	CWX	8	6550	12000	59
180	CW	8	12300	10000	68
215	CW	8	14500	10000	67
215	WW	8	13600	10000	63
Mercury-vapor					
40	Deluxe		1140	24000	28
50	Deluxe		1575	24000	31
75	Deluxe		2800	24000	37
100	Deluxe		4300	24000	43
175	Deluxe		8500	24000	48
175	Clear		7900	24000	45
250	Deluxe		13000	24000	52
250	Clear		12100	24000	48
400	Deluxe		23000	24000	57

Table 5-4 (*continued*)

Watts	Color	Length (feet)†	Initial Lumens	Life (hours)	Efficacy (lumens/watt)
400	Clear		21000	24000	52
700	Deluxe		43000	24000	61
1000	Deluxe		63000	24000	63
Metal-halide					
175			14000	7500	80
250			20500	7500	82
400			34000	15000	85
1000			105000	10000	105
High-pressure sodium					
50			3300	20000	66
70			5800	20000	82
100			9500	20000	95
150			16000	24000	106
150*			12000	12000	80
200			22000	24000	110
215*			19000	12000	88
250			27500	24000	110
400			50000	24000	125
1000			140000	24000	140
Low-pressure sodium					
65			4800	18000	137
65			8000	18000	145
90			13500	18000	150
135			22500	18000	166
180			33000	18000	183

†1ft = 0.3048 mm.
*Suitable for use in a mercury-vapor fixture.

CW—cool white.
WW—warm white.
CWX—deluxe cool white.

Deluxe—produces extra red light.
Clear—produces very little red light.

lamp (in the case of fluorescent lamps), the initial lumens, the life expectancy, and the efficacy of different types of lamps. This information is essential for anyone working with lighting in any situation.

LIGHTING CONTROLS

To make maximum use of available light it is necessary to control it. Light may be controlled by turning lights on and off, either manually or through the use of time clocks or photoelectric switches, and by controlling the level of illumination in a given area with dimmers and similar devices.

DIMMERS

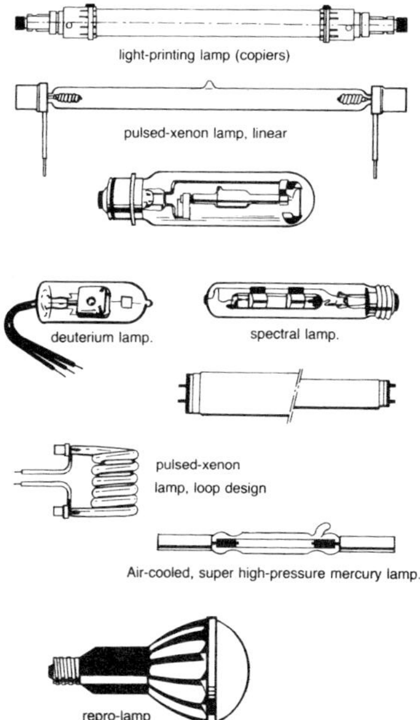

Figure 5-29. Some other types of lamps.

Solid-state circuitry dimmer switches for incandescent lighting adjust lights from bright to candle-glow without causing a flicker. In some dimmers a knob is turned to switch on the light and then rotated until the brightness level desired is reached. Turning off a light with such a dimmer involves rotating the knob back and clicking the switch off. In other types of dimmers, the knob is pushed in to turn the lights on, and then the knob is rotated to the brightness level desired. The lights are turned off by pushing in on the knob. Push the knob again to turn the lights back on at the exact brightness level that was set when it was turned off. This type of dimmer has to have a filter to prevent problems with television and radio reception within the house, especially if they are all on the same circuit.

The 1000-, 1500-, and 2000-watt models of incandescent dimmers have breakoff fins that permit close ganging of the switches, using a minimum of wall space. The 1500- and 2000-watt models have mounting holes for two-gang boxes.

Light control equipment comes in all sizes. Commercial-size units are usually transformers, and in some cases, such as stage lighting, saturable reactors are used. These reactors use a low voltage DC to control the saturation level of the core of a transformer that, in turn, decreases or increases the available current from the secondary winding of the transformer that supplies the lights.

In most cases a wall-mounted controller, mounted in a standard 4-inch wall box, is a continuously adjustable transformer. The rotating knob of the dimmer produces any light intensity from full bright to off or any level in between. It is a smooth, flickerless, stepless control process, suitable for either incandescent or 40-watt rapid-start fluorescent lamps.

Most newer models of dimmers are made with radio frequency interference suppression (RFI) to make sure that the dimmer does not interfere with the operation of a radio or television or any other electronic device on the same circuit or in the building. Triacs are made that can handle up to 4,000 amperes. The high currents are controlled or triggered by diacs that are in turn controlled by low voltage.

Motor-driven units, composed of electronic circuits with RFI suppression, permit remote control. The dimmer can be installed in an out-of-the-way spot and the control station (knob) at a convenient location.

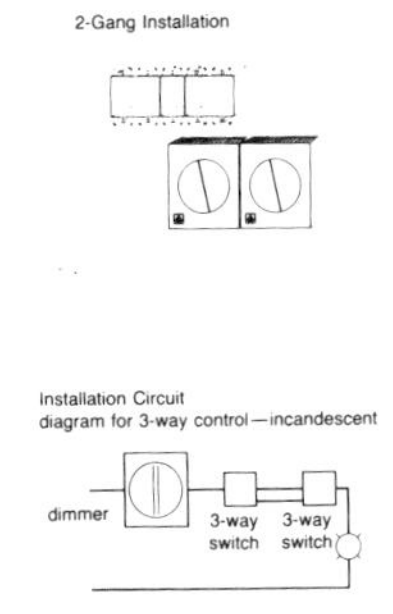

Figure 5-30. Dimmer installation for an incandescent lamp.

Dimmers are also available for use with fluorescent lamps. However, inasmuch as fluorescent lamps usually need about 100 volts to produce light, it is rare that anyone needs to dim them. The electronic dimmers for fluorescent lamps that are now available permit gradual control of the entire dimming range from full intensity to completely off.

The dimmer can be mounted in a standard double-gang sectional switch box with a depth of at least 2 1/2 inches. The dimmer comes with pigtails. When ganging dimmers horizontally, a single gang box is placed between double gang boxes. (See Figure

5-30.) Derating is not required when specifier series dimmers are ganged. Some dimmers handle from 4 to 20 40-watt lamps; others will handle from 4 to 40 40-watt lamps, or 16.66 amperes. Figure 5-31 shows the installation circuits for fluorescent dimmers.

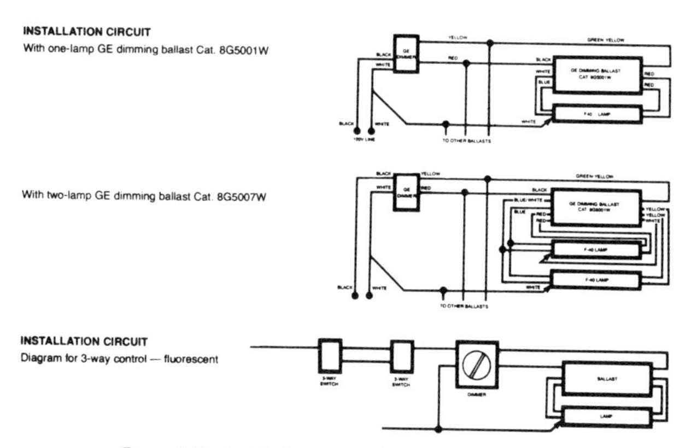

Figure 5-31. Installation circuits for fluorescent dimmers.

PHOTOELECTRIC SWITCHES

Lighting can also be controlled by using photoelectric switches that detect the available sunlight and turn lamps on and off accordingly. There are many different types of photoelectric control available—some to control indoor lighting, some to control outdoor lighting. They differ also in the sunlight level, measured in footcandles, required to trigger lamp turn-on.

TIME CLOCKS

Time clocks can also be used to turn lights on and off. Time clocks can be obtained with a number of features. Some can be set to turn lights on and off at certain times each day; others permit

TABLE 5–5[1]

Typical Characteristics of AC Motors

Motor Types	Spilt-Phase General-purpose	Split-Phase Special Service	Capacitor-Start Special Service	Capacitor-Start General-purpose	Polyphase 1 HP & Below
Starting Torque (% Full Load Torque	130%	175%	250%	350%	275%
Starting Current	Normal	High	Normal	Normal	Normal
Service Factor (% of Rated Load)	135%	100%	100%	135%	135%
Comparative Price Estimate (Based on 100% for Lowest Cost Motor)	110%	100%	135%	150%	150%

Remarks	Low starting torque. High service factor permits continuous loading—up to 35% over nameplate rating. Ideal for applications of medium starting duty.	Moderate starting torque, but has service factor of 1.0. Apply where load will not exceed nameplate rating for any extended duration of time. Because of higher starting current, use where starting is infrequent.	High starting torque but 1.0 service factor. Use only where load will not exceed nameplate rating for any extended duration of time. Starting current is normal.	Very high starting torque. High service factor permits continuous loading up to 35% over nameplate rating. Ideal for powering devices with heavy loads, such as conveyors.	Normal start current for polyphase is low compared to single-phase motors. High starting ability. High service factor permits continuous loading up to 35% over nameplate rating. Direct companion to general-purpose capacitor-start motor.

[1]Courtesy General Electric.

longer or shorter operation periods for each individual day of the week as well as the possibility of omitting any day. This feature is necessary in many commercial lighting operations such as stores and shopping centers, where working hours vary from day to day. In general, time clocks that can handle up to 40 amperes per contact in SPST, DPDT, 3PST, and SPDT configurations are available.

MOTORS, ELECTRIC

Electric motors are available in many sizes, shapes, and types for every job imaginable. Characteristics, installation procedures, uses, and troubleshooting techniques differ somewhat among these motors. However, there is some general information and basic characteristics of motors that every electrician should be thoroughly familiar with. Then, when called upon for installation, maintenance, and simple troubleshooting, the electrician will be able to use this basic information to interpret the specific details and manufacturer's instructions for the particular type of motor.

GENERAL CHARACTERISTICS AND NAMEPLATE INFORMATION

Fractional-horsepower motors may be single-phase or polyphase. These motors differ in starting torque, starting current, service factor, purpose (general or special), duty, reversibility, price, and other factors. These factors must be considered in choosing the right motor for a particular job and in maintaining and troubleshooting motors. Tables 5-5 and 5-6 list some important characteristics of fractional-horsepower AC motors.

TABLE 5–6[1]
Motor Characteristics

		Duty	Typical Reversibility	Speed Character	Typical Start Torque*
Polyphase	AC	Continuous	Rest/Rot.	Relatively Constant	175% & up
Split Phase Synch.	AC	Continuous	Rest Only	Relatively Constant	125–200%
Split Phase Nonsynchronous	AC	Continuous	Rest Only	Relatively Constant	175% & up
PSC Nonsynchronous High Slip	AC	Continuous	Rest/Rot.†	Varying	175% & up
PSC Nonsynchronous Norm. Slip	AC	Continuous	Rest/Rot.†	Relatively Constant	75–150%
PSC Reluctance Synch.	AC	Continuous	Rest/Rot.†	Constant	125–200%
PSC Hysteresis Synch.	AC	Continuous	Rest/Rot.†	Constant	125–200%
Shaded Pole	AC	Continuous	Uni-Directional	Constant	75–150%
Series	AC/DC	Int./Cont.	Uni-Directional•	Varying‡	175% & up
Permanent Magnet	DC	Continuous	Rest/Rot.§	Adjustable	175% & up
Shunt	DC	Continuous	Rest/Rot.	Adjustable	125–200%
Compound	DC	Continuous	Rest/Rot.	Adjustable	175% & up
Shell Arm	DC	Continuous	Rest/Rot.	Adjustable	175% & up
Printed Circuit	DC	Continuous	Rest/Rot.	Adjustable	175% & up
Brushless D-C	DC	Continuous	Rest/Rot.	Adjustable	75–150%
D-C Stepper	DC	Continuous	Rest/Rot.	Adjustable	■

[1]Courtesy of Bodine Electric Co.
* Percentages are relative to full-load rated torque. Categorizations are general and apply to small motors.
† Reversible while rotating under favorable conditions: generally when inertia of the driven load is not excessive.
• Usually unidirectional—can be manufactured bidirectional.
‡ Can be adjusted but varies with load.
§ Reversible down to 0°C after passing through rest.
■ Dependent upon load inertia and electronic driving circuitry.

NAMEPLATE READINGS

Each motor has a nameplate on which the manufacturer provides information to aid in motor selection, operation, maintenance, and troubleshooting. Although each manufacturer arranges the information in a different way, all provide the same data: the horsepower of the motor; the power required to operate the motor and the frequency of that power as well as its voltage and current; the frame, serial number, and model number of the motor; the capacitor to be used (if needed); and the temperature required for normal operation.

Operating According to Nameplate Directions

1. Do not operate motors at other than ±10% of the nameplate voltage.
2. Do not operate motors on nominal power source frequencies other than that indicated on the nameplate.
3. Do not overload the motor in excess of nameplate output rating.
4. Do not exceed the temperature of the nameplate insulation class.
5. Do not change the value of capacitance indiscriminately.
6. Do not subject the motor to duty cycles for which it was not designed.

Motor performance is adversely affected if it is operated under conditions that deviate from nameplate specifications. Table 5-7 lists some performance parameters adversely affected.

FACTORS TO CONSIDER IN SELECTING, OPERATING, AND MAINTAINING A MOTOR

Among the factors to consider in selecting a motor are the purpose to which the motor is to be used, whether some overload capacity is desired in the motor, the system voltage available, the power output needed, and the ease of reversibility in the motor. The electrician and apprentice must be familiar with all of these concepts to recognize the different types of motors available, understand nameplate data, and work efficiently on motors.

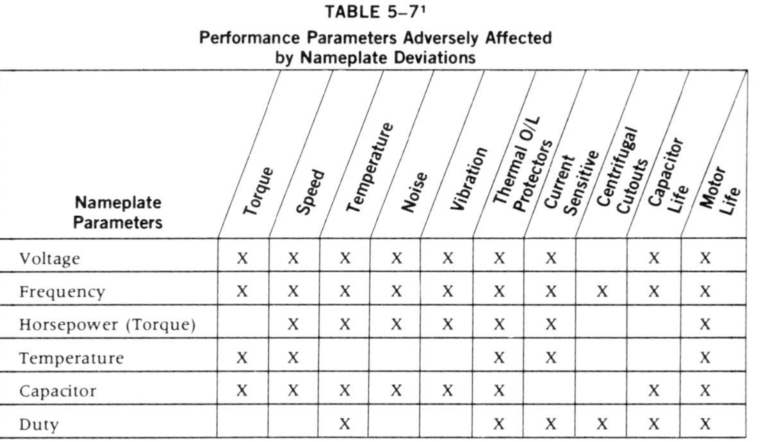

TABLE 5–7[1]

Performance Parameters Adversely Affected by Nameplate Deviations

Nameplate Parameters	Torque	Speed	Temperature	Noise	Vibration	Thermal O/L Protectors	Current Sensitive	Centrifugal Cutouts	Capacitor Life	Motor Life
Voltage	X	X	X	X	X	X	X		X	X
Frequency	X	X	X	X	X	X	X	X	X	X
Horsepower (Torque)		X	X	X	X	X	X			X
Temperature	X	X				X	X			X
Capacitor	X	X	X	X	X	X			X	X
Duty			X			X	X	X	X	X

[1]Courtesy Bodine Electric Co.

PURPOSE

Single-phase AC motors are available in two types: general purpose and special purpose. General purpose motors can be used in many and widely varying situations. The choice of which motor to use depends on the particular situation.

Special-purpose motors include heavy-duty motors designed especially for severe farm duty. These are capacitor-start induction-run motors that furnish high starting torque with normal current. These motors are gasketed throughout for protection from environmental hazards and have double-sealed ball bearings with a water flinger on the shaft end to protect the motor and bearings from contaminants. An oversize conduit box makes wiring these types of motors easy. Grounding provision is usually included. Low-temperature thermal-overload protectors can also be used, but a manual-reset overload button (with a rubber weather boot) is standard on these motors, so there is maximum operator safety. Table 5-8 provides power ratings and other information about farm-duty motors.

TABLE 5–8[1]

Farm-Duty Motors (Single-Phase)

HP	Speed (rpm)	Volts	NEMA Frame	Bearings	Therm. Prot.	Full-Load Amps
1/4	1725	115	48	Ball	None	4.5
		115	48	Ball	Auto	4.5
	1140	115	56	Ball	None	5.3
		115	56	Ball	Auto	5.3
1/3	1725	115	48	Ball	None	5.3
		115	48	Ball	Auto	5.3
		230	48	Ball	None	2.7
		230	48	Ball	Auto	2.7
	1140	115	56	Ball	None	7.0
1/2	1725	115	56	Ball	None	8.0

[1]Courtesy General Electric.

TABLE 5-9[1]
Service Factors

Horsepower	Service Factor Synchronous Speed			
	3600	1800	1200	900
1/20, 1/12, 1/8	1.40	1.40	1.40	1.40
1/6, 1/4, 1/3	1.35	1.35	1.35	1.35
1/2	1.25	1.25	1.25	1.15
3/4	1.25	1.25	1.15	1.15
1	1.25	1.15	1.15	1.15
1 1/2 and up	1.15	1.15	1.15	1.15

[1]Courtesy General Electric.

SERVICE FACTOR

Service factor (SF) is a measure of the overload capacity designed into a motor. It varies for motors of different horsepower and for different speeds. A service factor rating of 1.0 means that the motor should not be overloaded beyond its rated horsepower. A 1.15 SF means that the motor can deliver 15% more than the rated horsepower without injurious overheating. Motors used on farms should have a service factor of at least 1.35 in order to take the abuse (overload) they often receive under various work conditions. Table 5-9 gives standard NEMA service factors for various horsepower motors and motor speeds.

POWER SUPPLY

The system voltage must be known in order to select the proper motor or check its operation. Usually the motor nameplate lists a

TABLE 5–10[1]

Power System Voltage Standards

Polyphase 60 Hertz	
Nominal Power System Volts	Motor Nameplate Volts
208	200
240	230
480	440
600	575
Single-Phase 60 Hertz	
Nominal Power System Volts	Motor Nameplate Volts
120	115
240	230

[1]Courtesy General Electric.

voltage less than the nominal power system voltage. For example, the 120-volt line usually delivers between 110 and 120 volts depending on the location. The motor nameplate will usually list 115 volts, which is halfway between 110 and 120. A 240-volt system will usually have 230 volts on the nameplate. A joint committee of the Edison Electric Institute and the National Electrical Manufacturer's Association (NEMA) has recommended standards for both power system voltage and motor nameplate voltage. (See Table 5-10.)

POWER OUTPUT

In some cases it is necessary to compare the power output of various motors. Power output can be expressed in horsepower (English, or customary, measurement system) or in watts (SI system). To make comparisons the electrician may need to convert from one system of measurement to another. In converting, remember:

1 horsepower	= 746 watts
1000 watts	= 1 kilowatt
746 watts	= 0.746 kilowatts
1 millihorsepower	= 0.001 horsepower

Table 5-11 lists the horsepower, millihorsepower, and watt equivalents for common sizes of motors. Table 5-12 provides horsepower-kilowatt equivalents. Table 5-13 provides horsepower-watt equivalents for different torques.

TABLE 5–11[1]
Motor Power Output Comparison

Watts Output	*MHP	**HP
.746	1	1/1000
1.492	2	1/500
2.94	4	1/250
4.48	6	1/170
5.97	8	1/125
7.46	10	1/100
9.33	12.5	1/80
10.68	14.3	1/70
11.19	15	1/65
11.94	16	1/60
14.92	20	1/50
18.65	25	1/40
22.38	30	1/35
24.90	33	1/30
26.11	35	
29.80	40	1/25
37.30	50	1/20
49.70		1/15
60.17		1/12
74.60		1/10

[1]Courtesy General Electric.
*Millihorsepower
**Fractional h.p.

Note: Watts output is the driving force of the motor as calculated by the formula: $\frac{TN}{112.7} = WO$
where T = torque (oz. ft.), N = speed (rpm).

REVERSIBILITY

AC motors are not easily reversed. An AC induction motor will not always reverse while running. It may continue to run in the same direction but at a reduced efficiency. Most motors that are classified as reversible while running will reverse with a non-inertial-type load. They may not reverse if they are under no-load conditions or if they have a light load or an inertial load. A permanent-split capacitor motor that has insufficient torque to reverse a given load may just continue to run in the same direction.

TABLE 5–12[1]
Metric Motor Ratings

Horsepower	Kilowatt (kW)
1/20	0.025
1/16	0.05
1/8	0.1
1/6	0.14
1/4	0.2
1/3	0.28
1/2	0.4
1	0.8
1 1/2	1.1
2	1.6
3	2.5
5	4.0
7.5	5.6
10	8.0

[1]Courtesy Bodine Electric Co.

One of the problems related to reversing a motor while it is running is the damage done to the transmission system connected to the load or to the load itself. One of the ways to avoid this is to make sure the right motor is connected to a load.

To reverse a squirrel-cage three-phase motor it is necessary to interchange only two leads—any two leads. On single-phase motors the connections between the power source and the run winding and start winding have to be reversed.

Wiring Diagrams for Reversing of Motors. To reverse electric motors you need to know how the windings are wired in relation to one another and the color code of the leads that are brought out to the terminals for connection to a power source. Figure 5-32 shows the wiring schematics and how to reverse the direction of rotation of the various types of motors.

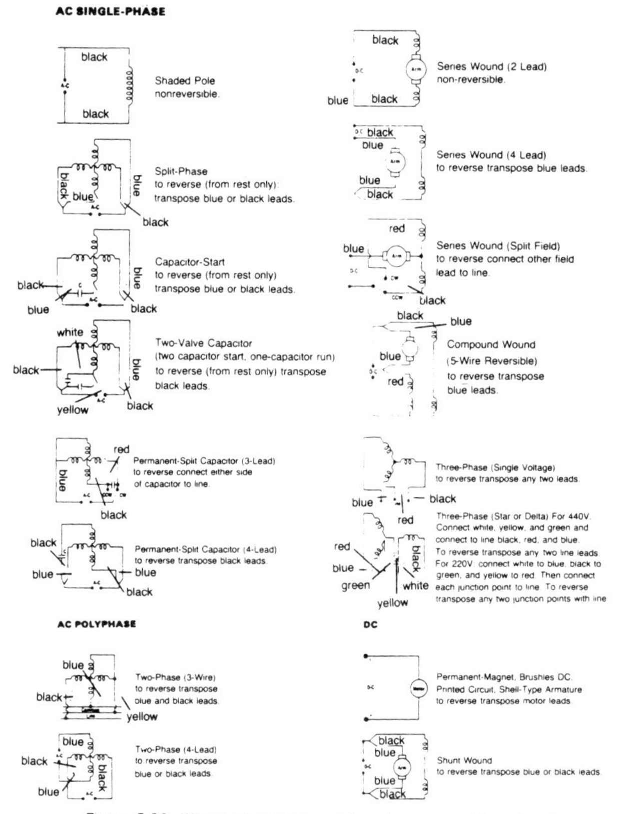

Figure 5-32. Wiring schematics and how to reverse different motors.

TABLE 5–13

Horsepower/Watts vs. Torque Conversion Chart

		@ 1125 r/min		@ 1200 r/min		@ 1425 r/min	
hp	watts	Oz.-in.	mN-m	Oz.-in.	mN-m	Oz.-in.	mN-m
1/2000	0.373	0.4482	3.1649	0.4202	2.9670	0.3538	2.4986
1/1500	0.497	0.5976	4.2198	0.5602	3.9561	0.4718	3.3314
1/1000	0.746	0.8964	6.3297	0.8403	5.9341	0.7077	4.9971
1/750	0.994	1.1951	8.4396	1.1205	7.9121	0.9435	6.6628
1/500	1.49	1.7927	12.6594	1.6807	11.8682	1.4153	9.9943
1/200	3.73	4.4818	31.6485	4.2017	29.6705	3.5383	24.9857
1/150	4.97	5.9757	42.1980	5.6023	39.5606	4.7177	33.3142
1/100	7.46	8.9636	63.2970	8.4034	59.3409	7.0765	49.9713
1/75	9.94	11.9515	84.3960	11.2045	79.1212	9.4354	66.6284
1/70	10.70	12.8052	90.4243	12.0048	84.7727	10.1093	71.3876
1/60	12.40	14.9393	105.4950	14.0056	98.9015	11.7942	83.2855
1/50	14.90	17.9272	126.5940	16.8068	118.6818	14.1531	99.9426
1/40	18.60	22.4090	158.2425	21.0085	148.3523	17.6913	124.9283
1/30	24.90	29.8787	210.9899	28.0113	197.8031	23.5884	166.5710
1/25	29.80	35.8544	253.1879	33.6135	237.3637	28.3061	199.8852
1/20	37.30	44.8180	316.4849	42.0169	296.7046	35.3827	249.8565
1/15	49.70	59.7574	421.9799	56.0225	395.6061	47.1769	333.1420
1/12	62.10	74.6967	527.4748	70.0282	494.5077	58.9711	416.4275
1/10	74.6	89.6361	632.9698	84.0338	593.4092	70.7653	499.7130
1/8	93.2	112.0451	791.2123	105.0423	741.7615	88.4566	624.6413
1/6	124.0	149.3934	1054.9497	140.0563	989.0153	117.9422	832.8550

1/4	186.0	224.0902	1582.4245	210.0845	1483.5230	176.9133	1249.2825
1/3	249.0	298.7869	2109.8994	280.1127	1978.0307	235.8844	1665.7101

		@ 1500 r/min		@ 1725 r/min		@ 1800 r/min	
hp	watts	Oz.-in.	mN-m	Oz.-in.	mN-m	Oz.-in.	mN-m
1/2000	0.373	0.3361	2.3736	0.2923	2.0640	0.2801	1.9780
1/1500	0.497	0.4482	3.1648	0.3897	2.7520	0.3735	2.6374
1/1000	0.746	0.6723	4.7473	0.5846	4.1281	0.5602	3.9561
1/750	0.994	0.8964	6.3297	0.7794	5.5041	0.7470	5.2747
1/500	1.490	1.3445	9.4945	1.1692	8.2561	1.1205	7.9121
1/200	3.730	3.3614	23.7364	2.9229	20.6403	2.8011	19.7803
1/150	4.97	4.4818	31.6485	3.8972	27.5204	3.7348	26.3737
1/100	7.46	6.7227	47.4727	5.8458	41.2806	5.6023	39.5606
1/75	9.94	8.9636	63.2970	7.7944	55.0409	7.4697	52.7475
1/70	10.70	9.6039	67.8182	8.3512	58.9723	8.0032	56.5152
1/60	12.40	11.2045	79.1212	9.7431	68.8011	9.3371	65.9344
1/50	14.90	13.4454	94.9455	11.6917	82.5613	11.2045	79.1212
1/40	18.60	16.8068	118.6818	14.6146	103.2016	14.0056	98.9015
1/30	24.90	22.4090	158.2425	19.4861	137.6021	18.6742	131.8687
1/25	29.80	26.8908	185.8909	23.3833	165.1226	22.4090	158.2425
1/20	37.3	33.6135	237.3637	29.2292	206.4032	28.0113	197.8031
1/15	49.7	44.8180	316.4849	38.9722	275.2043	37.3484	263.7374
1/12	62.1	56.0225	395.6061	48.7153	344.0053	46.6854	329.6718
1/10	74.6	67.2270	474.7274	58.4583	412.8064	56.0025	395.6061
1/8	93.2	84.0338	593.4092	73.0729	516.0080	70.0282	494.5077
1/6	124.0	112.0451	791.2123	97.4305	688.0107	93.3709	659.3436

TABLE 5–13 (continued)

1/4	186.0	168.0676	1186.8184	146.1458	1032.0160	140.0563	989.0153
1/3	249.0	224.0902	1582.4245	194.8610	1376.0213	186.7418	1318.6871
		@ 3000 r/min		@ 3450 r/min		@ 3600 r/min	
hp	watts	Oz.-in.	mN-m	Oz.-in.	mN-m	Oz.-in.	mN-m
1/2000	0.373	0.1681	1.1868	0.1461	1.0320	0.1401	0.9890
1/1500	0.497	0.2241	1.5824	0.1949	1.3760	0.1867	1.3187
1/1000	0.746	0.3361	2.3736	0.2923	2.0640	0.2801	1.9780
1/750	0.994	0.4482	3.1648	0.3897	2.7520	0.3735	2.6374
1/500	1.490	0.6723	4.7473	0.5846	4.1281	0.5602	3.9561
1/200	3.730	1.6807	11.8682	1.4615	10.3202	1.4006	9.8902
1/150	4.97	2.2409	15.8242	1.9486	13.7602	1.8674	13.1869
1/100	7.46	3.3614	23.7364	2.9229	20.6403	2.8011	19.7803
1/75	9.94	4.4818	31.6485	3.8972	27.5204	3.7348	26.3737
1/70	10.70	4.8019	33.9091	4.1756	29.4862	4.0016	28.2576
1/60	12.40	5.6023	39.5606	4.8715	34.4005	4.6685	32.9672
1/50	14.90	6.7227	47.4727	5.8458	41.2806	5.6023	39.5606
1/40	18.6	8.4034	59.3409	7.3073	51.6008	7.0028	49.4508
1/30	24.9	11.2045	79.1212	9.7431	68.8011	9.3371	65.9344
1/25	29.8	13.4454	94.9455	11.6917	82.5613	11.2045	79.1212
1/20	37.3	16.8068	118.6818	14.6146	103.2016	14.0056	98.9015
1/15	49.7	22.4090	158.2425	19.4861	137.6021	18.6742	131.8687
1/12	62.1	28.0113	197.8031	24.3576	172.0027	23.3427	164.8359
1/10	74.6	33.6135	237.3637	29.2292	206.4032	28.0113	197.8031
1/8	93.2	42.0169	296.7046	36.5364	258.0040	35.0141	247.2538
1/6	124.0	56.0225	395.6061	48.7153	344.0053	46.6854	329.6718
1/4	186.0	84.0338	593.4092	73.0729	516.0080	70.0282	494.5077
1/3	249.0	112.0451	791.2123	97.4305	688.0107	93.3709	659.3436

		@ 5000 r/min		@ 7500 r/min		@ 10,000 r/min	
hp	watts	Oz.-in.	mN-m	Oz.-in.	mN-m	Oz.-in.	mN-m
1/2000	0.373	0.1008	0.7121	0.0672	0.4747	0.0504	0.3560
1/1500	0.497	0.1345	0.9495	0.0896	0.6330	0.0672	0.4747
1/1000	0.746	0.2017	1.4242	0.1345	0.9495	0.1008	0.7121
1/750	0.994	0.2689	1.8989	0.1793	1.2659	0.1345	0.9495
1/500	1.490	0.4034	2.8484	0.2689	1.8989	0.2017	1.4242
1/200	3.730	1.0084	7.1209	0.6723	4.7473	0.5042	3.5605
1/150	4.97	1.3445	9.4945	0.8964	6.3297	0.6723	4.7473
1/100	7.46	2.0168	14.2418	1.3445	9.4945	1.0084	7.1209
1/75	9.94	2.6891	18.9891	1.7927	12.6594	1.3445	9.4945
1/70	10.70	2.8812	20.3455	1.9208	13.5636	1.4406	10.1727
1/60	12.40	3.3614	23.7364	2.2409	15.8242	1.6807	11.8682
1/50	14.90	4.0336	28.4836	2.6891	18.9891	2.0168	14.2418
1/40	18.60	5.0420	35.6046	3.3614	23.7364	2.5210	17.8023
1/30	24.90	6.7227	47.4727	4.4818	31.6485	3.3614	23.7364
1/25	29.80	8.0672	56.9673	5.3782	37.9782	4.0336	28.4836
1/20	37.30	10.0841	71.2091	6.7227	47.4727	5.0420	35.6046
1/15	49.70	13.4454	94.9455	8.9636	63.2970	6.7227	47.4727
1/12	62.10	16.8068	118.6818	11.2045	79.1212	8.4034	59.3409
1/10	74.6	20.1681	142.4182	13.4454	94.9455	10.0841	71.2091
1/8	93.2	25.2101	178.0228	16.8068	118.6818	12.6051	89.0114
1/6	124.0	33.6135	237.3637	22.4090	158.2425	16.8068	118.6818
1/4	186.0	50.4203	356.0455	33.6135	237.3637	25.2101	178.0228
1/3	249.0	67.2270	474.7274	44.8180	316.4849	33.6135	237.3637

[1]Courtesy Bodine Electric Co.

CAPACITOR-START MOTORS

The capacitor-start motor is a modified form of the split-phase motor. It has a capacitor in series with the start winding. It uses a centrifugal switch to remove the start winding after it has come up to about 75% of rated speed. One advantage of the capacitor-start motor is its ability to start under load and develop high starting torque.

The capacitor-start motor can be reversed at rest or while rotating. The speed is relatively constant, while the starting torque is 75 to 150% of rated torque. The starting current is normal.

Duty cycle for motor-starting capacitors is rated on the basis of 20 3-second periods per hour. Sixty 1-second periods per hour should be one equivalent duty cycle. Table 5-14 lists the ratings and test limits for AC electrolytic capcitors.

WARNING

When you replace a defective capacitor, it is imperative that the new capacitor be of the same voltage and microfarad rating.

MOTOR ENCLOSURES

Enclosure is the term used to describe the motor housing, shell, or case. There are several common types of enclosures.

Drip-proof (DP) Enclosures. Drip-proof enclosures are usually used indoors in fairly clean, dry locations. The ventilation openings in the end shields or bells and in the housing or shell are placed so that drops of liquid falling within an angle of 15° from the vertical will not affect performance.

Explosion-proof (EXP-PRF) Enclosures. An EXP-PRF enclosure is designed to withstand an internal explosion of specified

TABLE 5–14[1]

Ratings and Test Limits for AC Electrolytic Capacitors

Capacity Rating, Microfarads			110-Volt Ratings		125-Volt Ratings		220-Volt Ratings	
Nominal	Limits	Average	Amps. at Rated Voltage, 60 Hz	Approx. Max. Watts	Amps. at Rated Voltage, 60 Hz	Approx. Max. Watts	Amps. at Rated Voltage, 60 Hz	Approx. Max. Watts
	25– 30	27.5	1.04– 1.24	10.9	1.18– 1.41	14.1	2.07–2.49	43.8
	32– 36	34	1.33– 1.49	13.1	1.51– 1.70	17	2.65–2.99	52.6
	38– 42	40	1.56– 1.74	15.3	1.79– 1.98	19.8	3.15–3.48	61.2
	43– 48	45.5	1.78– 1.99	17.5	2.03– 2.26	22.6	3.57–3.98	70
50	53– 60	56.5	2.20– 2.49	21.9	2.50– 2.83	28.3	4.40–4.98	87.6
60	64– 72	68	2.65– 2.99	26.3	3.02– 3.39	33.9	5.31–5.97	118.2
65	70– 78	74	2.90– 3.23	28.4	3.30– 3.68	36.8	5.81–6.47	128.1
70	75– 84	79.5	3.11– 3.48	30.6	3.53– 3.96	39.6	6.22–6.97	138
80	86– 96	91	3.57– 3.98	35	4.05– 4.52	45.2	7.13–7.96	157.6
90	97–107	102	4.02– 4.44	39.1	4.57– 5.04	50.4	8.05–8.87	175.6
100	108–120	114	4.48– 4.98	43.8	5.09– 5.65	56.5	8.96–9.95	197
115	124–138	131	5.14– 5.72	50.3	5.84– 6.50	65		
135	145–162	154	6.01– 6.72	62.8	6.83– 7.63	85.8		
150	161–180	170	6.68– 7.46	69.8	7.59– 8.48	95.4		
175	189–210	200	7.84– 8.71	81.4	8.91– 9.90	111.4		
180	194–216	205	8.05– 8.96	83.8	9.14–10.18	114.5		
200	216–240	228	8.96– 9.95	93	10.18–11.31	127.2		
215	233–260	247	9.66–10.78	106.7	10.98–12.25	145.5		
225	243–270	257	10.08–11.20	110.9	11.45–12.72	151		
250	270–300	285	11.20–12.44	123.2	12.72–14.14	167.9		
300	324–360	342	13.44–14.93	147.8	15.27–16.96	201.4		
315	340–380	360	14.10–15.76	156				
350	378–420	399	15.68–17.42	172.5				
400	430–480	455	17.83–19.91	197.1				

[1]Courtesy General Electric.

gases or vapors and not allow the internal flame or explosion to escape. If the motor is below 1/3 horsepower, the enclosure may be nonventilated (EPNV); if the motor is larger, a fan-cooled enclosure (EPFC) is usually used.

Open (OP) Enclosure. Open enclosures are used indoors in fairly clean locations. Ventilation openings in end shields and/or in the shell permit the passage of cooling air over and around the windings. The location of the openings is not restricted.

Totally Enclosed (TE). In a totally enclosed housing, there are no openings in the motor housing, but it is not airtight. This type of housing is used in locations that are dirty, oily, and the like. There are two types of totally enclosed housings: *fan-cooled* (TEFC), which have a fan to blow cooling air over the motor, and *non-ventilated* (TENV), which are not equipped with a fan and depend on convection air for cooling.

POWER SOURCES

Most electrical power is generated commercially as three-phase. It is distributed from the generating plant to the substation near the community that uses it as three-phase. Then it is taken from the substation as single-phase and distributed locally to homes and farms. It is generated at 13,800 volts at the generating plant and then stepped up from 138,000 to as high as 750,000 volts to be distributed long distances. (See Figure 5-33.)

THREE-PHASE (3ϕ) Power

Three-phase power is available from electrical utilities. It can be brought to the farm, office, school, or industrial plant and used as three-phase, or it can be delivered as such and then broken down into single-phase for the equipment located on the premises. Three-phase power is expensive to install inasmuch as it uses different transformers and more wires than single-phase. Most farms use three-phase only if the farm is located near a school or factory with that service. They are able to use single-phase since they don't usually have motors rated over 5 horsepower.

Polyphase generators—3-phase generators—are sometimes employed by hospitals and industries for emergency power. These generators are large units pulled by engines generating up to 750 horsepower. They produce voltages from three different coils that are displaced 120° electrically. The output of the sine wave is like that shown in Figure 5-34. The three windings are placed on the armature 120° apart. As the armature is rotated, the outputs of the three windings are equal but out of phase by 120°. Three-phase windings are usually connected in delta or wye configuration. Each of these connections has definite electrical characteristics from which the designation "delta" and "wye" are derived. Figure 5-35 shows how the currents flow in the windings of a delta- and a wye-connected coil.

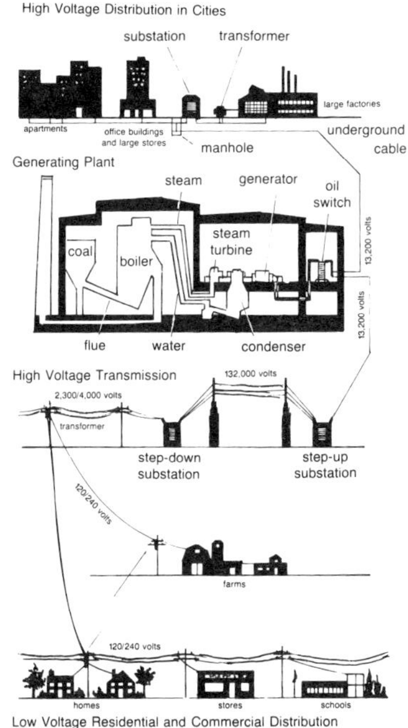

Figure 5-33. Generation and distribution of electricity.

VOLTAGES AND CURRENTS

To troubleshoot this type of service you have to understand some of the workings of the system.

In a balanced circuit, when the generators are connected in delta, the voltage between any two lines is equal to that of a single phase. The line voltage and voltage across any windings are in phase, but the line current is 30° or 150° out of phase with the current in any of the other windings.

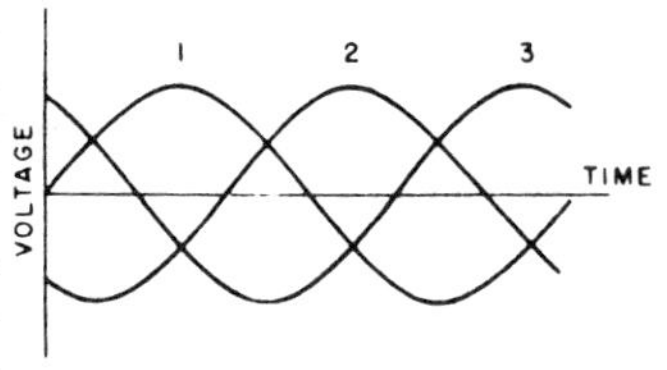

Figure 5-34. Output of the sine wave.

(Figure 5-35). In the *delta*-connected generator, the *line current* from any one of the windings is found by multiplying the phase current by $\sqrt{3}$, or 1.73.

In the *wye* connection, the current in the line is in phase with the current in the winding. The voltage between any two lines is not equal to the voltage of a single phase, but it is equal to the vector sum of the two windings between the lines. The current in line A of Figure 5-35, for instance, is

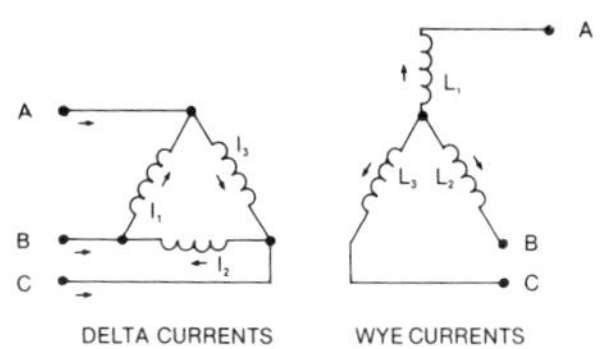

Figure 5-35. Current in delta-connected and wye-connected circuits.

current flowing through the winding L_1; that in line B is the current flowing through the winding L_2; and the current flowing in line C is that of the winding L_3. Therefore, the current in any line is in phase with the current in the winding that it feeds. Since the line voltage is the vector sum of the voltages across any two coils, the line voltage E_L and the voltage across the windings $E\phi$ are 30° out of phase. The line voltage may be found by multiplying the voltage of any winding by the individual voltage across any winding by 1.73.

TRANSFORMERS

A transformer is used to step up or step down voltage. The three-phase transformer is used when large power outputs are required. Either a single transformer or three separate transformers may be used, generally connected in delta or wye. Commercial three-phase voltage from power lines is usually 208 volts. The standard values of single-phase voltage can be supplied for the line shown in Figure 5-36. This is a wye-connected transformer. The various types or combinations of three-phase transformer connections are shown in Figure 5-37. The main reason for selecting one type over another is the need for various voltages and currents. Delta has an advantage of greater current output. Figure 5-38 shows the voltages available from the two types of transformers.

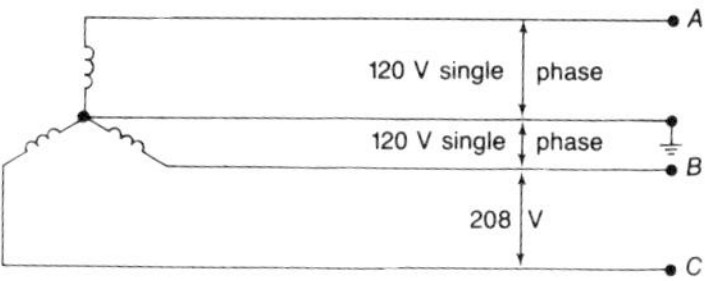

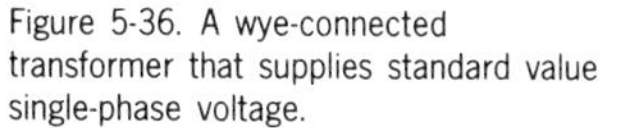
Figure 5-36. A wye-connected transformer that supplies standard value single-phase voltage.

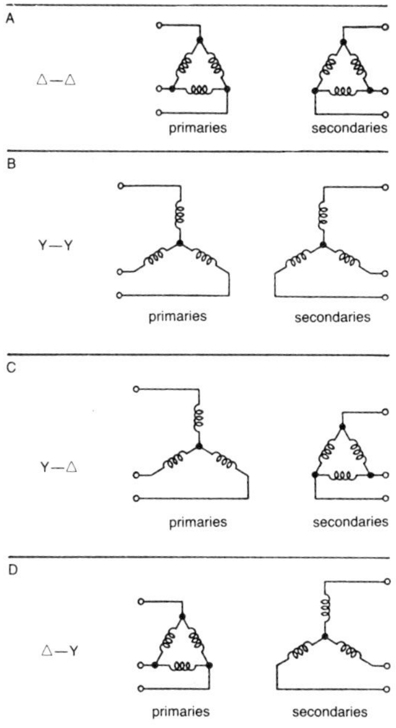

Figure 5-37. Methods of connecting three-phase transformer.

WIRING

One of the electrician's important considerations when it comes to three-phase current is the grounding of the system. This is very important if the system is in a damp or wet location. Farms are usually considered wet and damp, and the grounding has to be very well done for the sake of animals and humans.

The National Electrical Code has definite suggestions for the grounding of electrical systems. The various systems with voltages of 50 to 1000 volts are described in Section 250-5(b). Figure 5-39 describes the various types of electrical service and shows which of the transformer's secondary wires has to be grounded.

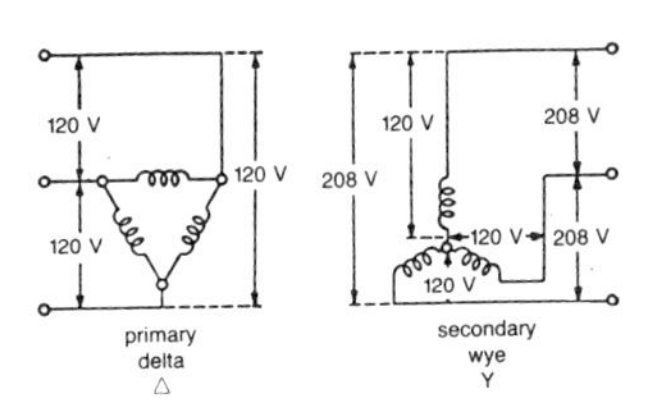

Figure 5-38. Voltages available from two types of transformers.

SINGLE-PHASE (1ϕ) POWER

If power is being generated by a utility company, single-phase power is generated as three-phase and then broken down. If generated by local generators driven by diesel engines or gasoline engines, it is usually generated as single-phase since that is what is required for the equipment to be connected to it.

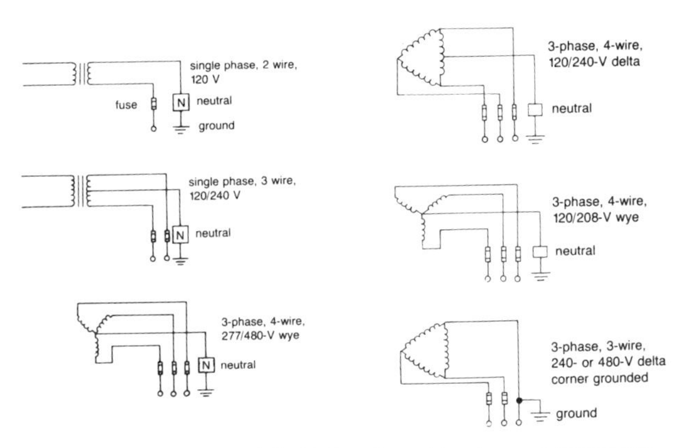

Figure 5-39. Various types of electrical service and transformer secondary wires to be grounded.

POWER DISTRIBUTION

Residential areas are served almost entirely by single-phase systems. Some utilities use radial primary systems, while others employ loop primaries. Some companies have both, the choice for a given area depending primarily on load density. In some instances, radial primaries are changed to loop when the load reaches a higher level.

On some systems a maximum number of 12 homes are served from a single transformer. There is a trend, however, toward fewer homes per transformer, especially in developments where homes are built with all-electric utilities and appliances. At present there is no strong trend concerning the location of service connections. Connections at the transformer compartment and in junction boxes below grade are methods preferred by utility companies. The use of pedestal and dome-type connections appears to be diminishing. Some companies are experimenting with a single transformer per house, with the service connection extending directly from a transformer that is either pad- or wall-mounted or installed in an underground vault.

Lightning Arresters. A majority of the utilities install *lightning arresters* on riser poles. In fact, many install arresters both on the poles and at the open point of a loop. Transformer primary protection is mostly by internal "weak links" or fuses; some companies use both. Various types of transformer switches are being used, but the use of "load-break elbows" for both transformer switching and sectionalizing is increasing.

Lighting arresters used on the line after it leaves the transformer and reaches the house are very useful when the distance from the transformer to the house is long enough so that a high-voltage surge may be set up in the power line when it is struck by lightning or when the strike is near the wires. The lightning arrester shown in Figure 5-40 can be installed outside at the drip loop, or it can be installed on the load center or disconnect. The

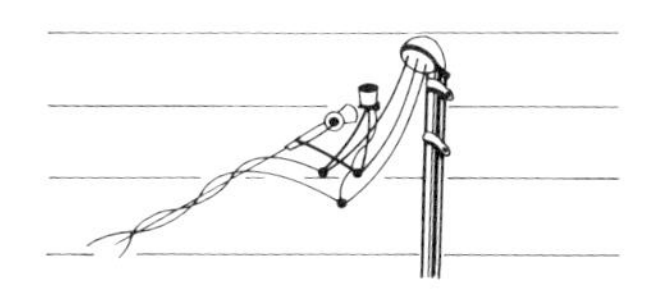

Figure 5-40. A lightning arrester connected to hot wires and the ground wire.

arrester is connected to each hot wire entering the building and the ground wire. If a surge is detected, it will be diverted to the ground through the arrester. Surge protection is very important today with all of the surge-sensitive electronic equipment (e.g., computers) installed in the typical home.

Surge Suppressors. Transient suppressors are available to protect electronic equipment. The larger devices can dissipate 8000 watts at one millisecond. Smaller diode-size devices have peak power dissipation of 600 watts. (See Figure 5-41.)

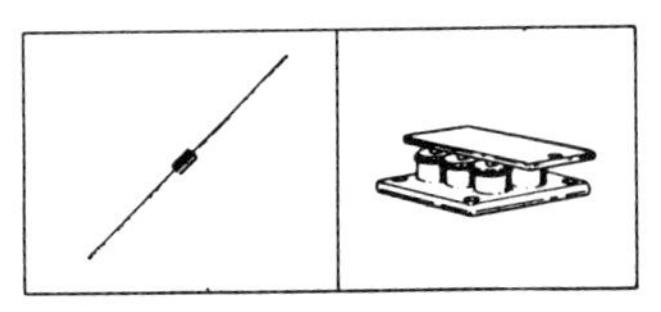

Figure 5-41. Two types of surge suppressors.

METER INSTALLATION

The electrician is usually required to place the meter socket trough for the kWh meter. It is usually supplied by the power company and installed by the customer. Figure 5-42 shows a typical one-meter installation with a socket for plugging in the meter once the service has been connected. In the case of apartment houses it is sometimes necessary to install two or more meters. Figure 5-43 shows the proper installation of two to six meters for a single-phase 3-wire, 120/240-volt, 150-ampere minimum service entrance.

ENTRANCE INSTALLATION

The service entrance is very important inasmuch as it is the end of the distribution line for the utility. Once the power has reached the user and enters the property, it is up to the owner to see to it that the wiring inside meets code requirements.

Figure 5-44 shows the attention given to the service entrance riser support on a low building. Note how far the meter is mounted above the ground. This work is usually done by the electrician; at least 24-inch service conductors are left for connection by the power company when it brings power up to the house or building.

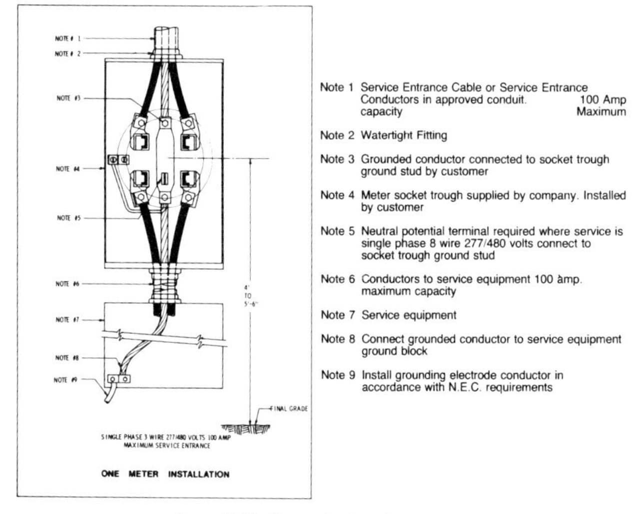

Figure 5-42. One-meter installation.

In most suburban developments the electrical service is brought in from the line to the house by underground cables. Figure 5-45 shows the requirements for such an installation. This service shows the pole in the rear of the house. Some localities also require that the entire electrical service to an area be underground. This eliminates poles in the rear of the house.

The standards for a farm meter pole by a local power company are shown in Figure 5-46. Note the division of ownership. Also note the 5/8-inch by 8-foot ground rods at least 3 feet from the pole. This is a single-phase, 3-wire 120/240-volt installation for loads exceeding 40kW demand.

LOW-VOLTAGE POWER

Most low-voltage systems use transformers from the 120-volt or 240-volt line and step down the voltage to either 16 volts or 24

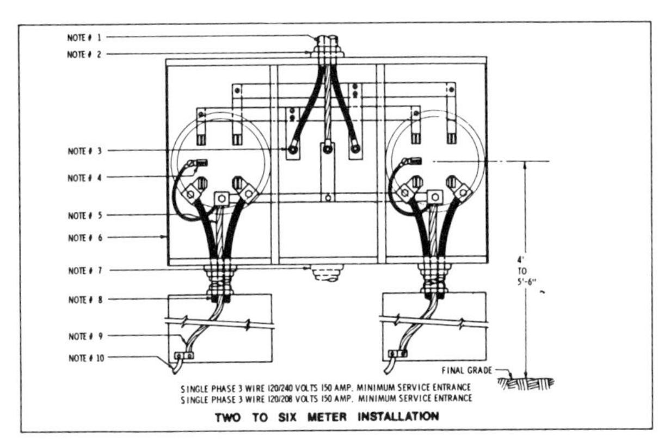

Note 1 Service entrance cable or service entrance conductors in approved capacity – 150 Ampere minimum capacity

Note 2 Watertight Fitting

Note 3 Preloaded compression for 3/8" stud size to be furnished and installed by contractors

Note 4 Grounded terminal for potential code required only where service is single phase 8 wire 120/208 volts. Connected to socket trough ground stud by customer.

Note 5 Grounded conductor. Connected to socket trough ground stub by customer.

Note 6 Meter socket troughs supplied by company. Installed by customer. Multimeter channel cannot be modified for additinal positions.

Note 7 Alternate service entrance location.

Note 8 Conductors to service equipment. 150 ampere maximum per position

Note 9 Connect grounded conductor to service equipment ground block

Note 10 Install grounding electrode conductor in accordance with N.E.C. requirements.

Figure 5-43. Installation of two to six meters.

volts to be used within a building as signaling or remote control circuits.

Chimes or door bells use a step-down transformer to take 120 volts down to 16. The wiring is usually a two-conductor or three-conductor # 18. The low-voltage switching may be a number of pushbutton types. This same voltage may be used in control circuits for garage door openers and certain burglar alarm systems.

A low-voltage transformer is used to obtain 24 volts for use in controlling the hot-air furnace that heats the house or, in some

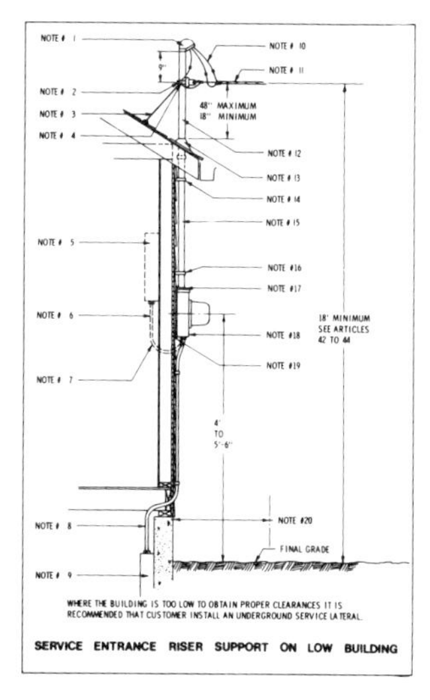

Figure 5-44. Service entrance riser support.

Note 1 Raintight service head

Note 2 Service bracket and mounting strap furnished by company. Installed by customer

Note 3 Back brace required for dimension greater than 24"

Note 4 Bond riser to service neutral

Note 5 Alternate service equipment

Note 6 Alternate service entrance

Note 7 Maximum length of unguarded service entrance conductors within wall shall be 12"

Note 8 Service entrance

Note 9 Service equipment

Note 10 Leave 24" of service entrance conductors for service drop connection by company

Note 11 Company's triplex service drop
150 A—1000 lbs. strain
200 A—1500 lbs. strain

Note 12 The riser shall be capable of withstanding a horizontal pull at the service drop attached. Provide back brace where necessary.

Note 13 Approved vent pipe flange

Note 14 3/8" U bolt as close to roof as possible

Note 15 2" or 2½" galvanized rigid steel conduit

Note 16 3/8" U bolt riser support

Note 17 Watertight fitting

Note 18 Meter socket trough furnished by company. Installed by customer.

Note 19 Provide treated wood backboard for mounting meter socket trough in a true vertical position. Fasten backboard to a structural member

Note 20 8' clear space to property line

cases, stores and other commercial and industrial plants. The main advantage of these control circuits is that they can use #18 wire for short runs and then move up to #16 or #14 for longer distances.

Low-voltage signaling and remote control circuits are covered by the NEC Article 725. Class 1, Class 2, and Class 3 circuits are detailed as to voltages and currents allowed.

Low voltage is also used in modern construction and home lighting systems. General Electric makes a system that has found wide use. Inasmuch as it is low-voltage and low-current, the wiring requirements for relay switching circuits are small, flexible wires that may be snaked in thinwall or steel partitions without the protection of metal raceways (except where local codes prohibit such installation). Wires can be dropped through hung ceilings into movable partitions as easily as rewiring telephones. In fact, wiring can be placed under rugs to switches located on desks

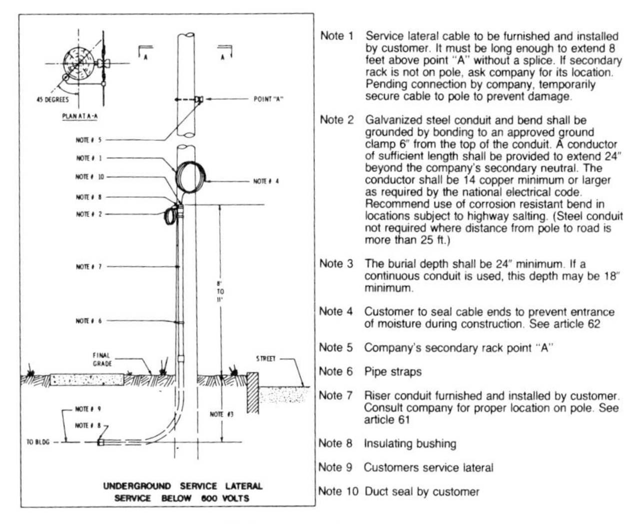

Figure 5-45. Underground service entrance.

and easily changed when required. The switches are modern in appearance and very compact.

The low voltage *circuits* are simple, and one that can be easily followed is seen in Figure 5-47. It shows the basic circuit, one transformer, one relay, and one switch controlling a load. The red, white, and black wire from the switch to the relay and transformer is usually # 20 AWG. This means it can be run longer distances for less cost than conventional wiring.

Figure 5-48 shows the basic circuitry for a remote control system. The rectifier (diode) is used to change the 24 volts from AC to DC. This produces less noisy relays—no AC hum.

The power supply for the GE system (used here as an example—there are others) is a 24-volt transformer connected to a 120-volt line. It can also be obtained to operate from 277 volts.

The *relays* (see Figure 5-49) are the mechanical latching type. That means the switching circuits require only a momentary im-

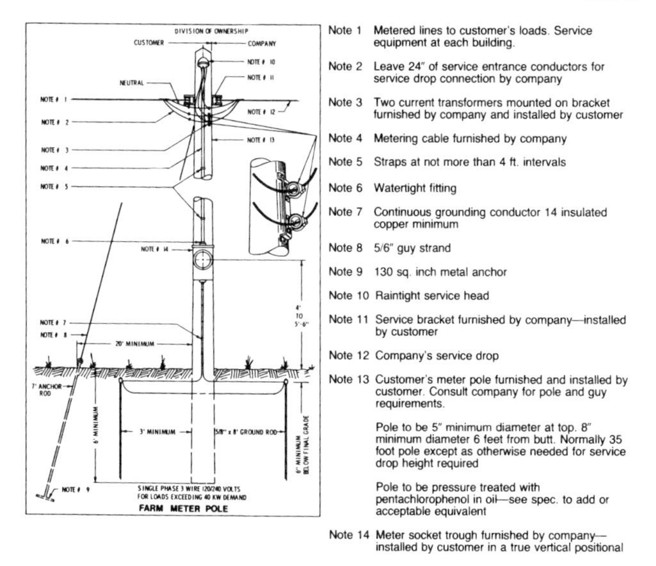

Figure 5-46. Standards for a farm meter.

pulse. These relays have a coil design that resists burnout due to equipment or operational failure. They can control 20 amperes of tungsten, fluorescent, or inductive loads at 125 volts AC and 277 volts AC. They can also be used for 1/2-horsepower, 240-volt AC and 1 1/2-horsepower, 125-volt AC motors. Some relays are available with a pilot light switch to indicate the on position in remote locations on a master panel.

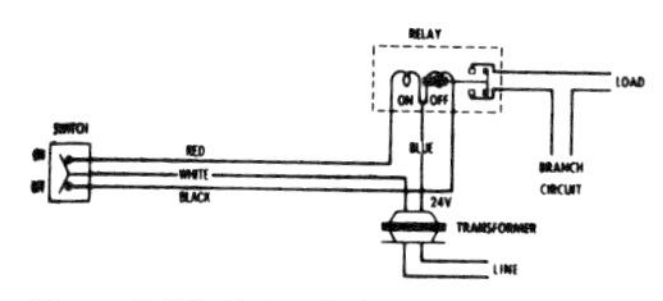

Figure 5-47. A simple low-voltage circuit.

Wiring can become a little more complicated, and the diagram has to be followed more closely when more stories are added to the system (see Figure 5-50). The best way to troubleshoot such a system is to follow the manufacturer's checklist of possible troubles and their wiring diagrams, usually on file in the building where the wiring is installed.

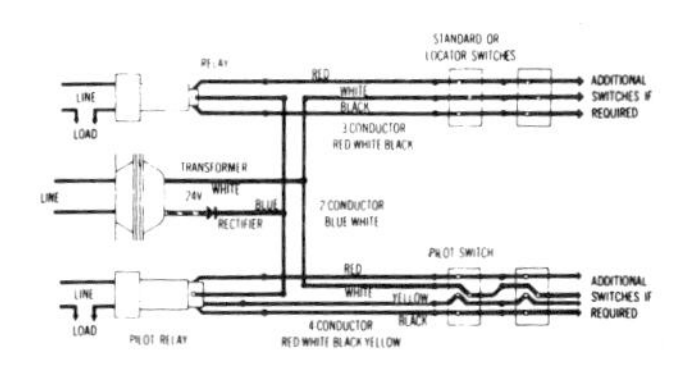

Figure 5-48. Basic circuitry for remote-control system.

EMERGENCY POWER

Emergency power has always been needed in hospitals, public buildings, subways, schools, manufacturing plants, and on farms, not only for emergency lighting but also to run refrigerators, typewriters, kidney machines, manufacturing processes, cash registers, elevators, heating equipment, fire pumps, telephones, computers, alarms, and other applications. The need for emergency power sources has increased in recent years. More electrical loads are considered essential, and therefore additional backup power is required.

Figure 5-51 shows a basic system for automatic emergency power transfer. The accessory group (AG) monitors the level of voltage from the normal source (utility). If the normal source fails or drops below acceptable levels, the AG signals the automatic engine starting controls (AESC) to start and monitor the engine generator set. When the generator-set voltage and frequency are adequate, the AG causes the automatic transfer switch (ATS) to transfer the load to the generator. When the normal source is restored, the load is retransferred, and the engine generator is shut down after a cooling off period. The controls then reset. The entire operation is automatic.

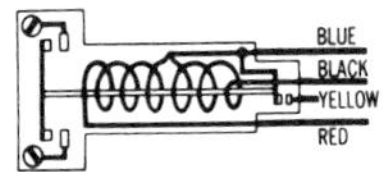

Figure 5-49. Internal wiring of a relay.

When normal power fails and emergency power is being used, it is important to make the most use of that power. Selective load transfer systems are effective, low-cost solutions to operating loads where only one or two out of a number of loads are operated from the emergency source at any one time. A bank of elevators can be used as an example. A selective load transfer system allows one

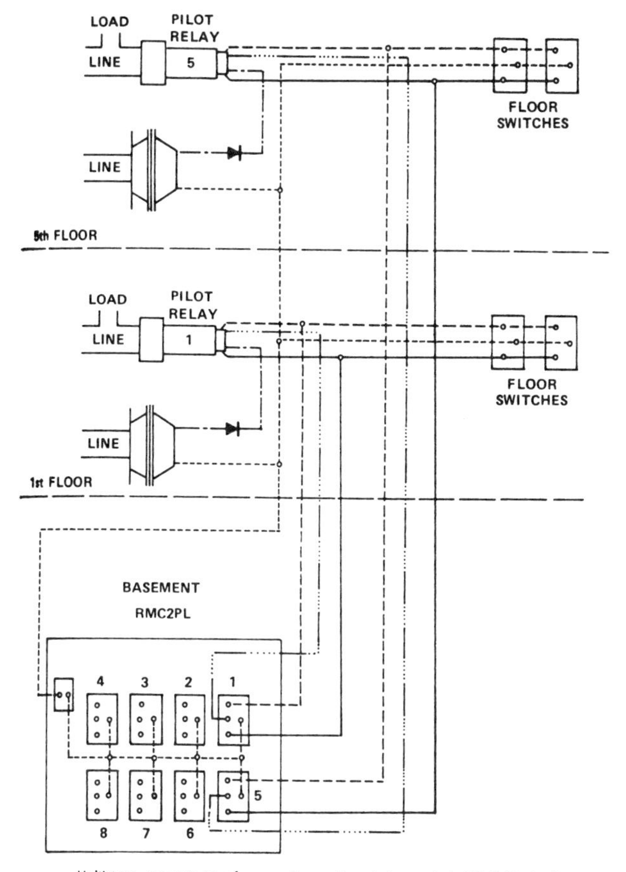

Multistory, separate transformers. For master-selector control of individual relays on different floors, where separate floor transformers and floor switches are also used, this circuit diagram explains the necessary wiring requirements. This circuit is especially useful for lights controlled from a watchman's station, or for control of corridor lights from a superintendent's office.

Figure 5-50. Wiring on more than one story.

elevator at a time to be operated when normal power fails. This uses the minimum amount of auxiliary power for operation because the standby generator can be sized for the necessary emergency load plus only one elevator. Any number of elevators in any combination of sizes to suit the needs of the application can be accommodated. Moreover, the system can be interfaced with other emergency loads for expanded operation if it is found that certain loads no longer require emergency power.

PROGRAMMABLE CONTROLLERS (PC)

Electricians who work in factories are called on to install and service programmable controllers. Programmable controllers are electronic devices using integrated circuits and semiconductors for remote switching. Since they are made of solid-state electronics, they have no moving parts. These units have been very successful in automating machinery and working for long periods without maintenance or attention of any kind. They have largely replaced relays as the workhorses of industrial controls.

In many locations and for many jobs the PC is preferred over the robot. It is less expensive than the robot and can be relied on to do a specific job accurately and repeatedly without attention.

The PC manufacturer supplies installation and programming instructions for each particular type of PC. Standard electrical wiring practices apply to the installation of these units. However, there are some particular concerns in troubleshooting these units. Since they use integrated circuits and semiconductors in their design, they are subject to the same problems as these devices.

The *output circuit* usually consists of a triac (in some instances, an SCR). The triac is sensitive to applied voltages, current, and internal power dissipation and is limited to a maximum peak off-state voltage. Exceeding this AC peak causes a dielectric-type breakdown that may result in a permanent short-circuit failure.

Often a semiconductor device, a varistor (THY), is placed across the triac to limit the peak voltage to some value below the maximum rating; in other cases, an RC snubber (made up of a resistor-capacitor filter circuit) alone may adequately protect the triac from excess voltage. The DC output consists of a power transistor protected against inductive loads by a diode.

Well-designed *input circuits* are less susceptible to damage than are output circuits using a triac. They do, however, respond to transients (line surges) and noise on the input line. Special precautions must be taken to reduce these sources of interference. If an input circuit is damaged, the circuits can fail, and the control circuit senses this. The various outputs then respond accordingly and, depending on the application of the system, could place the system in a hazardous condition. In situations involving such applications, an external means of monitoring the circuit inputs should be provided or redundant inputs should be used.

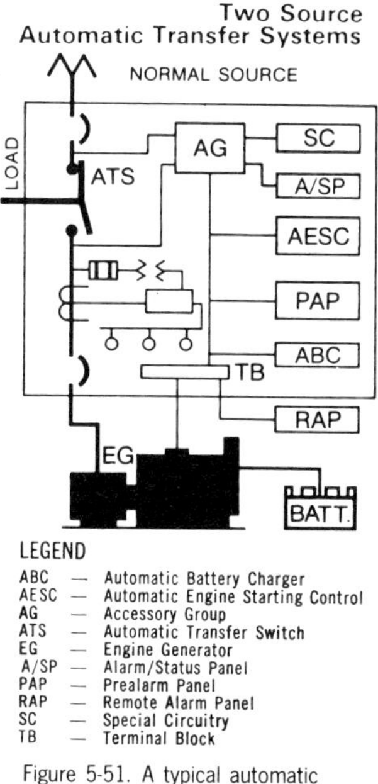

Figure 5-51. A typical automatic emergency power transfer system.

Figure 5-52 shows a block diagram of a typical programmable controller system with its inputs and outputs.

SOURCES OF DAMAGE TO SEMICONDUCTORS

Semiconductors can be damaged by temperature and certain atmospheric contaminants, by shock and vibration, and by noise.

TEMPERATURE

Excessive temperature can cause semiconductor materials to malfunction. The failure rate increases rapidly with increases in temperature. Even when stored, the devices are subject to problems caused by excessive temperature. Elevated ambient temperature can also cause intermittent problems that disappear when the temperature is lowered again. Air flow should be maintained around semiconductors to keep temperatures within the limits stated for the device.

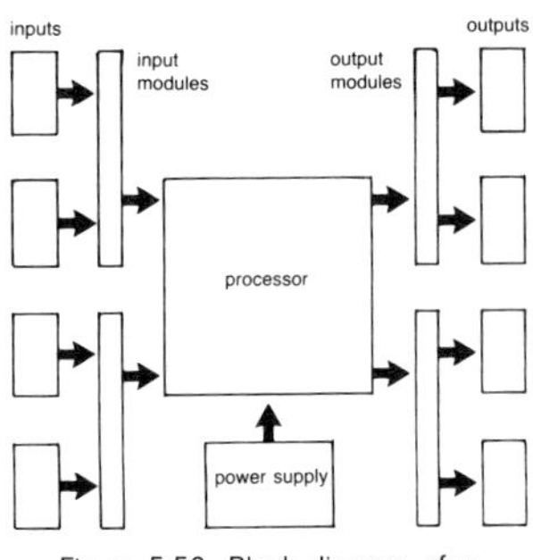

Figure 5-52. Block diagram of a programmable controller system.

SHOCK AND VIBRATION

Semiconductor-type circuits are generally not very susceptible to shock and vibration. However, it is possible that shock and vibration can cause problems if there are loose connections or worn insulation on wiring.

NOISE

Noise is generated by a number of sources in an industrial setting. Noise is defined here as electrical energy of random amounts and frequencies that adversely affects the functioning of electronic circuits. Most noise-caused malfunctions are of the nuisance type, causing operating errors, but some can result in hazardous machine operation.

Noise enters the control circuits by a number of different means—through the input lines, the output lines, or the power supply lines. It may be coupled into the lines electrostatically through capacitance between these lines and the lines carrying the noise signals. In most industrial areas a high potential is usually required, or long, closely spaced conductors are necessary. Noise caused by magnetic coupling is also quite common when control lines are close to lines carrying large current. The signals in this case are coupled through mutual inductances, such as in a transformer. Electrostatic and magnetic noise may also be directly coupled into the control logic circuit.

Programmable controllers have been designed to ignore problems created by noise. However, it is impossible to design for all situations. Therefore, filters, shielding circuitry, and insensitive circuitry have been placed in the design of the controllers for the purpose of eliminating noise problems. It is also good practice to make sure that no two PCs use the same grounding wire.

WIRES AND WIRING

Without wire there would be no electricity where we want and need it. The type, size, and insulation of wire affects the safety of the electrical operation for long periods of time. In addition to wire, it also takes boxes, fixtures, switches, plugs, and other devices to wire any facility. In this section we shall look at wire and some of the devices used in making an electrical wiring system operate without damage to people or buildings.

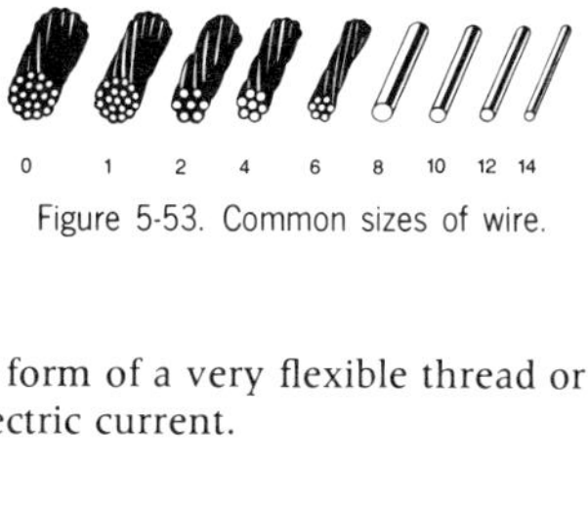

0 1 2 4 6 8 10 12 14

Figure 5-53. Common sizes of wire.

WIRE

A wire is a metal, usually in the form of a very flexible thread or slender rod, that conducts an electric current.

CONDUCTING MATERIALS

Although silver is the best conductor, its use is limited because of its high cost. Two commonly used conductors are aluminum and copper. Each has advantages and disadvantages. Copper has high conductivity and is more ductile (can be drawn out thinner). It is relatively high in tensile strength and can be soldered easily. However, it is more expensive than aluminum.

Aluminum wire has about 60% of the conductivity of copper. It is used in high-voltage transmission lines and sometimes in commercial and industrial wiring. Its use has increased in recent years. However, most electricians will *not* use it to wire a house today. There are a number of reasons for this, the most important being the safety of the installation and the possible deterioration of joints over the years due to expansion and contraction of the aluminum wire.

If copper and aluminum are twisted together in a wire nut, it is possible for moisture to get to the open metals over a period of time. Corrosion will take place, causing a high-resistance joint. This can result in a dimmer light or a malfunctioning motor.

WIRE SIZE

The size of wire is given in numbers. The size usually ranges from 0000 (referred to as 4-aught) to No. 40. *The larger the wire, the smaller its number.*

TABLE 5–15
Standard Annealed Solid Copper Wire

		Cross Section	
Gage Number[1]	Diameter (Mils)	Circular Mils	Square Inches
0000	460.0	212 000.0	0.166
000	410.0	168 000.0	0.132
00	365.0	133 000.0	0.105
0	325.0	106 000.0	0.829
1	289.0	83 700.0	0.0657
2	258.0	66 400.0	0.0521
3	229.0	52 600.0	0.0431
4	204.0	41 700.0	0.0328

TABLE 5–15 (*continued*)

Standard Annealed Solid Copper Wire			
		Cross Section	
Gage Number[1]	Diameter (Mils)	Circular Mils	Square Inches
5	182.0	33 100.0	0.0260
6	162.0	26 300.0	0.0206
7	144.0	20 800.0	0.0164
8	128.0	16 500.0	0.0130
9	114.0	13 100.0	0.0103
10	102.0	10 400.0	0.00615
11	91.0	8 230.0	0.00647
12	81.0	6 530.0	0.00513
13	72.0	5 180.0	0.00407
14	64.0	4 110.0	0.00323
15	57.0	3 260.0	0.00258
16	51.0	2 580.0	0.00203
17	45.0	2 050.0	0.00161
18	40.0	1 620.0	0.00128
19	36.0	1 290.0	0.00101
20	32.0	1 020.0	0.000802
21	28.5	810.0	0.000636
22	25.3	642.0	0.000505
23	22.6	509.0	0.000400
24	20.1	404.0	0.000317
25	17.9	320.0	0.000252
26	15.9	254.0	0.000200
27	14.2	202.0	0.000158
28	12.6	160.0	0.000126
29	11.3	127.0	0.0000995
30	10.0	101.0	0.0000789
31	8.9	79.7	0.0000626
32	8.0	63.2	0.0000496
33	7.1	50.1	0.0000394
34	6.3	39.8	0.0000312
35	5.6	31.5	0.0000248
36	5.0	25.0	0.0000196
37	4.5	19.8	0.0000156
38	4.0	15.7	0.0000123
39	3.5	12.5	0.0000098
40	3.1	9.9	0.0000078

[1]American wire gage—B&S

Table 5-15 lists the size of standard annealed solid copper wire in relationship to its number. Note the relationship of circular mils to square inches. A mil is 0.001 inch. Circular mils are often expressed by Roman numerals; 212,000 circular mils is 212 MCM. The first M stands for thousand (as with Roman numerals) and the CM stands for circular mils. Wire larger than 0000 is usually referred to according to MCM (thousands of circular mils). A frequently used wire is 750 MCM (750,000 circular mils).

Figure 5-53 shows the various sizes of wire most often encountered by electricians. Note that No. 6 and above consist of multistrands of wire. This allows the wire to be bent by hand. The larger sizes sometimes require a conduit bender to bend them to smaller radii.

TYPES OF WIRE

There are many types of wire manufactured. Three major types are discussed here.

Branch Circuit Wiring. General-purpose circuits should supply all lighting and all convenience outlets through the house, except those convenience outlets in the kitchen, dining room or dining area of other rooms, breakfast room or nook, family room, and laundry areas. Gen-

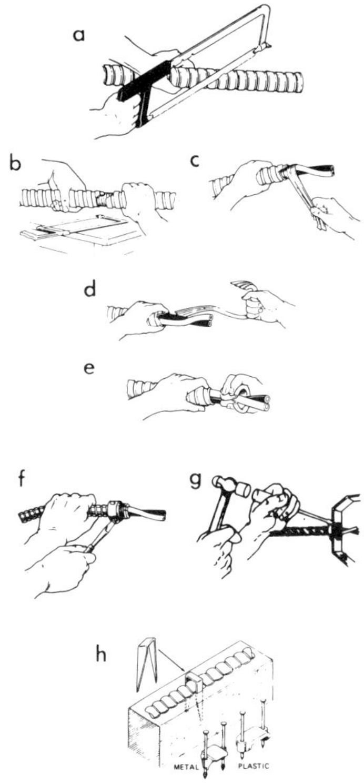

Figure 5-54. Installing armored cable.

eral-purpose circuits should be provided on the basis of one 20-ampere circuit for not more than every 500 square feet or one 15-ampere circuit for not more than every 375 square feet of floor area. Outlets supplied by these circuits should be equally divided among the circuits.

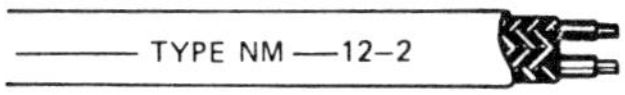

TYPE NM —12–2

Romex cable, 12/2, for use in home circuits.

Wrong way

Right way

The right and wrong way to cut insulation from a piece of wire.

Figure 5-55. Romex cable is often used for branch circuits.

Feeder Circuits should also be used. It is strongly recommended that consideration be given to the installation of branch circuit protective equipment served by appropriately-sized feeders located throughout the house, rather than at a single location.

Small Appliance Branch Circuits should also be installed. There should be in home wiring service at least one 3-wire, 120/240-volt, 20-ampere small-appliance branch circuit that is equipped with split-wired receptacles for all convenience outlets in the kitchen, breakfast and dining room, and family room. Two 2-wire, 120-volt, 20-ampere branch circuits are equally acceptable.

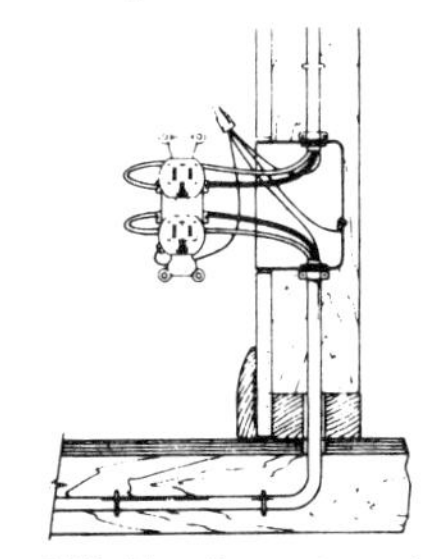

Figure 5-56. How Romex is used to wire a branch circuit with receptacle.

Cable of several different types can be used for branch wiring.

Armored cable, commonly referred to as BX, is available in 250-foot rolls for use where local codes permit. BX is hard to work with and needs some attention to details once the metal shield has been cut. (See Figure 5-54.)

Romex is used for branch circuits and is easier to work with. A cable stripper is used to strip off the first coating of insulation: then a knife or wire cutters is used to cut away any loose materials to expose the uninsulated copper wire and the black- and white-jacketed conductors. (In 3-wire cable there is also found a red-jacketed conductor.) (See Figure 5-55.)

Always strip wire far enough back so the wire will go at least three-quarters of the distance around a binding screw. When stripping be careful not to nick the conductor. Use correctly sized stripping tool or strip with a knife as if sharpening a pencil.

Wrap the wire at least ¾ of the distance around a wire binding screw without overlapping—then tighten the screw as securely as possible.

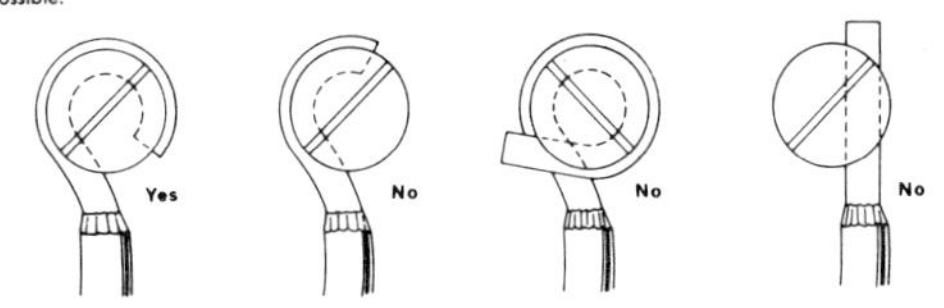

Do not use push-in terminals for aluminum conductor.

No

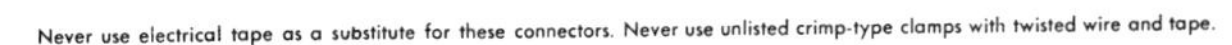

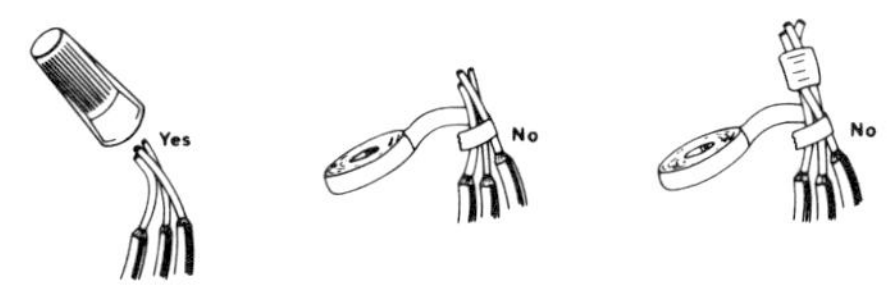

Figure 5-57. Proper wiring of receptacles and switches in branch circuits.

Figure 5-56 shows how Romex is used to wire a branch circuit with a receptacle. Figure 5-57 shows how to properly wire receptacles and switches in branch circuits. Figure 5-58 illustrates how branch circuits are loaded to balance the load on the different phases. Single-phase and three-phase loading is shown. Of course, the proper wire should be used for each of the circuits.

Service Entrance Cable. Power is brought from the pole or transformer into the rear of the house (in some cases the lines are underground) by means of three wires—one black, one red, and one white or uninsulated. These wires may be three separate ones, or they may be twisted together to look like one cable.

Once the cable is connected to the house, it is brought down to the meter by way of a sheathed cable with three wires: one red,

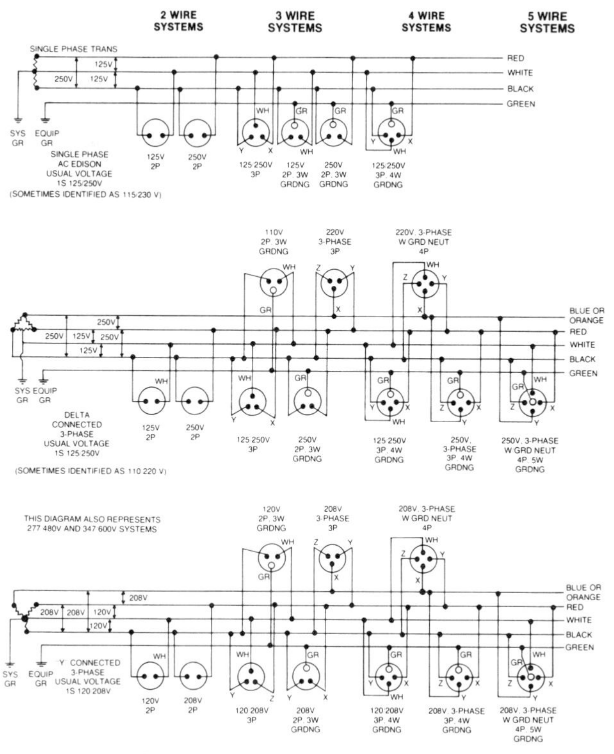

Figure 5-58. Loading of branch circuits.

one black, and one uninsulated (ground or neutral). The stranded, uninsulated wire is twisted at the end to make its connection. (See Figure 5-59.)

Wire size depends on the load to be applied to the line. The square footage of the house determines the amount of minimum service capacity needed (125 amperes, 150 amperes, or 200 amperes). From the outside the cable enters the house and is connected to the distribution box located somewhere easily accessible, such as in the basement. From the distribution box the branch circuits run to all parts of the house.

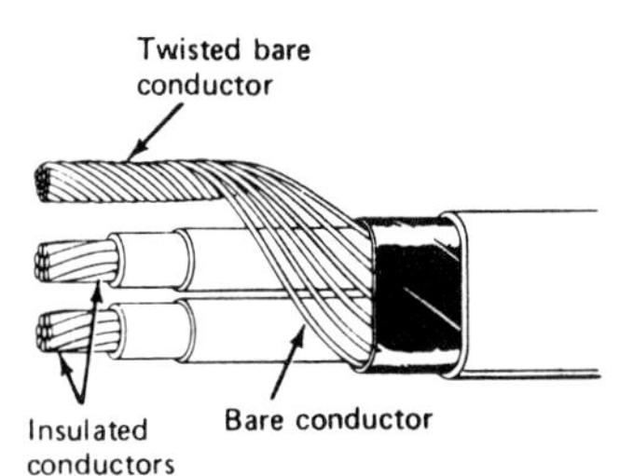

Figure 5-59. Connecting uninsulated wire in service entrance cable.

Underground Feeders and Branch Circuit Cable. This type of cable is for direct burial and is made with a polyvinyl chloride (PVC) insulation and jacket materials. (See Figure 5-60.) It is available in many sizes.

In multiple conductor cables the two- or three-circuit conductors are individually insulated and are laid parallel under a jacket to form a flat construction. A multiple-conductor cable may have in addition to the circuit conductors an uninsulated or bare conductor of the same wire size for grounding purposes.

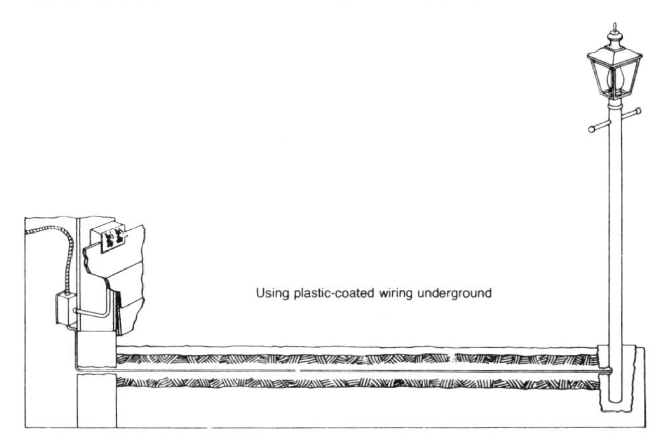

Figure 5-60. PVC-covered cable for underground feeders.

Type UF cables are designed for use in wiring methods recognized by the NEC in systems operating at potentials of 600 volts or less. The maximum conductor temperature in Type UF cables is 60°C. This type of wire may also be used for interior wiring in dry, wet, or corrosive locations. Multiple-conductor-type UF cables may be installed as nonmetallic sheathed cable. Once every 24 inches along the conductor jacket will appear *Sunlight Resistant UF, size of the wire, and the number of conductors plus 600 V UL.* (See Figure 5-61.)

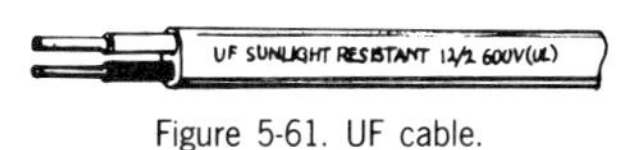

Figure 5-61. UF cable.

CABLE

Cable is a general term often applied to large conductors. They may be single-strand conductors or a combination of conductors (wires) insulated from one another but encased together. Many types of cable are used, two of the most common being Romex cable and BX cable.

ROMEX CABLE

Romex cable is used to carry power from a distribution panel box to the individual outlets within the house. This nonmetallic sheathed cable has plastic insulation covering the wires to insulate them from all types of environments. Some types of Romex cable may be buried underground; underground Romex has UF stamped on its outside covering.

BX CABLE

BX cable is the name applied to armored or metal-covered wiring. BX sometimes meets the need in home applications for flexible wiring. It is used to connect an appliance, such as a garbage disposal unit, which vibrates or moves a great deal. BX cable has to have special fittings to make sure its metal covering does not cut through the insulation of the wire it houses.

CONDUIT

Conduit protects and carries wires. Conduit comes in many types: thinwall metal, plastic, or rigid. Rigid conduit is like pipe with thick walls and ends with screw threads.

EMT (Thinwall Conduit)

EMT, electrical metallic tubing, also often called thinwall conduit, is commonly used. It is lightweight, thinner, and easier to bend than rigid conduit. Because it is metallic, it can handle physical abuse and is used wherever physical protection is needed but PVC cannot be used because of the presence of steam pipes or other sources of heat.

Conduit usually comes in 10-foot lengths. Couplings are used to extend the overall length of a piece of conduit. Thinwall conduit couplings are electrical fittings used to attach or couple the length of one conduit to another. EMT does not require threads to be cut on the ends. It uses specially made connectors for the ends so that it can be attached to boxes, panels, and other devices. (Rain-tight fittings should be used outside.) Figure 5-62 shows how EMT is fitted into a conduit box, and Figure 5-63 shows a variety of fittings used on EMT.

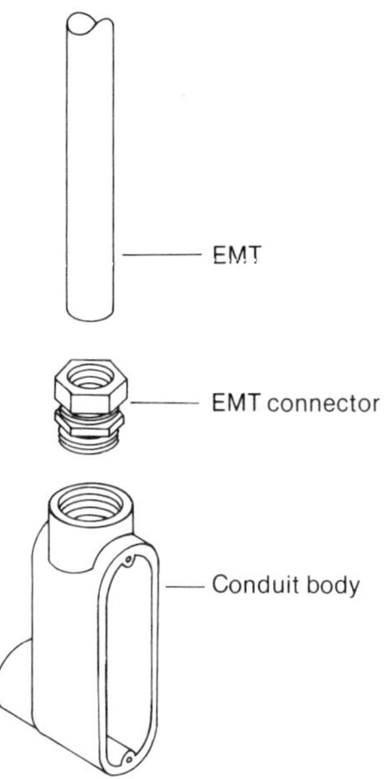

Figure 5-62. EMT fitting into a conduit box.

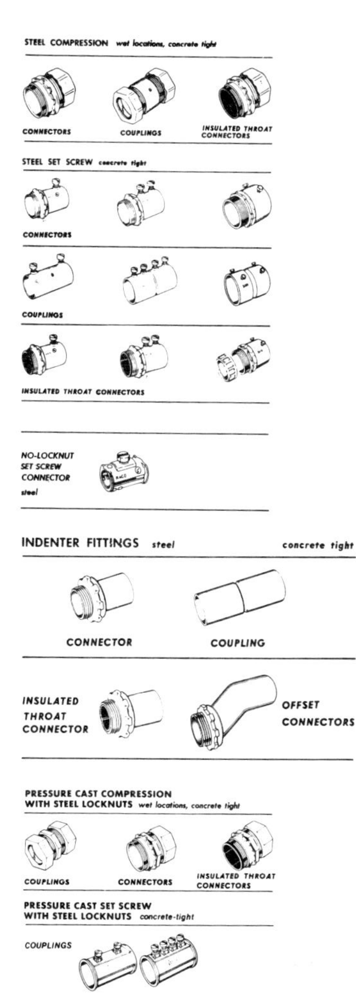

Figure 5-63. Variety of fittings used on EMT.

ENMT

Electrical nonmetallic tubing (ENMT) is also used as conduit. Made of the same material as PVC, ENMT is resistant to moisture and many atmospheric pollutants and is flame-retardant. Suitable for aboveground use, it is easily bent by hand but cannot be used where flexibility is needed, such as at motor terminations, to prevent noise and vibration. NEC Article 331 deals with ENMT and lists couplings, connectors, and fittings to be used with it.

RIGID CONDUIT

Industrial and commercial wiring must, because of the large amounts of current and high voltages required, be enclosed in large pipe to protect the wires from damage by equipment operating around them.

This large pipe, called rigid conduit, presents a number of problems, most of which arise whenever a bend has to be made. (For conduit of smaller diameter, special hand benders, or "hickeys," are used.) Dies are used to keep large pipe from collapsing as it is

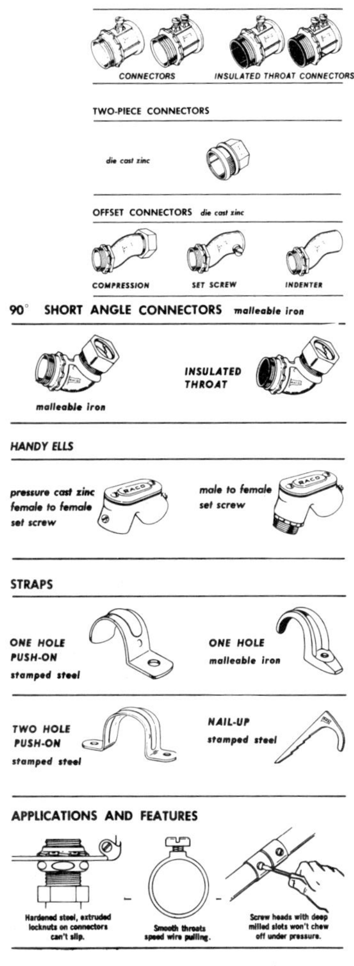

Figure 5–63 (*continued*)

bent. The inside part of the pipe is compressed as the outside portion is stretched. It is very easy to collapse the piece of pipe if proper care is not taken during the bending operation. That is why some very elaborate bending equipment is available. The skill associated with conduit bending comes with experience.

Threads have to be cut on the ends of rigid conduit for fittings. This can be done by hand tools in some cases, but in most instances a thread-cutting machine is used. Cutting, threading, and bending rigid conduit takes a number of years to master. Rigid conduit has special types of boxes for switches and receptacles, known as FS boxes. See Figure 5-64 for fittings used on rigid conduit.

When using rigid conduit, keep in mind that the entire conduit is not filled to capacity; the number of wires in the conduit is limited by NEC rules.

IMC (INTERMEDIATE METALLIC CONDUIT)

This is a relatively new type of conduit with wall thickness less than that of

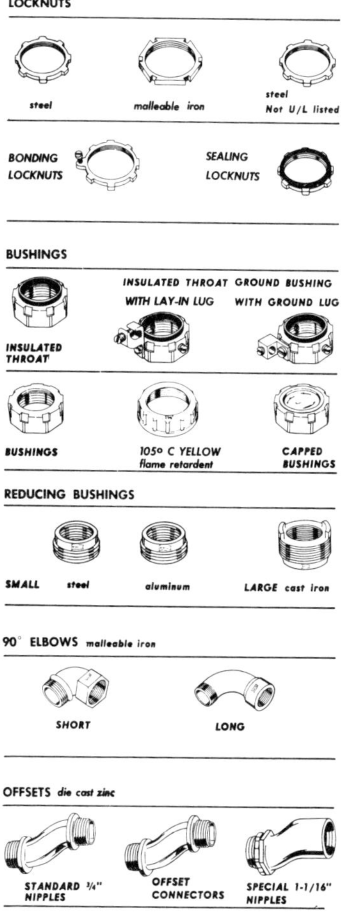

Figure 5-64. Fittings used on IMC and rigid conduit.

rigid metal conduit but greater than that of EMT. It uses the same threading methods and standard fittings as rigid metal conduit and has the same general application rules as rigid metal conduit. IMC is a lightweight rigid steel conduit that requires about 25% less steel than heavy-wall rigid conduit. Acceptance into the Code was based on a UL fact-finding report that showed through research and comparative tests that IMC performs as well as rigid steel conduit in many cases and surpasses rigid aluminum and EMT in more cases.

IMC may be used in any application for which rigid metal conduit is recognized by the NEC, including use in all classes and divisions of hazardous locations, as covered in Sections 501-4, 502-4, and 503-3. Its thinner wall makes it lighter and less expensive than standard rigid metal conduit. However it has physical properties that give outstanding strength. It has the same outside diameter as rigid conduit and the same trade sizes. The rules for number of wires in IMC are the same as for rigid metal conduit.

See Figure 5-64 for fittings used on IMC and rigid conduit.

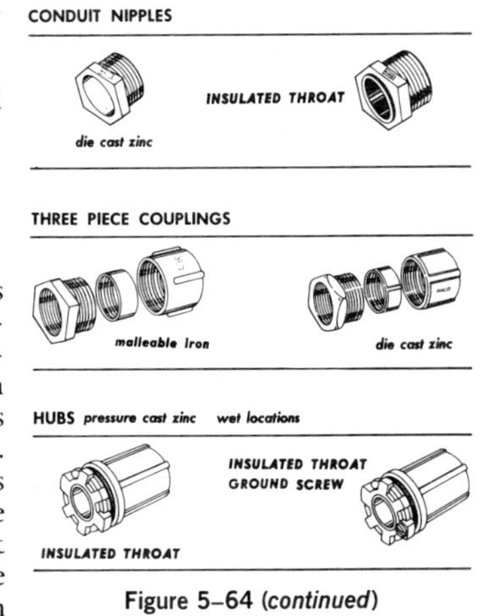

Figure 5–64 (*continued*)

PVC CONDUIT SYSTEMS

Polyvinyl chloride (PVC) is also used in conduit systems. The use of this plastic conduit (pipe) in electrical wiring systems has decided advantages. Nonmetallic conduits weigh one fourth to one fifth as much as metallic systems. They can also be easily installed in less than half the time and are easily fabricated on the job. (Nonmetallic conduit and raceway systems are covered by Article 347 of the National Electrical Code.)

PVC has high impact resistance to protect wiring systems from physical damage. It is resistant to sunlight and approved for outdoor usage.

The use of expansion fittings allows the system to expand and contract with temperature variations. PVC conduit will expand or contract approximately four to five times as much as steel and two and one-quarter times as much as aluminum. Installations where the expected temperature exceeds 14°C (25°F) should use expansion joints. The manufacturer furnishes the formulas for figuring out the expansion joint size and how often it is needed in any given installation.

Any plastic conduit should always be installed away from steam lines and other sources of heat. Support straps should be tightened only enough to allow for linear movement caused by expansion and contraction.

The PVC conduit that is widely used in the United States is called Plus 40. It is UL-listed for use underground, encased in concrete or direct burial, and for exposed or concealed use in most conduit applications above ground.

Plus 80 is designed for above-ground and underground applications where PVC conduit with extra heavy wall is needed. It is frequently used in situations where severe abuse may occur, such as for pole risers, bridge crossings, and heavy traffic areas. Typical applications are around loading docks, in high-traffic areas, and where threaded connections are required.

PVC conduit and fittings are similar to those used on rigid conduit. (See Figure 5-65.)

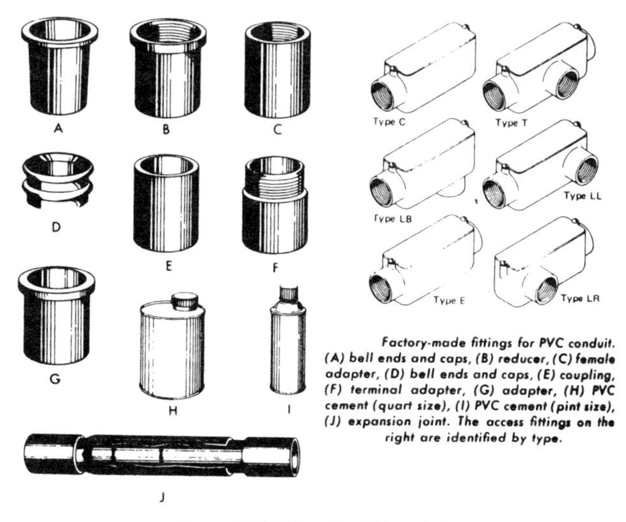

Factory-made fittings for PVC conduit. (A) bell ends and caps, (B) reducer, (C) female adapter, (D) bell ends and caps, (E) coupling, (F) terminal adapter, (G) adapter, (H) PVC cement (quart size), (I) PVC cement (pint size), (J) expansion joint. The access fittings on the right are identified by type.

Figure 5-65. Fittings for PVC conduit.

BOXES

Boxes are used to make connections of wiring or to mount switch boxes and fixtures. There are many types of boxes, metallic and nonmetallic.

Nonmetallic PVC switch boxes and receptacle boxes are used in residential wiring. The roughing-in time can be reduced by using these boxes inasmuch as they have their own connectors molded into the box. They can also be used in remodeling old work. (See Figure 5-66.)

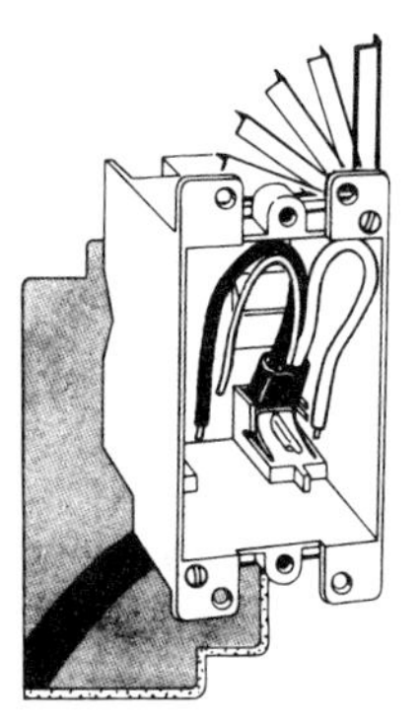

Figure 5-66. Nonmetallic box.

When you install an "old work" box into hollow walls, two arms swing out behind the wall. Tighten two screws and the arms draw tightly against the wall. Installation is completed with a few turns of the screwdriver. A template is included in the box to ensure an accurate cut in the wall. The swing-arm box can be used with wall material from 1/4-inch to 5/8-inch thick or from paneling to drywall. A cable clamp is molded into the box. Just push the wire in and pull back slightly to activate the clamping action.

PVC boxes are made for switches and ceiling installations. The ceiling boxes will support fixtures up to 50 pounds. Mounting posts have a pair of holes 2 3/4 inches and 3 1/2 inches on center to accept fixture canopies that require either of these spacings. Heavy-duty nails are already inserted. Just make sure you don't miss and hit the box with a hammer—the results can be disastrous. The plastic does crack when hit a hard blow by a hammer that missed its target. The PVC boxes are usually a bright color. They may be gold or blue, to name but two of the manufacturers' colors. However, the thermoset (hard plastic) boxes are black. They also have raised covers and reducers. Figure 5-67 shows some PVC boxes and thermoset boxes.

Boxes have been made of steel for years. There are many different manufacturers, but most of the boxes are made roughly to the same standards. Look for the CSA and UL labels before buying.

3-GANG SWITCH BOX

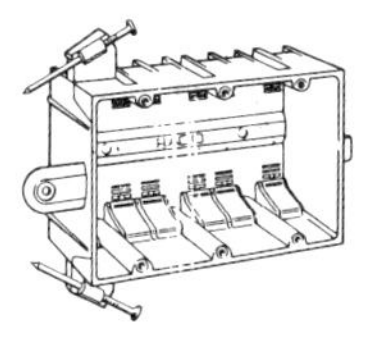

4" DIA. CEILING BOX

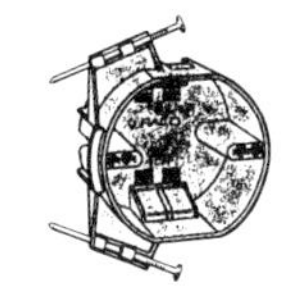

23.5 cubic inches

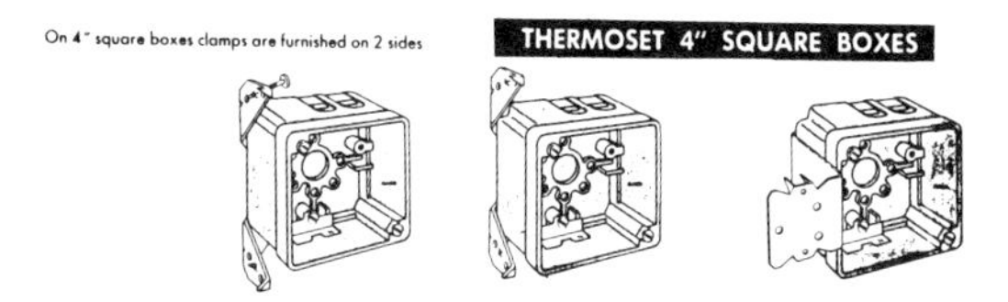

Figure 5-67. PVC boxes and thermoset boxes.

A wide variety of boxes is available. Each has its own identity and serves a specific purpose. Therefore, when looking for an easy way to mount a box, be sure to check all options. Many boxes come with brackets for specific types of mounting. (See Figure 5-68.) Plastic ears are included in many switch boxes. They are set forward 1/16 inch in the "old work" position. Two screw ears are supplied with shallow boxes and one screw ear on deep boxes. (See Figure 5-69.) Clamps inside boxes are made for armored cable or nonmetallic cable (Romex). (See Figure 5-70.)

Table 5-16 shows the volume needed per conductor so that the proper box can be chosen for a particular function. Remember that a clamp counts as a conductor.

BRACKETS

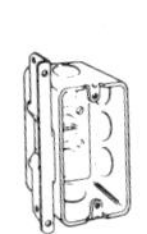

"A" **BRACKET**—Positions against side and face of stud. Bracket set back ⅝" on handy boxes.

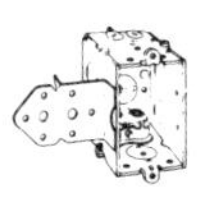

"B" **BRACKET**—Mounts on face of stud. Bracket set flush on 4" Square and set back ⅝" on handy and switch boxes.

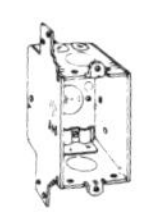

"D" **BRACKET**—Flat, bracket side on Non-gangable switch boxes. Gauging notches at ⅜", ½" and ⅝".

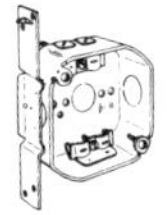

"FA" **BRACKET**— Side mount bracket with back up flange. Set back ½" on octagon and ⅝" on handy and switch boxes.

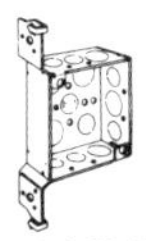

"FH" **BRACKET**—Side mount bracket with hooks that drive in face of stud. Flush on 4" square boxes and set back ½" on switch boxes.

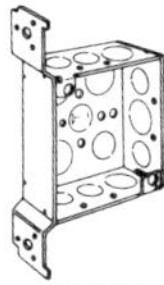

"FM" **BRACKET**—Side mount bracket for metal studs. Gauging notches set box ¼" behind front of stud.

"J" **BRACKET**—2 spurs, slotted holes for toe-nailing. Bracket set flush with gauging notches at ⅜" and ½".

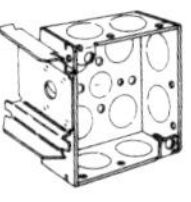

"MS" **BRACKET**— Snap-in bracket for metal studs. Mounts quickly without tools. Locks tightly in place for 1⅝", 2½" or 3⅝" studs.

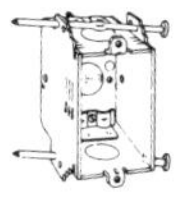

"S" **BRACKET**— Non-gangable switch box with sides extended. Option of staked, angled 16d nails. Gauging notches at ⅜", ½" and ⅝".

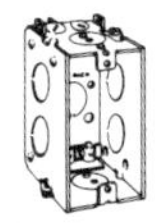

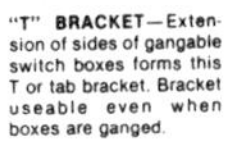

"T" **BRACKET**—Extension of sides of gangable switch boxes forms this T or tab bracket. Bracket useable even when boxes are ganged.

Figure 5-68. Boxes with brackets for different types of mounting.

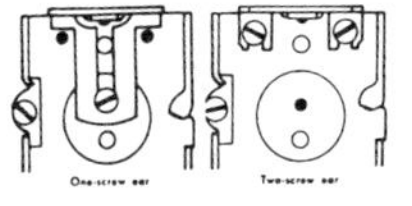

Figure 5-69. Boxes with screw ears for mounting.

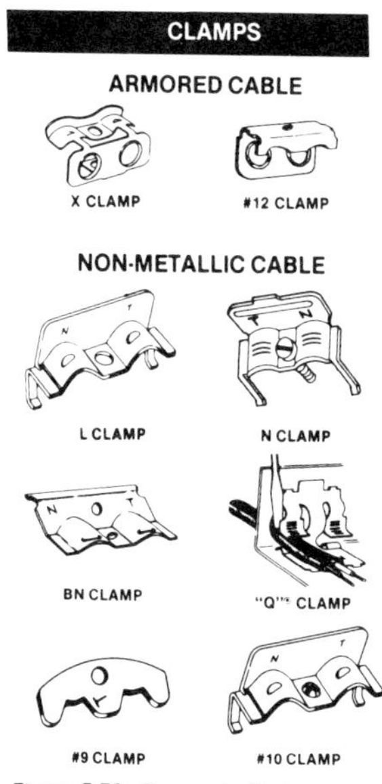

Figure 5-70. Clamps inside boxes.

TABLE 5–16

Volume Required per Conductor

Size of Conductor	Free Space Within Box for Each Conductor
No. 14	2. cubic inches
No. 12	2.25 cubic inches
No. 10	2.5 cubic inches
No. 8	3. cubic inches
No. 6	5. cubic inches

Equivalent NEMA Line	Wiring	Color Code	AC Voltage	100 Ampere Receptacle Connector	100 Ampere Plug Inlet	60 Ampere Receptacle Connector	60 Ampere Plug Inlet	30 Ampere Receptacle Connector	30 Ampere Plug Inlet	20 Ampere Receptacle Connector	20 Ampere Plug Inlet	AC Voltage	Color Code	Wiring	Equivalent NEMA Line
5	2 Pole 3 Wire	Yellow	125					GE 330 R 4 GE 330 C 4	GE 330 P 4	GE 320 R 4 GE 320 C 4	GE 320 P 4	125	Yellow	2 Pole 3 Wire	5
6	2 Pole 3 Wire	Blue	250			GE 360 R 6 GE 360 C 6	GE 360 P 6	GE 330 R 6 GE 330 C 6	GE 330 P 6	GE 320 R 6 GE 320 C 6	GE 320 P 6	250	Blue	2 Pole 3 Wire	6
7	2 Pole 3 Wire	Gray	277									277	Gray	2 Pole 3 Wire	7
8	2 Pole 3 Wire	Red	480			GE 360 R 7 GE 360 C 7	GE 360 P 7	GE 330 R 7 GE 330 C 7	GE 330 P 7			480	Red	2 Pole 3 Wire	8
9	2 Pole 3 Wire	Black	600					RESERVED FOR FUTURE CONFIGURATIONS				600	Black	2 Pole 3 Wire	9
14	3 Pole 4 Wire	Orange	125/250					GE 430 R 12 GE 430 C 12	GE 430 P 12	GE 420 R 12 GE 420 C 12	GE 420 P 12	125/250	Orange	3 Pole 4 Wire	14
15	3 Pole 4 Wire	Blue	250 3φ			GE 460 R 9 GE 460 C 9	GE 460 P 9	GE 430 R 9 GE 430 C 9	GE 430 P 9			250 3φ	Blue	3 Pole 4 Wire	15
16	3 Pole 4 Wire	Red	480 3φ			GE 460 R 7 GE 460 C 7	GE 460 P 7	GE 430 R 7 GE 430 C 7	GE 430 P 7	GE 420 R 7 GE 420 C 7	GE 420 P 7	480 3φ	Red	3 Pole 4 Wire	16
17	3 Pole 4 Wire	Black	600 3φ									600 3φ	Black	3 Pole 4 Wire	17
21	4 Pole 5 Wire	Blue	120/208 3φ									120/208 3φ	Blue	4 Pole 5 Wire	21
22	4 Pole 5 Wire	Red	277/480 3φ									277/480 3φ	Red	4 Pole 5 Wire	22
23	4 Pole 5 Wire	Black	347/600 3φ									347/600 3φ	Black	4 Pole 5 Wire	23

Figure 5-71. Plugs used with different current ratings.

		15 AMPERE		20 AMPERE		30 AMPERE		50 AMPERE		60 AMPERE	
		RECEPTACLE	PLUG	RECEPTACLE	PLUG	RECEPTACLE	PLUG	RECEPTACLE	PLUG	RECEPTACLE	PLUG
2-POLE/2-WIRE	125V	1–15R	1–15P								
	250V		2–15P	2–20R	2–20P	2–30R	2–30P				
	277V					(Reserved for Future Configurations)					
	600V					(Reserved for Future Configurations)					
2-POLE/3-WIRE GROUNDING	125V	5–15R	5–15P	5–20R	5–20P	5–30R	5–30P	5–50R	5–50P		
	Catalog No	5269-C • 8269-CHG	5266-P • 8266-PHG	5369-C • 8369-CHG	5366-P • 8366-PHG						
	250V	6–15R	6–15P	6–20R	6–20P	6–30R	6–30P	6–50R	6–50P		
	Catalog No	5669-C •	5666-P •	5469-C •	5466-P •						
	277V AC	7–15R	7–15P	7–20R	7–20P	7–30R	7–30P	7–50R	7–50P		
	Catalog No	5769-C •	5766-P •								
	347V AC	24–15R	24–15P	24–20R	24–20P	24–30R	24–30P	24–50R	24–50P		
	480V AC					(Reserved for Future Configurations)					
	600V AC					(Reserved for Future Configurations)					
3-POLE/3-WIRE	125/250V			10–20R	10–20P	10–30R	10–30P	10–50R	10–50P		
	3Ø 250V	11–15R	11–15P	11–20R	11–20P	11–30R	11–30P	11–50R	11–50P		
	3Ø 480V					(Reserved for Future Configurations)					
	3Ø 600V					(Reserved for Future Configurations)					
3-POLE/4-WIRE GROUNDING	125/250V	14–15R	14–15P	14–20R	14–20P	14–30R	14–30P	14–50R	14–50P	14–60R	14–60P
	3Ø 250V	15–15R	15–15P	15–20R	15–20P	15–30R	15–30P	15–50R	15–50P	15–60R	15–60P
	3Ø 480V					(Reserved for Future Configurations)					
	3Ø 600V					(Reserved for Future Configurations)					

Figure 5-72. General purpose nonlocking and locking plugs.

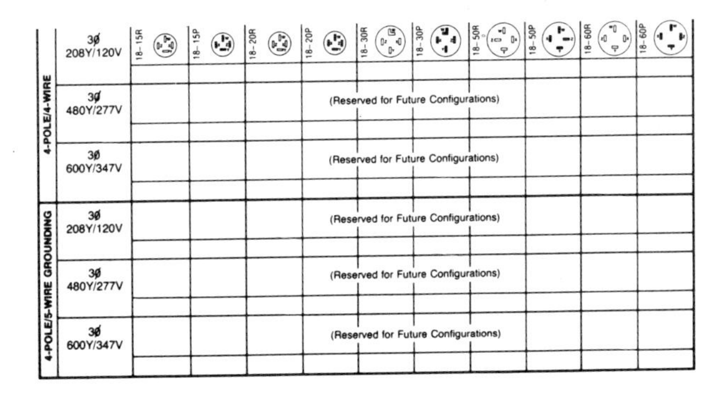

Figure 5-72 (continued).

PLUGS

Plugs are made in a number of sizes and shapes for use with 125-, 250-, 277-, 480-, and 600-volt power sources. They are also made to be used with three-phase as well as single-phase power. Current ratings are from 20 amperes up to 100 amperes. (See Figure 5-71.)

General-purpose nonlocking plugs and receptacles are shown according to NEMA configurations in Figure 5-72. The plugs are easily removed or inserted into the receptacle. They too run the range from 125 volts to 600 volts with currents of 15 amperes up to 60 amperes.

Locking plugs and receptacles slip into the receptacle and are twisted to lock into place. A variety of single- and three-phase configurations are available to fit most requirements.

Plugs for hazardous locations are specially made to extinguish any sparks that may occur before the plug is pulled from the receptacle. Figure 5-73 shows how this is done.

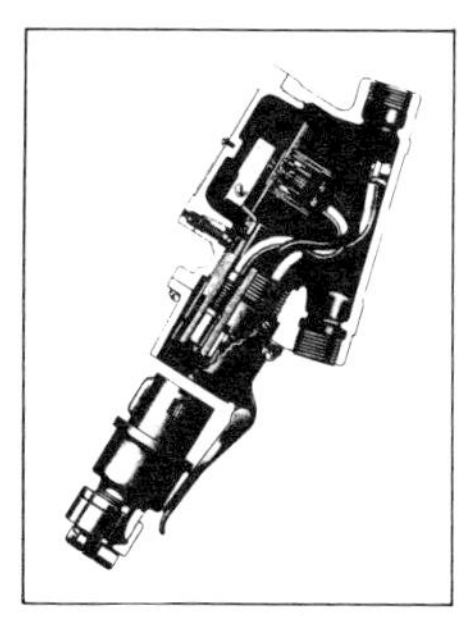

Receptacle constructed with an interlocked switch. Rotating the plug after insertion actuates this switch.

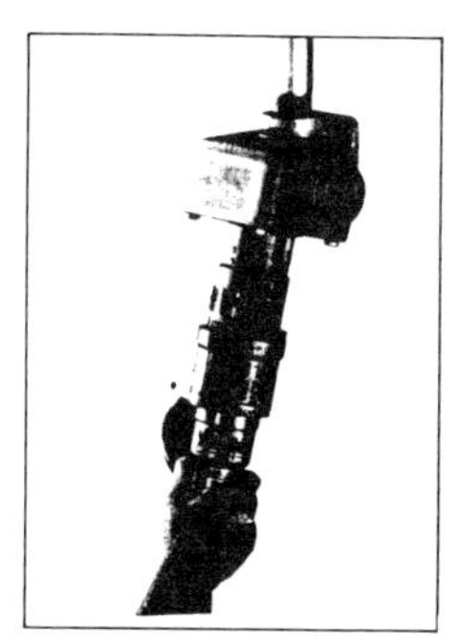

Plug partially withdrawn (delayed action stop in rotating sleeve prevents complete withdrawal). Contacts separated in explosionproof chambers.

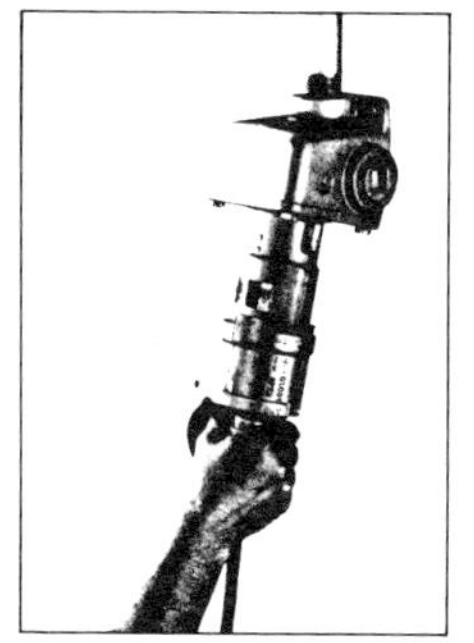

Plug about to be withdrawn.

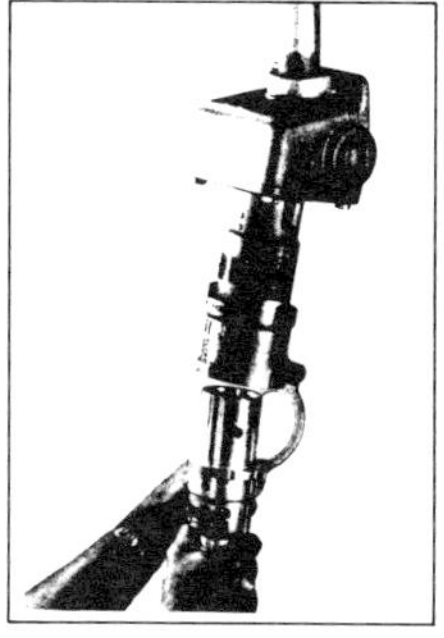

Plug completely withdrawn.

Figure 5-73. Specially made plugs for hazardous locations.

SWITCHES

Canopy switches are used for lamps or small devices that require less than 6 amperes at 125 volts. Figure 5-74 shows several types:

pushbutton (A), pullchain with a 3-foot cord (B), toggle with pigtails (C), a 4-position, 2-circuit rotary (D), and a type that will handle only 1 ampere at 125 volts in a rotary action (E).

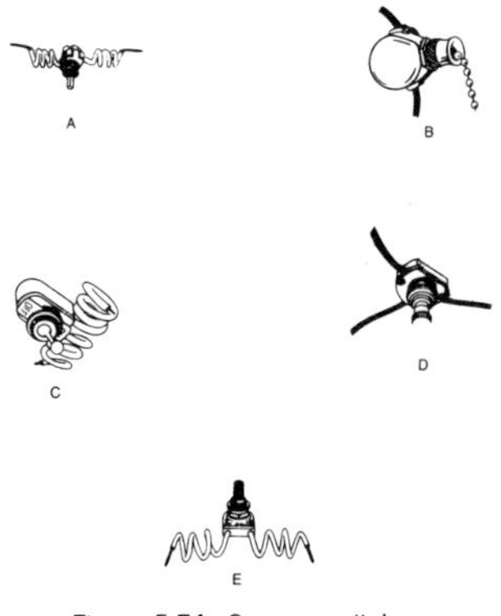

Figure 5-74. Canopy switches.

Cord switches are used to turn on lamps, heating pads, and other small portable equipment that usually demands no more than 6 amperes at 125 volts. The switch is designed to fit onto the cord by removing one of the twin leads and cutting it to make contact inside the disassembled switch.

Electronic switches are now part of home, office, and industrial plant wiring. An electronic switch may be single-pole or three-

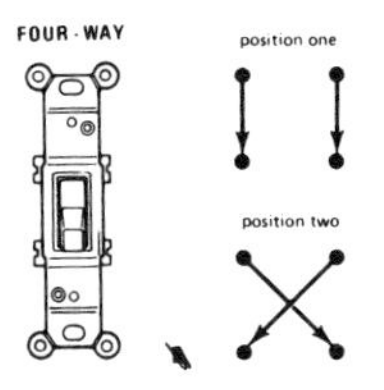

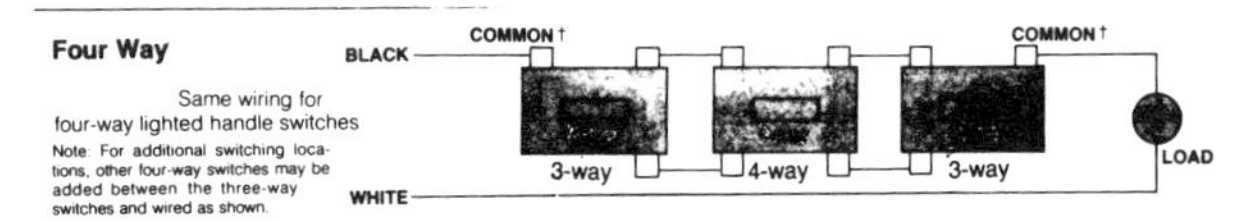

Figure 5-75. Four-way switches.

way, may allow manual override control at any time, can be installed in a single-gang switch box, may be programmed to turn lights on and off up to 8 times per day (with a minimum stay-on or -off time of 30 minutes), has pigtail terminals for wire nuts, and can handle up to 500 watts.

Four-way switches have four screws. It takes two three-way switches with one four-way to be able to operate a device from three locations. If more than three locations are needed, it still takes two three-way switches plus whatever number of four-way switches is needed after that. (See Figure 5-75.)

Snap switches with screw terminals are the most often used. They also come with a push-in connection and a green screw for a ground wire connection. The switch without the grounding screw has the ground wire from the Romex grounded to the metal box. Then, when the metal part of the switch is mounted to the box, it completes the grounding of the switch. The single-pole switch is used for on-off operations in 120-volt circuits in most residential circuits. (See Figure 5-76.)

SINGLE POLE

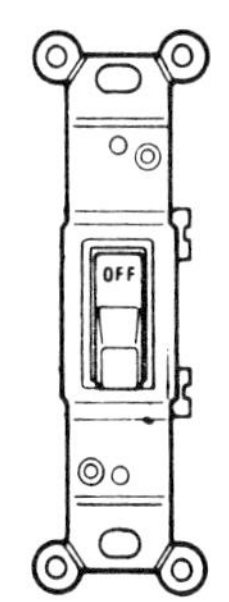

position two

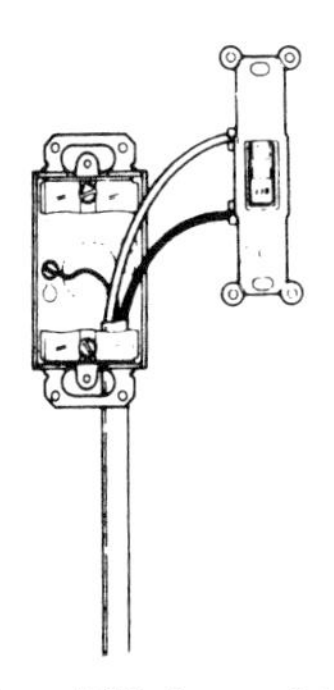

Figure 5-76. Snap switch.

Three-way switches have three screws. The three screws are connected so that the *C* is common and the other two are travelers. That means that as the switch is operated it completes the circuit from one traveler to the other. They have to be used in pairs in order to operate properly. Three-way switches are necessary in order to have four-ways operate. This type of switch is used to turn a light or device on or off from two different locations. (See Figure 5-77.)

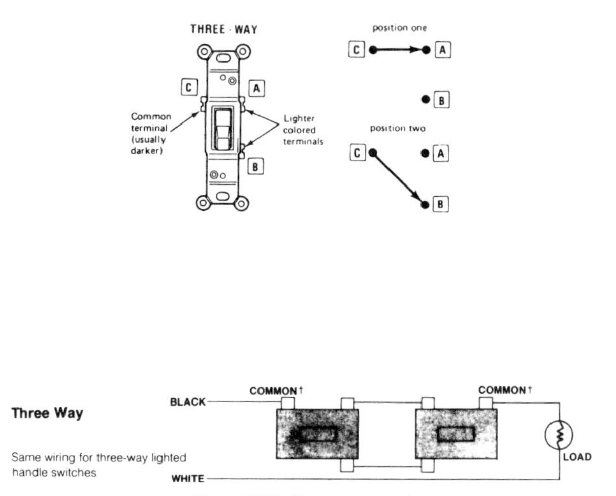

Figure 5-77. Three-way switch.

HAZARDOUS LOCATIONS EQUIPMENT

Special fixtures and devices are made for electrical wiring in hazardous locations. The NEC has extensive coverage of this type of wiring inasmuch as it is very dangerous. Only an experienced electrician should attempt wiring situations that are classified as hazardous. Check the NEC handbook for classes, groups, and divisions of hazardous materials (gases, vapors, dusts, etc.) and hazardous locations.

Sealed fittings, as specified in the code, are required in hazardous situations. Sealing fittings restrict the passage of gases, vapors, or flames from one portion of an electrical installation to another at atmospheric pressure and normal ambient temperatures. Sealed fittings limit explosions to the sealed-off enclosure and prevent precompression or "pressure piling" in conduit systems. Even though it is not a code requirement, many designers consider it good practice to minimize the effects of "pressure piling" through sectionalizing long conduit runs by inserting seals not more than 50 or 100 feet apart, depending on the conduit size.

Several electrical equipment makers offer a complete line of equipment for hazardous locations. The equipment is designed and produced in accordance with the requirements of the NEC. Some of this equipment needs conduit seals.

CONDUIT SEALS

The effectiveness of a conduit seal depends on properly filling the sealing fittings with a compound, following instructions furnished with the product. It should be noted that NEC Sections 501-5(c(4) prohibits splices or taps in sealing fittings and states that no compound is to be used to fill any fittings in which splices or taps are made.

NOTE

The importance of careful workmanship cannot be overemphasized. The safe operation of the entire explosion-proof electrical system depends on properly made, correctly located seals.

Chico A® Sealing Compound. One of the most commonly used sealing compounds is Chico A®. This sealing compound comes in a package with water in a plastic mixing pouch. The compound is easily mixed with water and then poured. It sets quickly. Setting takes place in 30 minutes, and it becomes hard within 60 to 70 minutes. It hardens to full compression strength within two days.

The compound is insoluble in water, is not attacked by petroleum products, and is not softened by heat. It is not injurious to any rubber or plastic wire insulation. In fact, continued cycling tests have shown that successive periods of heat, moisture, and freezing cold have no harmful effects on the seal after it has throughly hardened.

The compound expands slightly in setting, filling the sealing chamber completely. A properly made seal will withstand very high explosion pressure without crumbling or blowing out the seal fittings. (See Figure 5-78.)

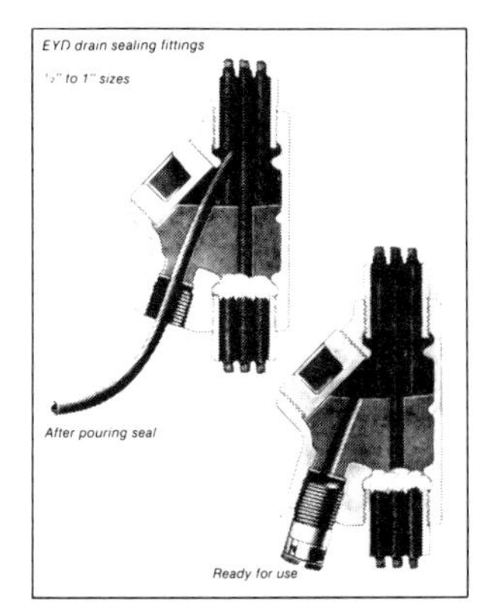

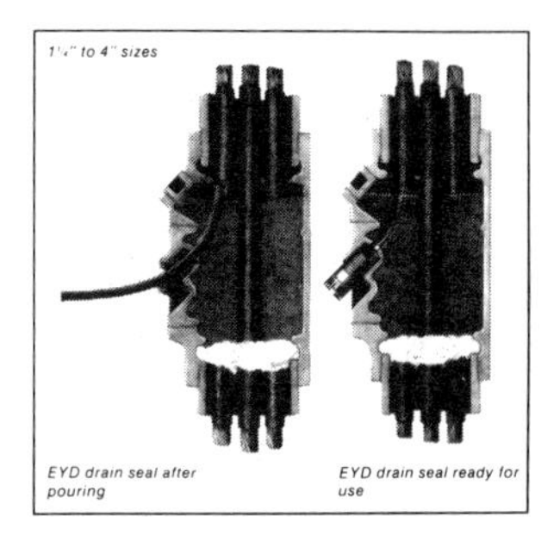

Figure 5-78. Sealing process.

The Intrapak®. The Intrapak® is an easy-pour, two-compartment plastic mixing pouch containing the Chico-A® compound and the precise amount of pure water needed for proper mixing. No mixing or measuring implements are required. A hard squeeze of the water compartment releases the water into the compound compartment. Mixing is completed by kneading the transparent pouch for 1 minute. A corner of the pouch is cut, and the mixture is then poured directly into the sealing fitting. No funnel is required.

How To Use Sealing Compound

Figure 5-79. Carefully pouring compound into sealing fitting.

1. Using a clean mixing vessel, mix two volumes of compound to each volume of clean, cold water. Warm water makes it set too quickly. Stir immediately and thoroughly.
2. Pour mixed compound carefully into the sealing fittings, using a funnel for best results. Underwriters' Laboratories' standard for safety requires the depth of the seal to be equal to or greater than the trade size of the conduit, with a minimum of 5/8 inch. (See Figure 5-79.)
3. Close the pouring opening immediately after pouring the compound.
4. Do not mix more than will be used within 15 minutes because once the compound has started to set, it cannot be thinned without destroying its effectiveness.
5. Do not mix and pour Chico-A® compound in freezing temperatures as the water in the mixture will freeze before setting and curing, resulting in an unsafe seal. In addition, freezing water may expand sufficiently to damage the seal fittings.

How To Use Sealing Fibers

To prevent sealing compound, in its fluid state from leaking out, use sealing fiber to dam each conduit hub or sealing fitting in horizontal conduit and only the bottom hub in vertical conduit. (See Figure 5-80.)

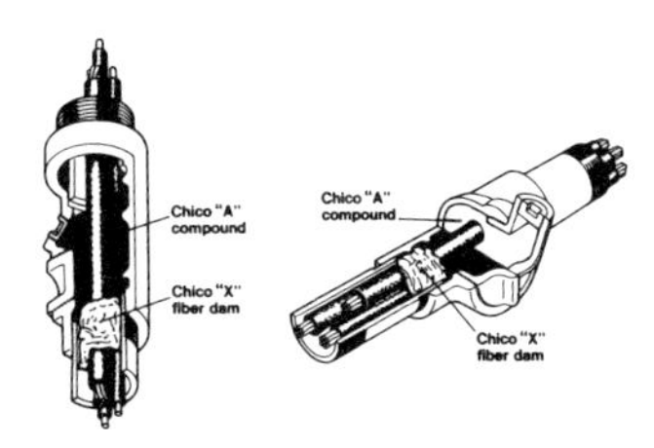

Figure 5-80. Sealing fiber used to dam sealing fittings.

1. Pack the sealing fiber between and around the conductors in each conduit hub. If the conductors are stiff, temporary wooden wedges inserted between the conductors will be helpful. It is important that the conductors be permanently separated from one another and from sealing fitting walls so that the sealing compound will surround each conductor.
2. Do not leave shreds of fiber clinging to the inside walls of the sealing chamber or to the conductors. Such shreds, when embedded in the compound, may form leakage channels. Dampening the fiber slightly will make its use easier and will help to prevent the shreds from clinging to the walls and wires.
3. Be sure completed dam is even with the integral bushing. Remember that the dam has to be strong enough and tight enough to prevent a considerable weight of fluid sealing compound from seeping out.
4. If the fitting has a separate damming opening, close the cover before pouring the seal.

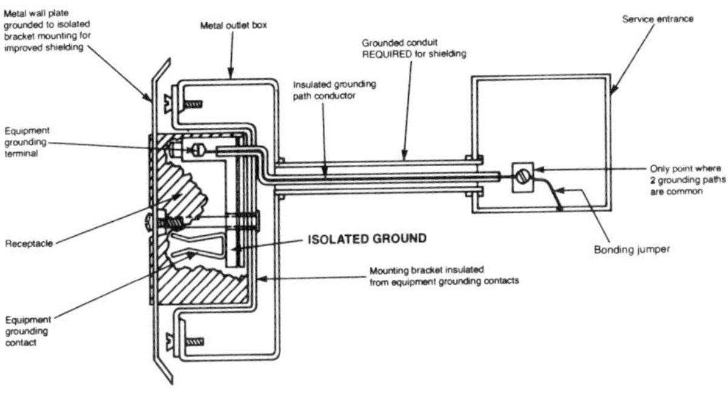

Figure 5-81. Isolated ground receptacle.

ISOLATED GROUND RECEPTACLES

With the increase in the number of pieces of sensitive electronic gear in the home, it has become necessary to improve on the quality of the electrical power these devices tap into. The computer and the videocassette recorder are very sensitive to variations in the power supply, especially to electromagnetic interference (EMI). One of the answers to this problem has been the isolated ground receptacle. (See Figure 5-81.)

Transmitted electromagnetic radiation waves from other electrical equipment would normally induce interfering ground path noise on the equipment grounding conductor, but the isolated grounding conduit provides an electric noise shield.

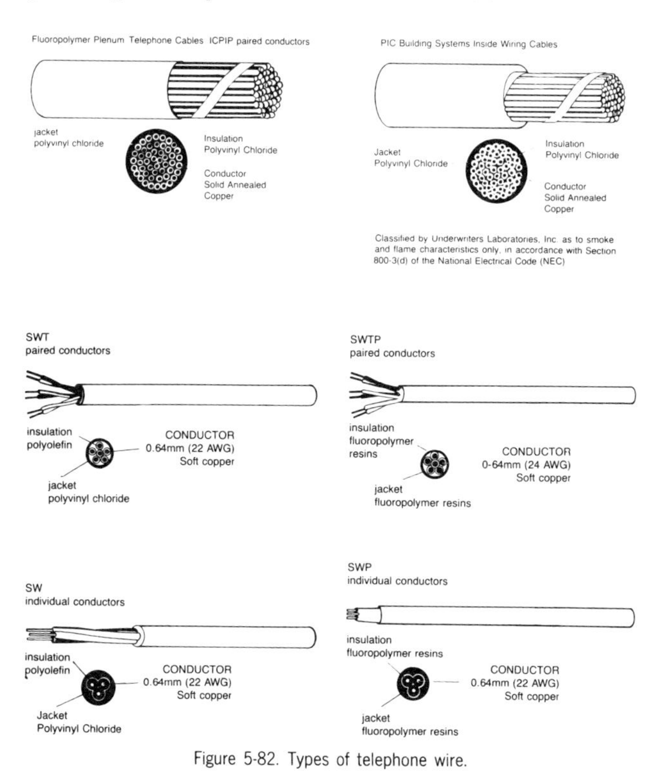

Figure 5-82. Types of telephone wire.

WIRING FOR TELEPHONE INSTALLATION

Now that the electrician is permitted to do telephone wiring of houses and other locations, the apprentice should be familiar with the types of wires used. Figure 5-82 shows different types of telephone wiring.

Station Wire. Station wire is designed for inside-outside use in station installations from the station protector to the telephone terminal block. It is designated SW with individual conductors and SWT (station wire twisted) with paired conductors that are twisted and then encased in PVC. The PVC jacket is tough and weather- and flame-resistant. It provides protection when exposed to weather conditions and inside cleaning products such as detergents, waxes, oils, and most solvents. The wire is installed with a stapling gun (see Figure 5-83).

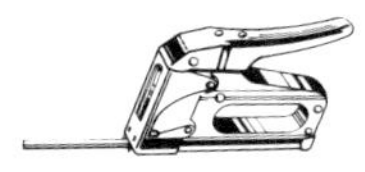

Figure 5-83. Staple gun used to attach station wire to wooden molding or walls.

Insulated conductors are twisted together in two-, three-, or four-conductor configurations and then jacketed. The PVC-jacketed wire is available in beige or olive. It is packaged in 500-foot rolls or coils in a box.

PIC Building Systems Inside Wiring Cables. Positive Identification Cables (PIC) are designed for inside use. The conductor is solid copper rated at #24 AWG. The main purpose of this type of semirigid polyvinyl cable is use in private branch exchange (PBX) and private attended branch exchange (PABX) systems. The conductor insulation is color-coded with the telephone industry's standard colors. Each insulated conductor is band-stripped at approximately 1-inch intervals with the color of the mating insulated conductor of the assembled pair. The individual conductors are mated in accordance with PI band stripes and twisted together into pairs. The pairs are assembled into a cylindrical core or into units and then formed into a core. If the cable contains more than 25 pairs, the pairs are molded into a 25-pair color group, each color group bound with a single color binder. (100-pair cable is available in either the standard make-up of four 25-pair units or five 20-pair units.) The jacket is flame-retardant and abrasion-resistant PVC. Jackets are available in olive or beige.

Station Wire Plenum (SWP), Individual Conductors. This type of wire is designed for use in air ducts and plenums without metal conduit in PBX and PABX, key system, and telephone instrument communications systems. (Key telephone systems [KTS], which can have 50 or more associated telephones, are similar to PBX systems. However the KTS has mostly outside calls and is little used for internal calls. A PBX has a larger share of internal calls.) SWP is used in installations that require less station wiring or shorter direct runs, and it reduces station installation costs. It has a fully annealed, solid bare-copper conductor that is #22 AWG in size. Insulation is a fluoropolymer resin.

Two-conductor station wire has one conductor insulated with red and the other with green. Three-conductor wire has the third conductor insulated with yellow; in four-conductor wire, the fourth conductor is black. The four-wire quad is made by spiraling the four wires together into the star quad configuration. The red and green conductors are placed diagonally across from each other to form pair 1, and the yellow and black conductors form pair 2. The insulated conductors are twisted in a two-, three- or four-conductor configuration and then jacketed. The jacket is made of fluoropolymer resin and is uncolored or olive in color. It is packaged in 500-foot coils in cartons. The cartons have a knock-out center to permit dispensing the wire directly from the carton.

Station Wire Twisted Plenum (SWTP), Paired Conductors. This station wire is designed for use in air ducts and plenums without conduit in PBX, PABX, key system, and telephone instrument communication systems. It too is made of #22 AWG fully annealed, solid bare-copper wire. It varies slightly from the individual-conductors cable inasmuch as it has a color code for the four pairs: pair 1 is formed by a white and blue (tip) and a blue (ring) wire; pair 2 by a white and orange (tip) and an orange (ring) wire; pair 3 by a white and green (tip) and a green (ring) wire; and pair 4 by a white and brown (tip) and a brown (ring) wire.

The tip (white) conductor insulation is band-stripped with the color of its mate to provide positive identification of each conductor.

ICPIP Pair Conductors. The individual conductors are mated in accordance with positive identification band stripes and then twisted together into pairs. The pairs are assembled into a cylindrical core or into units and then formed into a core. If the cable contains more than 25 pairs, the pairs are coded into a 25-pair color group, each color group bound with a unique color binder. The 100-pair cable is available in either the standard makeup of 25-pair units or five 20-pair units.

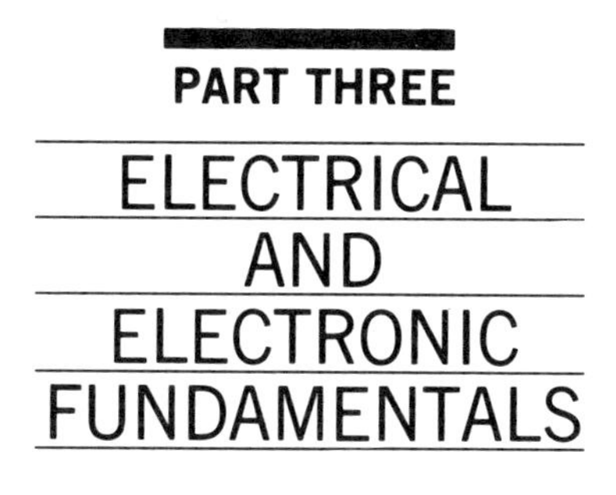

PART THREE

ELECTRICAL AND ELECTRONIC FUNDAMENTALS

6

Basic Mathematics and Measurement Conversion

Mathematics is a necessary tool for the electrician. Solving electrical formulas (see Chapter 7) involves the use of basic arithmetic (addition, subtraction, multiplication, and division), an understanding of squares and square roots, algebra, and trigonometry. Refer to a basic mathematics text if you need to review these subjects. In this chapter we will focus on several topics involving mathematics as it is encountered by the electrician and apprentice. Scientific notation is one of these topics, as is an understanding of the metric system of measurement and how to convert from U.S customary to the SI (metric) system. We also discuss how an electrician can estimate a job and, finally, how the use of a calculator can make computation and problem solving easier.

SCIENTIFIC NOTATION

The multiplication and division of large numbers can be simplified by using scientific notation. Notation by powers of 10 is used to indicate the position of the decimal point. Multiples of 10 from 1 to 1,000,000, with their equivalents in powers of 10, are given in Table 6-1.

This notation provides for both positive and negative powers of 10. When the power of 10 is positive, the decimal point of the number is moved to the right as many spaces as the exponent. When the power of 10 is negative, the decimal point is moved to the left.

Let's use the example 6.8759×10^4—a number written in scientific notation—to illustrate this. This number 6.8759×10^4 equals 68759 because the decimal point is moved to the right four spaces. The notation 6.8759^{-3} equals 0.0068759 because the decimal point was moved three spaces to the left.

TABLE 6–1
Powers of 10

$$1 = 10^0$$
$$10 = 10^1$$
$$100 = 10^2$$
$$1000 = 10^3$$
$$10000 = 10^4$$
$$100000 = 10^5$$
$$1000000 = 10^6$$

Likewise, powers of 10 can be used to simplify decimal expressions. The submultiples of 10 from 0.1 to 0.000001, with their equivalents in powers of 10, are:

$$0.1 = 10^{-1}$$
$$0.01 = 10^{-2}$$
$$0.001 = 10^{-3}$$
$$0.0001 = 10^{-4}$$
$$0.00001 = 10^{-5}$$
$$0.000001 = 10^{-6}$$

Scientific notation comes in handy for electricians when they need to solve problems dealing with frequency and capacitance.

THE SI SYSTEM

The metric system was originally based on a meter designed to be one ten-millionth of the earth's meridian at a particular spot in France. Later, this was found to be inaccurate and revisions were made.

In 1954 the *Conference Générale des Poids et Mesures* adopted a standard metric system of measure based on MKSA units: meter-kilogram-second-ampere. Later, the kelvin, candela, and mole were added for temperature, luminous intensity, and substance quality measures. This system of measurement was then named the *Système Internationale d'Unités,* abbreviated SI. It has been

TABLE 6–2
SI Prefixes

Multiplying Factor	Prefix	Symbol
1 000 000 000 000 = 10^{12}	tera	T
1 000 000 000 = 10^{9}	giga	G
1 000 000 = 10^{6}	mega	M
1 000 = 10^{3}	kilo	k
100 = 10^{2}	hekto	h
10 = 10^{1}	deka	da
0.1 = 10^{-1}	deci	d
0.01 = 10^{-2}	centi	c
1.001 = 10^{-3}	milli	m
0.000 001 = 10^{-6}	micro	μ
0.000 000 001 = 10^{-9}	nano	n
0.000 000 000 001 = 10^{-12}	pico	p
0.000 000 000 000 001 = 10^{-15}	femto	f
0.000 000 000 000 000 001 = 10^{-18}	atto	a

TABLE 6–3

Conversion Factors in Converting from Customary (U.S.) Units to Metric Units

To Find	Multiply	By
microns	mils	25.4
centimeters	inches	2.54
meters	feet	0.3048
meters	yards	0.9144
kilometers	miles	1.609344
grams	ounces	28.349523
kilograms	pounds	0.4539237
liters	gallons (US)	3.7854118
liters	gallons (Imperial)	4.546090
milliliters (cc)	fluid ounces	29.573530
milliliters (cc)	cubic inches	16.387064
square centimeters	square inches	6.4516
square meters	square feet	0.09290304
square meters	square yards	0.83612736
cubic meters	cubic feet	2.8316847×10^{-2}
cubic meters	cubic yards	0.76455486
joules	BTU	1054.3504
joules	foot-pounds	1.35582
kilowatts	BTU per minute	0.01757251
kilowatts	foot-pounds per minute	2.2597×10^{-5}
kilowatts	horsepower	0.7457
radians	degrees	0.017453293
watts	BTU per minute	17.5725

widely accepted throughout the world. Even in the United States and a few other nations where it is not commonly used, it is employed in many scientific and technical fields.

With the three principal SI units—the meter, the unit of length; the liter, the unit of capacity; and the gram, the unit of weight—prefixes are used to indicate multiples and divisions. Multiples of the units are expressed by adding the Greek prefixes *deka* (10), *hekto* (100), and *kilo* (1000). Divisions of the units are expressed

TABLE 6–4

Metric-Customary Equivalents

Measures of Length

1 meter = 39.3 inches; 3.28083 feet; 1.0936 yards

1 centimeter = .3937 inch

1 millimeter = .03937 inch, or nearly 1/25 inch

1 kilometer = 0.62137 mile

1 foot = .3048 meter

1 inch = 2.54 centimeters; 25.40 millimeters

Measures of Surface

1 square meter = 10.764 square feet; 1.196 square yards

1 square centimeter = .155 square inch.

1 square millimeter = .00155 square inch

1 square yard = .836 square meter

1 square foot = .0929 square meter

1 square inch = 6.452 square centimeters; 645.2 square millimeters

Measures of Volume and Capacity

1 cubic meter = 35.314 cubic feet; 1.308 cubic yards; 264.2 U. S. gallons (231 cubic inches)

1 cubic decimeter = 61.0230 cubic inches; .0353 cubic feet

1 cubic centimeter = .061 cubic inch

1 liter = 1 cubic decimeter; 61.0230 cubic inches; 0.0353 cubic foot; 1.0567 quarts (U. S.); 0.2642 gallon (U. S.); 2.2020 lb. of water at 62°F.

1 cubic yard = .7645 cubic meter

1 cubic foot = .02832 cubic meter; 28.317 cubic decimeters; 28.317 liters

1 cubic inch = 16.383 cubic centimeters

1 gallon (British) = 4.543 liters

1 gallon (U. S.) = 3.785 liters

TABLE 6–4 (*continued*)

Measures of Weight

1 gram = 15.432 grains
1 kilogram = 2.2046 pounds
1 metric ton = { .9842 ton of 2240 pounds; 19.68 cwts.; 2204.6 pounds }
1 grain = .0648 gram
1 ounce avoirdupois = 28.35 grams
1 pound = .4536 kilogram
1 ton of 2240 lb. = { 1.016 metric tons; 1016 kilograms }

by adding the Latin prefixes *deci* (0.1), *centi* (0.01), and *milli* (0.001). The prefixes are the same in all languages, thus improving communication. Table 6-2 lists the common prefixes, their symbols, and multiplying factors.

We should point out here that there are some exceptions to the SI system and prefixes that the electrician may encounter in ordering supplies. In their catalogs, most electrical supply companies use the letter "C" to mean 100 of an item and the letter "M" to mean 1,000 of an item.

TABLE 6–5

Horsepower-Kilowatt Equivalents[1]

Horsepower	Kilowatt (kW)
1/20	0.025
1/16	0.05
1/8	0.1
1/6	0.14
1/4	0.2
1/3	0.28
1/2	0.4
1	0.8
1 1/2	1.1
2	1.6
3	2.5
5	4.0
7.5	5.6
10	8.0

[1]Courtesy Bodine Electric Co.

In some cases, the electrician may need to convert from one system of measurement to another. Table 6-3 gives the conversion factors used to convert from the customary (U.S.) system of measurement to the metric system for measurement units that the electrician may encounter. Table 6-4 lists metric-U.S. equivalents for many common units. Table 6-5 gives common horsepower-kilowatt equivalents.

TEMPERATURE CONVERSION

Temperature may be expressed according to the Fahrenheit scale or the Celsius scale. Many motor and conductor temperatures are given in degrees Celsius and must be converted to the Fahrenheit scale if that is what the electrician is familiar with. The formulas for conversion are

$$°F = 1.8 \times °C + 32$$
$$°C = 0.55555555 \times °F - 32$$

Many times, however, the electrician may be able to refer to a table, such as Table 6-6, that provides the equivalent temperatures in both scales.

The numbers in italics in the center column refer to the temperature, either in degrees Celsius or degrees Fahrenheit, that is to be converted to the other scale. If converting a Fahrenheit temperature to Celsius, find the number in the center column and then look to the left column for the Celsius temperature. If converting a Celsius temperature, find the number in the center column and look to the right for the Fahrenheit equivalent.

ESTIMATING A JOB

All electricians have to do some estimating. People want to know what it will cost before contracting for a job. In electrical wiring, estimating requires a lot of time and effort. The electrician must charge the customer for the materials to be used (e.g., wires, switches, plugs), the cost of labor and supervision, overhead ex-

TABLE 6–6

Temperature Conversion

The numbers in italics in the center column refer to the temperature, either in degrees Celsius or degrees Fahrenheit, that is to be converted to the other scale. If converting a Fahrenheit temperature to Celsius, find the number in the center column and then look to the left column for the Celsius temperature. If converting a Celsius temperature, find the number in the center column and look to the right for the Fahrenheit equivalent.

−100 to 30			31 to 71			72 to 212			213 to 620			621 to 1000		
C		F	C		F	C		F	C		F	C		F
−73	*−100*	−148	−0.6	*31*	87.8	22.2	*72*	161.6	104	*220*	428	332	*630*	1166
−68	*−90*	−130	0	*32*	89.6	22.8	*73*	163.4	110	*230*	446	338	*640*	1184
−62	*−80*	−112	0.6	*33*	91.4	23.3	*74*	165.2	116	*240*	464	343	*650*	1202
−57	*−70*	−94	1.1	*34*	93.2	23.9	*75*	167.0	121	*250*	482	349	*660*	1220
−51	*−60*	−76	1.7	*35*	95.0	24.4	*76*	168.8	127	*260*	500	354	*670*	1238
−46	*−50*	−58	2.2	*36*	96.8	25.0	*77*	170.6	132	*270*	518	360	*680*	1256
−40	*−40*	−40	2.8	*37*	98.6	25.6	*78*	172.4	138	*280*	536	366	*690*	1274
−34.4	*−30*	−22	3.3	*38*	100.4	26.1	*79*	174.2	143	*290*	554	371	*700*	1292
−28.9	*−20*	−4	3.9	*39*	102.2	26.7	*80*	176.0	149	*300*	572	377	*710*	1310
−23.3	*−10*	14	4.4	*40*	104.0	27.2	*81*	177.8	154	*310*	590	382	*720*	1328
−17.8	*0*	32	5.0	*41*	105.8	27.8	*82*	179.6	160	*320*	608	388	*730*	1346
−17.2	*1*	33.8	5.6	*42*	107.6	28.3	*83*	181.4	166	*330*	626	393	*740*	1364
−16.7	*2*	35.6	6.1	*43*	109.4	28.9	*84*	183.2	171	*340*	644	399	*750*	1382
−16.1	*3*	37.4	6.7	*44*	111.2	29.4	*85*	185.0	177	*350*	662	404	*760*	1400
−15.6	*4*	39.2	7.2	*45*	113.0	30.0	*86*	186.8	182	*360*	680	410	*770*	1418
−15.0	*5*	41.0	7.8	*46*	114.8	30.6	*87*	188.6	188	*370*	698	416	*780*	1436
−14.4	*6*	42.8	8.3	*47*	116.6	31.1	*88*	190.4	193	*380*	716	421	*790*	1454
−13.9	*7*	44.6	8.9	*48*	118.4	31.7	*89*	192.2	199	*390*	734	427	*800*	1472
−13.3	*8*	46.4	9.4	*49*	120.0	32.2	*90*	194.0	204	*400*	752	432	*810*	1490
−12.8	*9*	48.2	10.0	*50*	122.0	32.8	*91*	195.8	210	*410*	770	438	*820*	1508
−12.2	*10*	50.0	10.6	*51*	123.8	33.3	*92*	197.6	216	*420*	788	443	*830*	1526

−11.7	*11*	51.8	11.1	*52*	125.6	33.9	*93*	199.4	221	*430*	806	449	*840*	1544
−11.1	*12*	53.6	11.7	*53*	127.4	34.4	*94*	201.2	227	*440*	824	454	*850*	1562
−10.6	*13*	55.4	12.2	*54*	129.2	35.0	*95*	203.0	232	*450*	842	460	*860*	1580
−10.0	*14*	57.2	12.8	*55*	131.0	35.6	*96*	204.8	238	*460*	860	466	*870*	1598
−9.4	*15*	59.0	13.3	*56*	132.8	36.1	*97*	206.6	243	*470*	878	471	*880*	1616
−8.9	*16*	60.8	13.9	*57*	134.6	36.7	*98*	208.4	249	*480*	896	477	*890*	1634
−8.3	*17*	62.6	14.4	*58*	136.4	37.2	*99*	210.2	254	*490*	914	482	*900*	1652
−7.8	*18*	64.4	15.0	*59*	138.2	37.8	*100*	212.0	260	*500*	932	488	*910*	1670
−7.2	*19*	66.2	15.6	*60*	140.0	43	*110*	230	266	*510*	950	493	*920*	1688
−6.7	*20*	68.0	16.1	*61*	141.8	49	*120*	248	271	*520*	968	499	*930*	1706
−6.1	*21*	69.8	16.7	*62*	143.6	54	*130*	266	277	*530*	986	504	*940*	1724
−5.6	*22*	71.6	17.2	*63*	145.4	60	*140*	284	282	*540*	1004	510	*950*	1742
−5.0	*23*	73.4	17.8	*64*	147.2	66	*150*	302	288	*550*	1022	516	*960*	1760
−4.4	*24*	75.2	18.3	*65*	149.0	71	*160*	320	293	*560*	1040	521	*970*	1778
−3.9	*25*	77.0	18.9	*66*	150.8	77	*170*	338	299	*570*	1058	527	*980*	1796
−3.3	*26*	78.8	19.4	*67*	152.6	82	*180*	356	304	*580*	1076	532	*990*	1814
−2.8	*27*	80.6	20.0	*68*	154.4	88	*190*	374	310	*590*	1094	538	*1000*	1832
−2.2	*28*	82.4	20.6	*69*	156.2	93	*200*	392	316	*600*	1112			
−1.7	*29*	84.2	21.1	*70*	158.0	99	*210*	410	321	*610*	1130			
−1.1	*30*	86.0	21.7	*71*	159.8	100	*212*	414	327	*620*	1148			

penses (e.g., the cost of maintaining an office and operating trucks)—plus enough to make a profit.

If you bid on a commercial or industrial job, as the electrical contractor you must provide a detailed listing of what it will cost to do the job. A bid on residential wiring does not require so much detail. However, before preparing any bid, you should have a set of drawings that specify all of the materials to be used and what is to be done.

CHECKLIST FOR ESTIMATING

- Receptacles
- Light fixtures
- Switches: 3-way, 4-way, single-pole
- Receptacle cover plates
- Switch cover plates
- Boxes: square, octagonal, switch, receptacle
- Wire nuts, hangers, connectors, ground clips, staples
- 240-volt service
  - wire
  - receptacles
  - oven
  - surface burners
  - dryer
  - other
- Circuit breakers
- Circuit breaker box
- Wire: signal and Romex
- Conduit
- Conduit fittings
- Labor
- Travel
- Overhead
- Miscellaneous

ESTIMATING THE COST OF MATERIALS

In determining the cost of materials to be figured in the overall estimate, one technique is to take the price paid for the material and double it. Most electrical supply houses have a list price and a net price. You, the electrician, will pay the net price, but you should use the list price in estimating. Estimates done this way leave enough leeway to allow you to offer the customer a discount for prompt payment—and still make a profit.

USING PAST EXPERIENCE AS A GUIDE IN ESTIMATING

After a job is complete, you can determine exactly what it cost to do it. You can figure what it cost to install a fixture, an outlet, a switch, or any other device. Then you can use this figure when preparing estimates for other, similar jobs. It will save much time if you have a reliable price to attach to specific jobs, such as the installation of a switch or dimmer. Then you simply count the number of switches, dimmers, outlets, or whatever is to be installed and multiply by the per-unit installation price you have determined based on past experience. This technique allows you to arrive at an accurate estimate for a job quickly.

TECHNIQUE FOR SAVING MONEY

One money-saving technique is buying in quantity. If, for example, you buy the supplies for two, three, or more jobs at one time, you can buy by the box or by 1,000 feet of wire and obtain a better price than if you were buying by the unit or by smaller rolls of wire. Unit cost is lower this way, and you benefit from this by being able to bid lower on a job, thus increasing your chances of getting the job, and/or by being able to please customers by offering a discount for prompt payment.

USING A CALCULATOR

The hand-held calculator has become so inexpensive that almost everyone now has one. It is a valuable tool for the electrician and electronics technician inasmuch as it can speed up the problem-solving and computation that is often a part of electrical and electronics work.

Many models of calculators are available. Some are complicated, with many formulas and other information stored on a chip. With simpler models, you must know the formula to be able to figure out the units you need for a particular problem. In general, electricians do not need complicated and expensive calculators. All that is needed is a basic a calculator with at least one level of memory capable of performing the following functions:

Square root	$\sqrt{x}$
Reciprocals	$1/x$
Squares	x^y or x^2
Trig functions, especially cosine	cos
Addition	$+$
Subtraction	$-$
Division	$\div$
Multiplication	$\times$

A pi (π) key is also useful in some instances. In most cases, the calculator will take π out to eight decimal places. If, however, π is rounded to 3.14 or 3.14159, a different answer will result.

Most calculators with memory and trigonometric tables also have the ability to do scientific notation. Many also are designed with U.S.–metric conversion formulas already stored for easy use.

When you purchase a calculator, read the instructions carefully and practice how to use the calculator to solve different types of problems. Then, when you need to use it on the job, you will be thoroughly familiar with its capabilities and be able to solve problems quickly.

7

Electrical and Electronic Formulas

The electrician or apprentice is often called on to calculate the current on a line, the resistance in a circuit, or other factors. Mathematical formulas provide the means for doing this. The formulas can be solved with a basic knowledge of mathematics (addition, subtraction, multiplication, and division), algebra, and, in some cases, trigonometry.

The formulas most often used by an electrician are given in Table 7-1.

USING OHM'S LAW

Ohm's law is one of the most important and frequently used formulas. Georg Simon Ohm (1787–1854), an early experimenter with electrical phenomena, formulated the law. Doing a mathematical analysis of circuits, he found that the voltage, current, and resistance in any circuit have a definite mathematical relationship to one another. Specifically, he found that the current or intensity of electron flow (I) in any circuit was equal to the voltage or electromotive force (E) divided by the resistance (R). Or, in mathematical terms

$$I = \frac{E}{R}$$

TABLE 7–1

Electrical and Electronic Formulas

Resistance		
Series Resistors	**Parallel Resistors**	
$R_T = R_1 + R_2 + R_3 + \ldots$	$R_T = \dfrac{R_1 \times R_2}{R_1 + R_2}$ (For TWO resistors *only*)	
$I_T = I_{R_1} = I_{R_2} = I_{R_3} \ldots$	$\dfrac{1}{R_T} = \dfrac{1}{R_1} + \dfrac{1}{R_2} + \dfrac{1}{R_3} + \ldots$	
$E_A = E_{R_1} + E_{R_2} + E_{R_3} + \ldots$	$I_T = I_{R_1} + I_{R_2} + I_{R_3} + \ldots$	
	$E_A = E_{R_1} = E_{R_2} = E_{R_3}$	
Ohm's Law		
$E = I \times R$	$I = \dfrac{E}{R}$	$R = \dfrac{E}{I}$
Power		
$P = E \times I$	$P = I^2R$	$P = \dfrac{E^2}{R}$

Time Constant

$$T = \frac{L}{R} \qquad T = R \times C$$

Inductive *Capacitive*

Alternating Current

Peak = $rms \times 1.414$ $\qquad$ rms = peak $\times$ 0.7071

Average = peak $\times$ 0.637 $\qquad$ Peak-to-peak = $rms \times 2.828$

Inductive Reactances $X_L = 2\pi FL$

In Series

$$X_{L_T} = X_{L_1} + X_{L_2} + \dots$$

In Parallel

$$X_{L_T} = \frac{X_{L_1} \times X_{L_2}}{X_{L_1} + X_{L_2}}$$

$$\frac{1}{X_{L_T}} = \frac{1}{X_{L_1}} + \frac{1}{X_{L_2}} + \frac{1}{X_{L_3}}$$

Transformers

$$\frac{E_p}{E_s} = \frac{I_s}{I_p} \qquad \frac{T_p}{T_s} = \frac{E_p}{E_s}$$

I_p = current, primary
I_s = current, secondary

T_p = turns, primary
T_s = turns, secondary

E_p = voltage, primary
E_s = voltage, secondary

Table 7–1 (*continued*)

Capacitances	
Capacitors in Series (All in the same unit of measurement) $\frac{1}{C_T} = \frac{1}{C_1} + \frac{1}{C_2} + \frac{1}{C_3} + \ldots$ $C_T = \frac{C_1 \times C_2}{C_1 + C_2}$ (2 only)	*Capacitors in Parallel* (All in the same unit of measurement) $C_T = C_1 + C_2 + C_3 + \ldots$
Capacitive Reactances	
$X_C = \frac{1}{2\pi FC}$	
In Series $X_{C_T} = X_{C_1} + X_{C_2} + X_{C_3}$	*In Parallel* $\frac{1}{X_{C_T}} = \frac{1}{X_{C_1}} + \frac{1}{X_{C_2}} + \frac{1}{X_{C_1}} + \ldots$ $X_{C_T} = \frac{X_{C_1} \times X_{C_2}}{X_{C_1} + X_{C_2}}$ (For TWO *only*)
Ohm's Law for Alternating Current (AC) Circuits	
$E_L = I_L \times X_L$ $I_L = \frac{E_L}{X_L}$ $X_L = \frac{E_L}{I_L}$	$E_C = I_C \times X_C$ $I_C = \frac{E_C}{X_C}$ $X_C = \frac{E_C}{I_C}$

Series *RL* Circuit	Series *RC* Circuit
$Z = \sqrt{R^2 + X_L^2}$	$Z = \sqrt{R^2 + X_C^2}$
$E_A = \sqrt{E_R^2 + E_L^2}$	$E_A = \sqrt{E_R^2 + F_C^2}$
$I_T = I_R = I_L$	$I_T = I_R = I_C$
$E_A = I_T Z$	$E_A = I_T Z$
*PF = Cos $\angle\theta$	PF = Cos $\angle\theta$
$Cos\ \angle\theta = \frac{R}{Z} = \frac{E_R}{E_A}$	$\text{Cos}\ \angle\theta = \frac{R}{Z} = \frac{E_R}{E_A}$
**TP = AP × PF	TP = AP × PF
TP = Watts	TP = Watts
***AP = Volt-Amperes	AP = Volt-Amperes
Phase $\angle$ = $\angle\theta$	Phase $\angle$ = $\angle\theta$
*PF = Power factor	
**TP = True power	
***AP = Apparent power	

Table 7–1 (*continued*)

Series *RCL* Circuits	Series *LC* Circuits
$Z = \sqrt{R^2 + (X_L - X_C)^2}$	$X_L = X_C$ Resonance
$E_A = \sqrt{E_R^2 + (E_L - E_C)^2}$	$E_L = E_C$ Resonance
$I_T = I_R = I_L = I_C$	$I_T =$ Maximum (∞)
$\text{Cos} \angle\theta =$ Power factor *(PF)*	$Z =$ Minimum (0)
$\text{Cos} \angle\theta = \frac{R}{Z} = \frac{E_R}{E_A}$	
$E_A = I_T Z$	
$\text{TP} = \text{AP} \times \text{Cos} \angle\theta$	
$\text{AP} = V \times A$	

Coil Merit [No Unit of Measurement]

$$Q = \frac{X_L}{R}$$

Resonance

$$F_r = \frac{1}{2\pi\sqrt{LC}} \quad \text{or,} \quad F_r = \frac{0.159}{\sqrt{LC}}$$

Parallel *RL* Circuits	Parallel *RC* Circuits
$Z = \frac{E_A}{I_T}$	$Z = \frac{E_A}{I_T}$
$E_A = E_R = E_L$	$E_A = E_R = E_C$
$I_T = \sqrt{I_R^2 + I_L^2}$	$I_T = \sqrt{I_R^2 + I_C^2}$
$\text{Cos} \angle\theta = \frac{I_R}{I_T}$	$\text{Cos} \angle\theta = \frac{I_R}{I_T}$
$\text{PF} = \text{Cos} \angle\theta$	$\text{PF} = \text{Cos} \angle\theta$
$\text{TP} = \text{AP} \times \text{PF}$	$\text{TP} = \text{AP} \times \text{PF}$
$\text{Phase} \angle = \angle\theta$	$\text{Phase} \angle = \angle\theta$

Parallel *RCL* Circuits

$Z = \frac{E_A}{I_T}$	$\text{Cos} \angle\theta = \frac{I_R}{I_T}$
$E_A = E_R = E_L = E_C$	$\text{TP} = \text{AP} \times \text{PF}$ (Watts)
$I_T = \sqrt{I_R^R + (I_L - I_C)^2}$	$\text{AP} = \text{V} \times \text{A}$ (Volt-Amperes)
$\text{Phase} \angle = \angle\theta$	$\text{PF} = \text{Cos} \angle\theta$

Antiresonance

Parallel *LC* Circuits	or	Tank Circuits
$X_L = X_C$	$I_{\text{circulating}} = I_C$ or I_L	
Z = maximum (∞)	$I_C = \frac{E_A}{X_C}$	$I_L = \frac{E_A}{X_L}$
I_{line} = minimum (0)		

Figure 7-1 shows how you can develop the formula for Ohm's law rather quickly and accurately. Let your finger cover the letter representing the value you are looking for, and the remaining two letters will show you the relationship between them by being situated either one on top of the other (division) or one next to the other (multiplication).

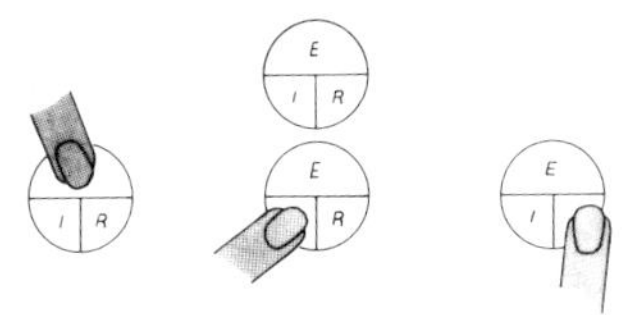

Figure 7-1. Graphic representation of Ohm's Law.

■ EXAMPLE 1

A resistor has a voltage of 120 volts applied to its 40 ohms of resistance. What is the current that will flow through the resistor?

Step 1. Decide which formula to use. Since the unknown is the current (I), the formula to use is

$$I = \frac{E}{R}$$

Step 2. Substitute the known values—voltage (E) and resistance (R)—in the formula.

$$I = \frac{120}{40}$$

Step 3. Perform the mathematical operation—in this case, division.

$$I = 3 \text{ amperes}$$

■ EXAMPLE 2

How much voltage do you need to cause 2 amperes to flow through a resistance of 20 ohms?

Step 1. Decide which formula to use. Since the unknown is the voltage (E), the formula to use is

$$E = I \times R$$

Step 2. Substitute the known values—the current (I) and resistance (R)—in the formula.

$$E = 2 \times 20$$

Step 3. Perform the mathematical operation—in this case, multiplication.

$$E = 40 \text{ volts}$$

■ EXAMPLE 3

What is the resistance of a circuit that has a voltage source of 24 volts pushing 6 amperes through a resistor?

Step 1. Determine the unknown and use the appropriate formula.

$$R = \frac{E}{I}$$

Step 2. Substitute the known values in the formula.

$$R = \frac{24}{6}$$

Step 3. Perform the mathematical operation.

$$R = 4 \text{ ohms}$$

USING POWER FORMULAS

A certain amount of power is expended whenever an electric current is passed through a load such as a resistor. This energy is dissipated in the form of heat, which must be dissipated into the surrounding medium. The amount of heat that can be dissipated from the surface of the resistor determines its wattage rating, or power rating.

The unit of measure for electric power is the watt (W). The watt was named for James Watt (1736–1819), a Scottish engineer and the inventor of the modern condensing steam engine, one of the first sources of mechanical power.

The watt is mathematically equal to the product of the voltage (E) and current (I),

$$P \text{ (power in watts)} = E \times I$$

Many relationships between P, E, R, and I can be obtained by using simple algebra.

$$P = E \times I \qquad P = \frac{E^2}{R}$$

$$I = \frac{P}{E} \qquad P = I^2R$$

EXAMPLE 1

What is the power consumed by an electric iron that has 120 volts at 10 amperes?

Step 1. Determine which formula to use. Since you know the voltage (E) and the current (I), the formula to use is

$$P = E \times I$$

Step 2. Substitute the known values in the formula.

$$P = 120 \times 10$$

Step 3. Perform the mathematical computation.

$$P = 1{,}200 \text{ watts}$$

EXAMPLE 2

How much current does a 120-volt, 60-watt light bulb pull?

Step 1. Determine which formula to use. Since you know the voltage (E) and the wattage, or power (P), the formula to use is

$$I = \frac{P}{E}$$

Step 2. Substitute the known values in the formula.

$$I = \frac{60\ \text{W}}{120\ \text{V}}$$

Step 3. Perform the mathematical computation.

$$I = 0.5\ \text{ampere}$$

EXAMPLE 3

What is the current required to light a 120-volt, 15-watt light bulb to full brilliance?

Step 1. Use the appropriate formula—in this case

$$I = \frac{P}{E}$$

Step 2. Substitute the known factors in the equation.

$$I = \frac{15}{120}$$

Step 3. Perform the mathematical computation.

$$I = 0.125\ \text{amperes}$$

Power formulas can be developed, and the current that each wattage of light bulb will draw on a 120-volt circuit can be determined. (See Table 7-2.) This type of table is handy when you need to compute how much current a circuit of so many light bulbs will draw, and it also helps in selecting the right size of wire for a job.

TABLE 7–2

Current Drawn by Different Wattage Bulbs on a 120-volt Circuit

Voltage 120	
Wattage	**Current (amperes)**
15	0.1250
25	0.2083
40	0.3330
60	0.5000
75	0.6250
100	0.8333
150	1.2500
200	1.6660
300	2.5000

POWER FACTOR

The power factor (PF) is the ratio of true power (TP) to apparent power (AP), or

$$PF = \frac{TP}{AP}$$

where true power is given in watts (W) and apparent power in volt-amperes (VA). Therefore, the power factor relationship can also be expressed as

$$PF = \frac{W}{VA}$$

EXAMPLE 1

If true power is 100 watts and apparent power is 120 volt-amperes, what is the power factor?

Step 1. Use the correct formula.

$$PF = \frac{TP}{AP} = \frac{W}{VA}$$

Step 2. Substitute the known factors in the equation.

$$PF = \frac{100}{120}$$

Step 3. Perform the mathematical computation.

$$PF = 0.8$$

Note: An important fact to remember with power factors is that the ratio is never greater than 1. For estimating purposes, the power factor in any circuit can be assumed to be as follows:

Incandescent light circuit	0.95 to 1.0 PF
Lighting and motors	0.85 PF
Motors only	0.80 PF

The power factor formula can also be used to show other relationships. If, for example, in a direct-current circuit, true power is expressed in kilowatts (kW) and apparent power as the product of kilovolts and amperes (kVA), then

$$\text{kW} = \frac{E \times I}{1{,}000} = \frac{\text{kVA}}{1{,}000}$$

For alternating current,

$$\text{kW} = \text{kVA} \times \text{PF}$$

DETERMINING DIFFERENT PHASE POWER

If the voltage (E) and current (I) are known, the power factor (PF) can be used to determine different phase power in kilowatts (kW).

To determine three-phase power (3ϕ) use the formula

$$\text{kW} = \frac{1.73 \times E \times I \times \text{PF}}{1{,}000}$$

To determine two-phase power (2ϕ), use the formula

$$\text{kW} = \frac{2 \times E \times I \times \text{PF}}{1{,}000}$$

To determine single-phase power (1ϕ), use the formula

$$\text{kW} = \frac{E \times I \times \text{PF}}{1{,}000}$$

If you wish to have the answer in wattage, eliminate dividing by 1,000, and then your answer will be in watts.

CONNECTIONS AND RESISTANCE IN THREE-PHASE POWER

Most commercial power is generated as three-phase. There are two types of three-phase hookups: delta and wye. Delta and wye connections are used in connecting transformers and in connecting the loads to these power sources. A three-phase delta resembles the Greek letter delta, Δ , which is triangular in shape, and the three-phase wye is shaped like the letter "Y."

WYE CONNECTION

The wye connection in Figure 7-2 has terminals labeled *a, b,* and *c,* and a common point called the neutral, shown at point 0. The terminal pairs, *a-b, b-c, c-a,* provide the three-phase supply.

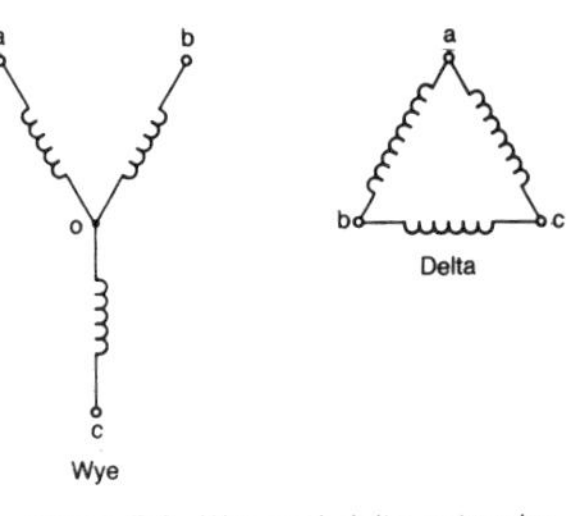

Figure 7-2. Wye and delta networks.

In this connection the line voltage is $\sqrt{3}$, which is 1.732050808 times the coil voltage, while the line current is the same as the coil current. The neutral point normally is grounded. It can be brought out to the power-consuming device by way of a four-wire power system for a dual voltage supply. This means it can produce a 120/208-volt system. If three wires are used and connected to points *a*, *b*, and *c*, then three-phase AC is available for use with motors and other loads. If you are using a 208-volt system, you know that you have a wye-connected transformer supplying the power. Single-phase AC is available if you connect from 0 to any one of the points *a*, *b*, or *c*. From 0 to *a* will produce 120 volts of single-phase AC. From 0 to *b* will produce 120 volts of single-phase AC, and from 0 to *c* will also produce 120 volts of single-phase AC.

DELTA CONNECTION

The delta connection has an advantage in current production because two of the coils are in series with each other and are parallel with the third. A parallel arrangement produces better current availability. The voltage available from any two terminals, *a* to *b* or *b* to *c* or *c* to *a*, is the same. This voltage is also single-phase. However if you wish to have 240 volts of three-phase AC, all you need to do is bring out three wires—one each from point *a*, point *b*, and point *c*. The current available in a three-phase delta connection is $\sqrt{3}$ times the current capability of any one coil.

DELTA AND WYE RESISTOR CIRCUITS

In the delta network (See Figure 7-3), the resistance between terminals may be determined by combining the formulas for series and parallel resistances:

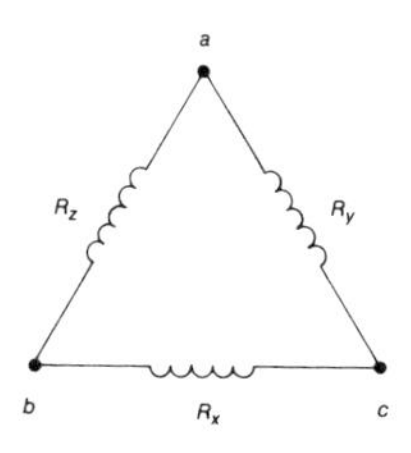

Figure 7-3. Delta network.

$$a \text{ to } b = \frac{R_z \cdot (R_x + R_y)}{R_z + (R_x + R_y)}$$

$$a \text{ to } c = \frac{R_y \cdot (R_x + R_z)}{R_y + (R_x + R_z)}$$

$$b \text{ to } c = \frac{R_x \cdot (R_z + R_y)}{R_x + (R_z + R_y)}$$

To simplify resistor networks for determining equivalent resistances, it is frequently necessary to convert a delta network to a wye (see Figure 7-4), or a wye to a delta. The mathematics for such conversions are simple and are based on the formulas for series and parallel resistance.

To convert from wye to delta

$$R_x = \frac{R_a \cdot R_b + R_b \cdot R_c + R_c \cdot R_a}{R_a}$$

$$R_y = \frac{R_a \cdot R_b + R_b \cdot R_c + R_c \cdot R_a}{R_b}$$

$$R_z = \frac{R_a \cdot R_b + R_b \cdot R_c + R_c \cdot R_a}{R_c}$$

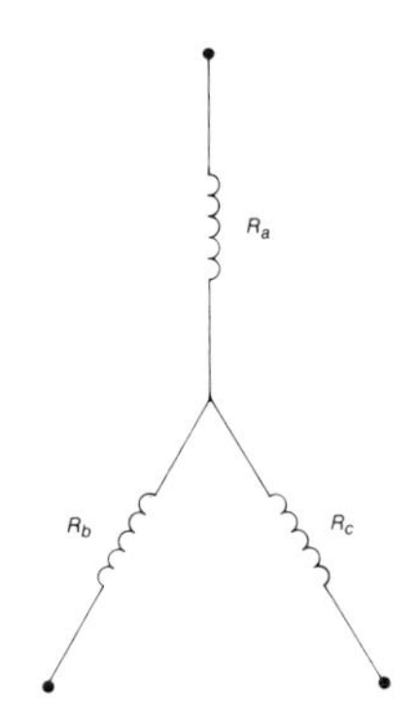

Figure 7-4. Wye network.

To convert from delta to wye, the formulas to use are:

$$R_a = \frac{R_y \times R_z}{R_x + R_y + R_z}$$

$$R_b = \frac{R_x + R_z}{R_x + R_y + R_z}$$

$$R_c = \frac{R_x + R_y}{R_x + R_y + R_z}$$

Simply put in the correct values and multiply and add to obtain the desired results.

USING TRANSFORMER FORMULAS

A transformer, you remember, is a device that increases or decreases voltage by inducing electrical energy from one coil to another through magnetic lines of force. The amount of voltage induced on either side (primary or secondary) of a transformer depends directly on the number of turns. The back electromotive force induced by the changing current in the primary is not equal to the electromotive force induced in the secondary unless the number of turns in the primary equals the number in the secondary. However, since the back electromotive force in the primary is equal to the applied voltage, a ratio may be set up to determine the electromotive force induced in the secondary in terms of applied voltage and the turns ratio of the two coils. The formula is

$$\frac{E_p}{T_p} = \frac{E_s}{T_s}$$

where

E_s is the voltage induced in the secondary

E_p is the voltage induced in the primary

T_p is the number of turns in the primary

T_s is the number of turns in the secondary

The formula may also be written as

$$E_pT_s = E_sT_p$$

Another form of the formula is

$$E_s = \frac{E_pT_s}{T_p}$$

■ EXAMPLE 1

A transformer with 1,000 turns in the secondary and 250 turns in the primary has a turn ratio of 4:1. If 120 volts AC is applied to the primary of this transfer, what is the voltage induced in the secondary?

Step 1. Decide which formula to use. Since you know the number of turns in the primary and secondary, the formula to use is

$$E_s = \frac{E_p T_s}{T_p}$$

Step 2. Substitute the known values and perform the mathematics computation.

$$E_s = \frac{120 \times 1{,}000}{250}$$

$$E_s = \frac{120{,}000}{250}$$

$$E_s = 480 \text{ volts}$$

This means that the transformer in question is a step-up transformer because the secondary voltage is greater than the primary voltage, or in other words, the voltage has been stepped up.

The number of volts per turn in a transformer can also be determined using the formula

$$\frac{\text{number of turns}}{\text{voltage}}$$

If you have 675 turns for a 115-volt primary, the 675 divided by 115 means you have 5 turns-per-volt in the primary. Then, if you want to make a 12-volt secondary, you simply multiply the 5 turns-per-volt times the 12 volts you need in the secondary. This means you have to put 60 turns in the secondary to produce

12 volts of output. The correct number of turns in the secondary can be figured rather quickly using this method.

If you need to put 900 turns in the primary to increase the X_L of the transformer primary and have it draw less current while sitting idle on the line, then you divide the 900 turns by the 120 volts of the line and you get 7.5 turns-per-volt. Then, if you need 12 volts for the secondary voltage, take the 12 and multiply it by 7.5 to get the 90 turns needed to provide the voltage. Don't forget to add about 5% to make up for the inefficiency of the transformer.

This is a step-down transformer, since the secondary voltage is lower than the primary voltage.

USING ALTERNATING CURRENT FORMULAS

A sine wave is the signature of alternating current. It has two alternations per hertz, or cycle. It is the alternation that is examined to obtain the rms, the peak value (E_m), and the average values (E_{av}) of AC. (See Figure 7-5.)

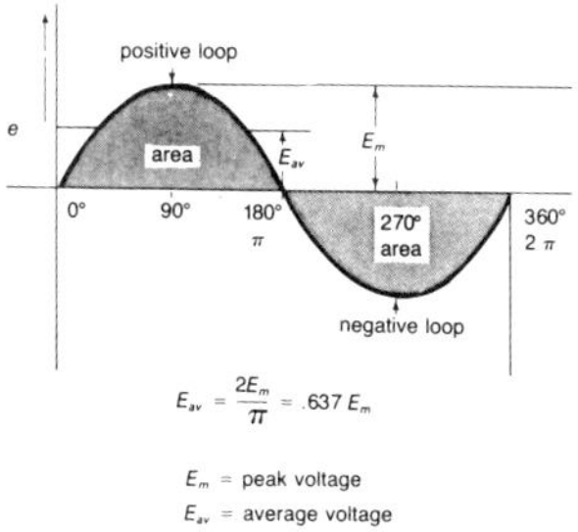

Figure 7-5. A sine wave is the signature of alternating current.

RMS

The abbreviation *rms* means root-mean-square, the value used when a meter is calibrated to read voltage or current. When we say 120-volts AC, we really mean it is 120-volts rms.

The rms is a way of equating AC and DC. It is the heating effect of AC compared to DC. A DC power source provides a steady voltage that rises almost instantaneously and stays at that level with no cooling period. AC varies, causing the resistor to heat up slowly and then cool down. This is why AC has only 70.71% of the heating value of an equal amount of DC. Remember this when comparing AC and DC systems.

PEAK VALUE

Peak is the maximum or highest point on the sine wave. To convert rms to peak, multiply by 1.414. A 120-volt rms thus has a 169.68-volt peak.

Peak-to-peak (p-p) is a term used to describe the signal voltage in television receivers. It is a term used more with electronics than with electrical work.

AVERAGE VALUE

The average value of a sine wave function is defined as the area under one loop divided by the base of the loop. It is the average value of AC being delivered. The base of one loop is 180° or pi (π) radians in length. However, in order to find the area of such an irregular surface as a sine loop, the figure must be broken down into a series of small rectangles whose areas can be easily determined. The sum of all these small areas will then be the area of the loop.

To convert from peak to average, multiply by 0.637. The 169.68-volt peak (see above) would then have an average of 108.

FORMULAS USED WITH MOTORS

Working with motors sometimes requires the use of special formulas—for example, to figure the horsepower of a motor or to determine the power needed to drive a pump or a fan. Table 7-3 lists abbreviations used in dealing with motors and gives useful formulas.

TABLE 7–3

Formulas for Motor Applications

T = torque or twisting moment (Force × moment arm length)
π = 3.1416
N = revolutions per minute
HP = horsepower (33,000 ft.-lbs. per min.) applies to power output
R = radius of pulley, in feet
E = input voltage
I = current in amperes
P = power input in watts

$$HP = \frac{T(\text{lb-in.}) \times N(\text{rpm})}{63{,}025}$$

$$HP = T(\text{oz-in.}) \times N \times 9.917 \times 10^{-7}$$

$$= \text{approx. } T(\text{oz-in.}) \times N \times 10^{-6}$$

$$P = EI \times \text{power factor} = \frac{HP \times 746}{\text{motor efficiency}}$$

Power to Drive Pumps:

$$HP = \frac{\text{Gal. per min.} \times \text{Total Head (inc. friction)}}{3{,}960 \times \text{eff. of pump}}$$

Where:

Approx. friction head (ft.) =

$$\frac{\text{pipe length (ft.)} \times [\text{velocity of flow (fps)}]^2 \times 0.02}{5.367 \times \text{diameter (in.)}}$$

Eff. = Approx. 0.50 to 0.85

Time to Change Speed of Rotating Mass:

$$\text{Time (sec.)} = \frac{WR^2 \times \text{change in rpm}}{308 \times \text{torque (ft-lb.)}}$$

$$\text{Where: } WR^2 \text{ (disc)} = \frac{\text{Weight (lbs.)} \times [\text{radius (ft.)}]^2}{2}$$

WR^2 (rim) =

$$\frac{\text{Wt. (lbs.)} \times [(\text{outer radius in ft.})^2 + \text{inner radius in ft.})^2]}{2}$$

Power to Drive Fans:

$$HP = \frac{\text{Cu. ft. air per min.} \times \text{water gage pressure (in.)}}{6.350 \times \text{Eff.}}$$

8

Basic Electronics

Electronics got its start with the invention of the vacuum tube. A vacuum tube has the ability to rectify and to amplify. Edison was experimenting with his first electric lamp when he discovered that the frequent burning out of the carbon filament near the positive end produced a black deposit inside the bulb and a shadow on the positive leg of the lamp filament. He tried to remedy the condition by placing a metal plate inside the lamp near the filament and connecting it through a galvanometer to the filament battery. When the positive terminal was connected to the metal plate, a deflection of the galvanometer was observed. When the plate was connected to the negative terminal, there was no deflection of the galvanometer. Edison recorded these effects in his notebook but did not pursue the phenomenon.

Sir J. J. Thomson, an English physicist, discovered the electron around the turn of the century. Once the electron was discovered and defined, it was an easy task to describe what happened during the "Edison effect." Electrons escaped from the heated carbon filament and, being negatively charged, moved freely through the vacuum to the plate when it was connected to the positive battery terminal. However, when the plate was negative, the electrons were repelled, and no deflection was observed on the galvanometer.

The vacuum tube diode was the first device to take advantage of the Edison effect. It was made to *rectify* alternating current—that is, to change the alternating current to direct current.

DIODES

There are two types of diodes. One is the vacuum tube diode, and the other is the semiconductor diode. Both types possess two elements: a cathode and an anode. These elements have polarity. A cathode in a vacuum tube is labeled negative (−), and the anode is positive (+). In a semiconductor device, the cathode is positive, which is indicated by a ring around one end of the diode. Both types of diode serve the same functions: they rectify current and are used for switching operations.

Diodes can also be broken down into two other groups—power diodes and signal diodes. Power diodes handle large power requirements in power supplies and are used to change AC to pulsating DC. They can be either vacuum tubes or semiconductors. Signal diodes are smaller than power diodes and may be either vacuum tubes or semiconductors. As semiconductors, they are very small and encapsulated in plastic or glass.

RECTIFIER DIODES

The actual size of these semiconductor diodes is small, but they can handle up to 6 amperes at 400 volts and are very inexpensive. When properly used in circuits that do not exceed their ratings, they will last for many years. Rectifier diodes are identified by three or four numbers with a 1N or MR prefix. See Figure 8-1 for the wide range of currents and voltages available, and see how the characteristic numbers are used to identify them.

Description	Industry Part No.
1 AMP 50V	1N4001
1 AMP 500V	1N4002
1 AMP 200V	1N4003
1 AMP 400V	1N4004
1 AMP 400V	1N4004
1 AMP 600V	1N4005
1 AMP 100V	1N4007
1 AMP 100V	1N4007
3 AMPS 100V	1N5401
3 AMPS 200V	1N5402
3 AMPS 300V	1N5403
3 AMPS 400V	1N5404
3 AMPS 600V	1N5406
6 AMPS 200V	MR751
6 AMPS 200V	MR752
6 AMPS 400V	MR754
3 AMPS 100V FAST RECOVERY	MR851
3 AMPS 400V FAST RECOVERY	MR854
1 AMP 50V FAST RECOVERY	1N4933
1 AMP 100V FAST RECOVERY	1N4934
1 AMP 600V FAST RECOVERY	1N4937

Figure 8-1. Rectifier diodes are available in a wide range of currents and voltages.

BRIDGE RECTIFIERS

Bridge rectifiers are made up of four diodes. They produce a full-wave rectification so that both halves of the wave are utilized in power supplies. They are found in many devices from battery chargers to meters. Many electrical power supplies have bridge rectifiers. They are packaged in a variety of forms. (See Figure 8-2.)

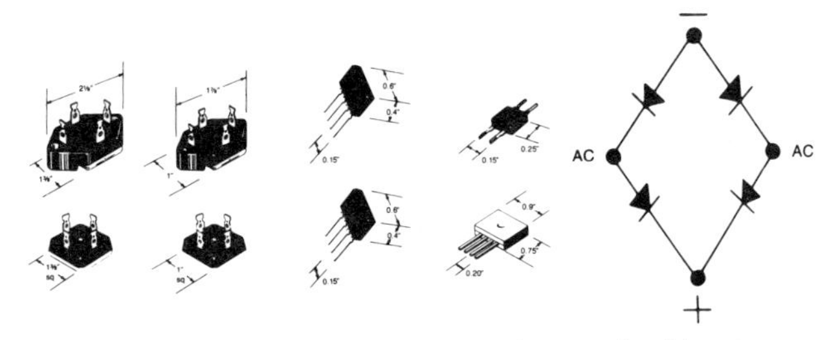

Figure 8-2. Bridge rectifiers, packaged in many forms, are found in many devices.

DIACS

This device (diode [two-element] AC switch) is used in switching-amplifier control systems. It is used in combination with a triac to serve as a light dimmer and to regulate the speed of electric motors. The diac is made specifically to trigger triacs but is also useful in conjunction with silicon controlled rectifiers. It will trigger with either positive or negative pulses. (See Figure 8-3.)

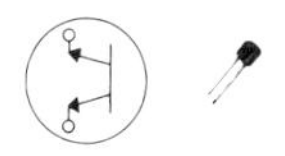

Figure 8-3. A diac, or diode AC switch, is used with triacs and SCRs.

SILICON CONTROLLED RECTIFIERS (SCRs)

These are also called thyristers. They have many uses, such as for speed controls, light dimmers, and many other control circuits. (See Figure 8-4.)

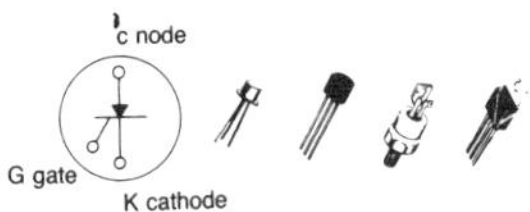

Figure 8-4. A silicon controlled rectifier (SCR), also called a thyrister, is used in many control circuits.

TRIACS

The triac (standing for *tri*ode [three-element] AC switch) can be triggered by positive and negative gate signals and conducts during both halves of the AC cycle. This device is used in level controls, light dimmers, and speed controls for motors. (See Figure 8-5.)

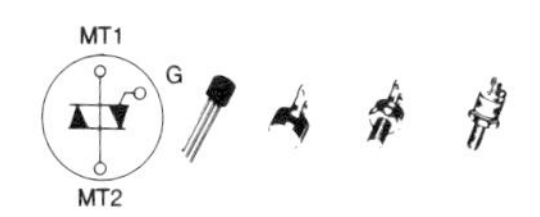

Figure 8-5. A triac, or triode AC switch, is used in level controls and speed control devices.

SWITCHING AND SIGNAL DIODES

These diodes are for small currents and voltages of less than 100. They can be found in many older devices of the germanium type. Switching diodes are used for computer and control circuits. (See Table 8-1.)

ZENER DIODES

This type of diode is used for voltage regulation and is found in almost every piece of industrial and commercial electronics equipment or control device. The zener is also prefixed with a 1N designation. Wattage ratings are from 400 milliwatts to 5 watts. Be sure to obtain the right wattage rating if you want the diode to operate without burning up. Zener diodes resemble the standard rectifier diode in appearance. However, the symbol for the zener is slightly different, so it can be spotted on a schematic. (See Table 8-2.)

TABLE 8–1

Switching and Signal Diodes

Description	Industry Part No.	Description	Industry Part No.
5MA 60V Germanium Signal	1N34A	200MA 75V Switching	1N4148
200 MA 100V Switching	1N270	200MA 50V Switching	1N4150
200MA 75V Switching	1N914B	100MA 25V Switching	1N4154
200MA 75V Switching	1N914B	200MA 70V Switching	1N4446
200MA 75V Switching	1N4148	200MA 75V Switching	1N4448

TABLE 8–2
ZENER DIODES QUICK REFERENCE GUIDE

Low Power 500 MW

NOMINAL VOLTAGE V_2	400mW DO-35	500mW DO-35	400mW DO-7		500mW DO-7	250 & 400mW DO-7	250mW DO-7
1.8						1N4614	1N4678
2.0						1N4615	1N4679
2.2						1N4616	1N4780
2.4		1N5985	1N4370		1N5221	1N4617	1N4681
2.5					1N5222		
2.7		1N5986	1N4371		1N5223	1N4618	1N4682
2.8					1N5224		1N4683
3.0		1N5987	1N4372		1N5225	1N4619	1N4683
3.3		1N5988	1N746		1N5226	1N4620	1N4684
3.6		1N5989	1N747		1N5277	1N4621	1N4685
3.9		1N5990	1N748		1N5228	1N4622	1N4686
4.3		1N5991	1N749		1N5229	1N4623	1N4687
4.7	1N5728B	1N5992	1N750		1N5230	1N4624	1N4688
5.1	1N4729B	1N5993	1N751		1N5231	1N4625	1N4689
5.6	1N5730B	1N5994	1N752		1N5232	1N4626	1N4690
6.0					1N5233		
6.2	1N5731B	1N5995	1N753		1N5234	1N4627	1N4691
6.8	1N5732B	1N5996	1N754	1N957	1N5235	1N4099	1N4692
7.5	1N5733B	1N5997	1N755	1N958	1N5236	1N4100	1N4693
8.2	1N5734B	1N5998	1N756	1N959	1N5237	1N4101	1N4694
8.7					1N5238	1N4102	1N4695
9.1	1N5735B	1N5999	1N757	1N960	1N5239	1N4103	1N4696
10	1N5736B	1N6000	1N759	1N961	1N5240	1N4104	1N4697
11	1N5737B	1N6001		1N962	1N5243	1N4105	1N4698
12	1N5738B	1N6002	1N758	1N963	1N5242	1N4106	1N4699
13	1N5739B	1N6003		1N964	1N5243	1N4107	1N4700
14					1N5244	1N4108	1N4701
15	1N5740B	1N6004		1N965	1N5245	1N4109	1N4702
16	1N5741B	1N6005		1N966	1N5246	1N4110	1N4703
17					1N5247	1N4111	1N4704
18	1N5742B	1N6006		1N967	1N5248	1N4112	1N4705
19					1N5249	1N4113	1N4706
20	1N5743B	1N6007		1N968	1N5250	1N4114	1N4707
22	1N5744B	1N6008		1N969	1N5251	1N4115	1N4708
21	1N5745B	1N6009		1N970	1N5252	1N4116	1N4709

Table 8–2 (*continued*)

NOMINAL VOLTAGE V_2	400MW DO-35	500MW DO-35	400MW DO-7	500MW DO-7	250 & 400MW DO-7	250MW DO-7
25				1N5253	1N4117	1N4710
27	1N5746B	1N6010	1N971	1N5254	1N4118	1N4711
28				1N5255	1N4119	1N4712
30	1N5747B	1N6011	1N972	1N5256	1N4120	1N4713
33	1N5748B	1N6012	1N973	1N5257	1N4121	1N4714
36	1N5749B	1N6013	1N974	1N5258	1N4122	1N4715
39	1N5750B	1N6014	1N975	1N5259	1N4123	1N4716
43	1N5751B	1N6015	1N976	1N5260	1N4124	1N4717
47	1N5752B	1N6016	1N977	1N5261	1N4125	1N4718
51	1N5753B		1N978	1N5262	1N4126	
56			1N979	1N5263	1N4127	
60				1N5264	1N4128	
62			1N980	1N5265	1N4129	
68			1N981	1N5266	1N4130	
75			1N982	1N5267	1N4131	
82			1N983	1N5268	1N4132	
87				1N5269	1N4133	
91			1N984	1N5270	1N4134	
100			1N985	1N5271	1N4135	
110			1N986	1N5272		
120			1N987	1N5273		
130			1N988	1N5274		
140				1N5275		
150			1N989	1N5276		
160			1N990	1N5277		
170				1N5278		
180			1N991	1N5279		
190				1N5280		
200			1N992	1N5281		

Table 8–2 (*continued*)

Medium Power 1 Through 5 Watt

NOMINAL VOLTAGE V_z	1 WATT DO-41	1 WATT DO-13	1 WATT CASE J	2 WATT CASE J	3 WATT CASE J	5 WATT CASE T-18
3.3	1N4728A	1N3821	1N4728			1N5333
3.6	1N4729A	1N3822	1N4729	2EZ3.6D5		1N5334
3.9	1N4730A	1N3823	1N4730	2EZ3.9D5	3EZ3.9D5	1N5335
4.3	1N4731A	1N3824	1N4731	2EZ4.3D5	3EZ4.3D5	IN5336
4.7	1N4732A	1N3825	1N4732	2EZ4.7D5	3EZ4.7D5	IN5337
5.1	1N4733A	1N3826	1N4733	2EZ5.1D5	3EZ5.1D5	IN5338
5.6	1N4734A	1N3827	1N4734	2EZ5.6D5	3EZ5.6D5	1N5339
6.0						1N5340
6.2	1N4735A	1N3828	1N4735	2EZ6.2D5	3EZ6.2D5	1N5341
6.8	1N4736A	1N3829 1N3016	1N4736	2EZ6.8D5	3EZ6.8D5	1N5342
7.5	1N4737A	1N3830 1N3017	1N4737	2EZ7.5D5	3EZ7.5D5	1N5343
8.2	1N4738A	1N3018	1N4738	2EZ8.2D5	3EZ8.2D5	1N5344
8.7						1N5345
9.1	1N4739A	1N3019	1N4739	2EZ9.1D5	3EZ9.1D5	IN5346
10	1N4740A	1N3020	1N4740	2EZ10D5	3EZ10D5	1N5347
11	1N4741A	1N3021	1N4741	2EZ11D5	3EZ11D5	1N5348
12	1N4742A	1N3022	1N4742	2EZ12D5	3EZ12D5	1N5349
13	1N4743A	1N3023	1N4743	2EZ13D5	3EZ13D5	1N5350
14				2EZ14D5	3EZ14D5	1N5351
15	1N4744A	1N3024	1N4744	2EZ15D5	3EZ15D5	1N5352
16	1N4745A	1N3025	1N4745	2EZ16D5	3EZ16D5	1N5353
17				2EZ17D5	2EZ17D5	1N5354
18	1N4746A	1N3026	1N4746	2EZ18D5	3EZ18D5	1N5355
19				2EZ19D5	3EZ19D5	1N5356
20	1N4747A	1N3027	1N4747	2EZ20D5	3EZ20D5	1N5357
22	1N4748A	1N3028	1N4748	2EZ22D5	3EZ22D5	1N5358
24	1N4749A	1N3029	1N4749	2EZ24D5	3EZ24D5	1N5359
25						1N5360
27	1N4750A	1N3030	1N4750	2EZ27D5	3EZ27D5	1N5361
28						1N5362
30	1N4751A	1N3031	1N4751	2EZ30D5	3EZ30D5	1N5363
33	1N4752A	1N3032	1N4752	2EZ33D5	3EZ33D5	1N5364
36	1N4753A	1N3033	1N4753	2EZ36D5	3EZ36D5	1N5365
39	1N4554A	1N3034	1N4754	2EZ39D5	3EZ39D5	1N5366
43	1N4555A	1N3035	1N4755	2EZ43D5	3EZ43D5	1N5367

Table 8–2 (*continued*)

NOMINAL VOLTAGE V_Z	1 WATT DO-41	1 WATT DO-13	1 WATT CASE J	2 WATT CASE J	3 WATT CASE J	5 WATT CASE T-18
47	1N4756A	1N3036	1N4756	2EZ47D5	3EZ47D5	1N5368
51	1N4757A	1N3037	1N4757	2EZ51D5	3EZ51D5	1N5369
56		1N3038	1N4758	2EZ56D5	3EZ56D5	1N5370
60					3EZ60D5	1N5371
62		1N3039	1N4759	2EZ62D5	3EZ62D5	1N5372
68		1N3040	1N4760	2EZ68D5	3EZ68D5	1N5373
75		1N3041	1N4761	2EZ75D5	3EZ75D5	1N5374
82		1N3042	1N4762	2EZ82D5	3EZ82D5	1N5375
87						1N5376
91		1N3043	1N4763	2EZ91D5	3EZ91D5	1N5377
100		1N3044	1N4764	2EZ100D5	3EZ100D5	1N5378
110		1N3045	1EZ110D5	2EZ110D5	3EZ110D5	1N5379
120		1N3046	1EZ120D5	2EZ120D5	3EZ120D5	1N5380
130		1N3047	1EZ130D5	2EZ130D5	3EZ130D5	1N5381
140			1EZ140D5	2EZ140D5	3EZ140D5	1N5382
150		1N3048	1EZ150D5	2EZ150D5	3EZ150D5	1N5383
160		1N3049	1EZ160D5	2EZ160D5	3EZ160D5	1N5384
170			1EZ170D5	2EZ170D5	3EZ170D5	1N5385
180		1N3050	1EZ180D5	2EZ180D5	3EZ180D5	1N5386
190			1EZ190D5	2EZ190D5	3EZ190D5	1N5387
200		1N3051	1EZ200D5	2EZ200D5	3EZ200D5	1N5388

INTEGRATED CIRCUITS (ICs)

A single, monolithic chip of semiconductor was developed in 1958. J. S. Kilby was responsible for its fabrication. Active and passive circuit elements were successively diffused and deposited on a single chip. Shortly afterward, Robert Noyce made a complete circuit in a single chip. This led the way to the modern, inexpensive integrated circuits.

Resistors, capacitors, and transistors, as well as diodes, can be placed on a chip. Diodes can be arranged in many groups to do different things. Photolithography, a combination of photographic and printing techniques, has been used to aid in the layout of the diode arrangement and has made it possible to mass-produce sophisticated devices with high reliability.

Integrated circuits are packaged in a number of types of ceramic and plastic cases. They may have four or as many as 48 pins. The identification of the pins starts with the notch in the top of the case. Usually it will also have a dot on top to indicate the No. 1 pin. (See Figure 8-6.)

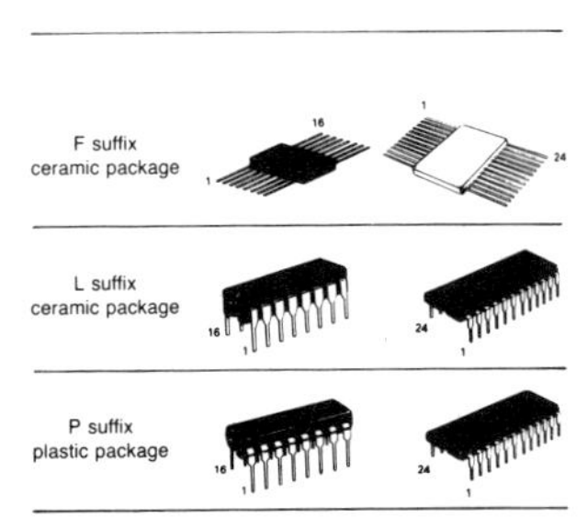

Figure 8-6. Integrated circuits with from 4 to as many as 48 pins are packaged in a variety of ceramic and plastic cases.

Integrated circuits are relatively standardized. This is partly to take advantage of mass-production techniques. Standard components and connections are needed if the finished electronic equipment is to be low in cost.

Large numbers of ICs are used in computers and calculators, as well as in television and other communications equipment. They are small and very reliable, and they consume little power. It is possible to obtain ICs with large memory capacities. Thousands of electronic elements may be placed on a single chip; this is done through a method known as large-scale integration (LSI).

Robotics uses electronics. Electricians wire the installations and are responsible for getting power to the control consoles. Inside the consoles are many types of ICs. Their identification may be very useful. (See Figure 8-6.) Check catalogs for the right ones for the particular job. In some cases, an IC has been created and manufactured for a specific piece of equipment. Manufacturers' numbers have to be obtained from the top of the chip in order to replace it with the correct substitute.

OPTOELECTRONIC DEVICES

Optoelectronics has been used in many industrial and commercial operations. Displays are used everywhere to indicate calculator results, cash register amounts, weights on digital scales, and readouts on digital meters. Photo diodes, infrared-emitting diodes, and photo detectors, as well as photo transistors, reflective object sensors, and infrared light-emitting diodes (LEDs) are used in industry in control devices. Many of these are also used in fiber optics through which telephone messages are transmitted and received.

DISPLAYS

The seven-element LED display comes in five sizes. This type of display produces light and can be used where there is no ambient light. They usually glow red. Optoelectronic displays are used everywhere a sturdy readout is needed. They can be replaced if soldered in place or in sockets. The right size is important when trying to locate a replacement.

Liquid crystal displays (LCDs) need ambient light in order to be seen. They do not use much energy, working with microwatts of energy, whereas the LED display that glows brightly enough to be seen in the dark or daylight draws more energy in terms of milliwatts.

Light-emitting diodes are available with either common cathodes or common anodes. In the former (common cathode) case, particular segments are energized when a positive voltage is applied to them, with the remainder "down." In the latter (common anode) case, energized portions are "high," and those portions to be energized are "low" that is, at 0 volts. A 7447 decoder is needed to illuminate the appropriate segments of the display.

LIGHT-EMITTING DIODES (LEDs)

The light-emitting diode is used for indicator lights, calculator numerical displays, and computer display systems. The LED is basically a PN junction diode, operating in a forward-biased mode. Recombinations of electrons and holes release the stored energy in the excited electrons. The P-type gallium arsenide LED must be made very thin for efficient light output. Electrons are pumped up by an externally supplied direct current.

Light-emitting diodes are used as infrared sources for many sensing operations in industry. Along with reflective object sensors, they can be used to count and do quality control operations. The photo sensor responds to infrared radiation from the LED only when a reflective object is within its field of view.

OPTICAL COUPLERS

The triac-driven opto-couplers each contain a gallium arsenide or gallium aluminum arsenide infrared-emitting diode coupled to a photodiode and a zero-voltage bidirectional triac driver mounted in a standard six-pin DIP, or dual inline package, the type ICs use. These devices are intended to be used for low-power DC control of power triacs, which in turn control resistive, inductive, or capacitive loads powered from the 120-volt AC or 220-volt AC with required LED drive currents of 30, 15, 10, or 5 milliamperes. Zero-voltage crossing ensures that the device will not turn on until the line voltage reduces to 15 volts for the 120-volt devices and 25 volts for the 220-volt devices.

TRANSISTORS

Transistors are semiconductor devices made from N- and P-type crystals. Once joined, the two different types of crystal produce junctions. Transistors are identified according to emitter junction and collector junction.

A PNP transistor is formed by a thin N region between two P regions. The center N region is called the base. This base is usually 0.001 inch thick. A collector junction and an emitter junction are also formed.

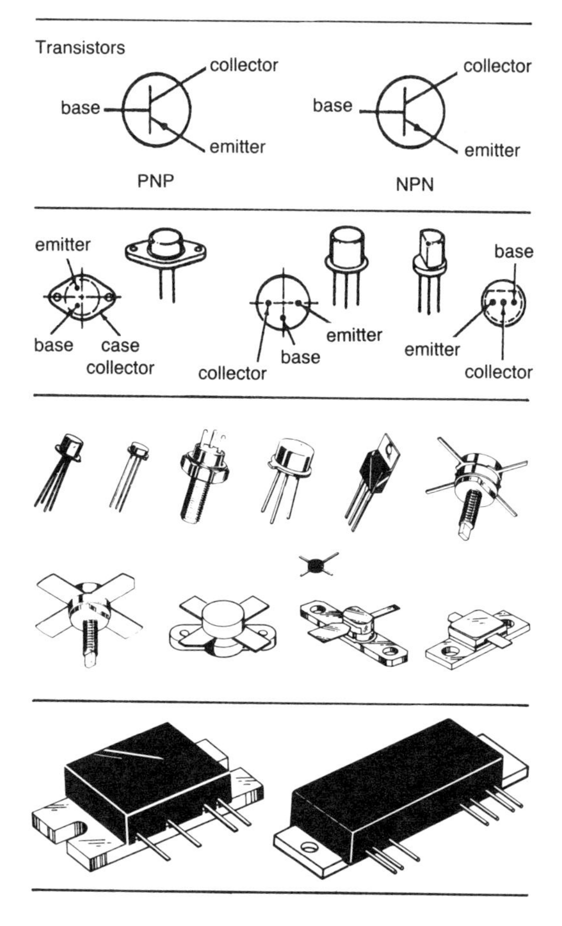

Figure 8-7. The two most popular types of transistors, PNP and NPN, come in a wide variety of pin designations.

NPN and PNP transistors (see Figure 8-7) are the two most popular types. The main difference between the two transistors is in polarity. Polarity can be recognized by pin locations on transistors. Pin designations for specific transistors are found in a transistor handbook.

The important thing to remember in transistor circuits is polarity and voltage. The polarity of the voltage is very important in the proper operation of the transistor. NPN and PNP types differ only in their polarity.

TUBES

Electron tubes are devices in which electrons are generated within a glass envelope. Vacuum tubes are units from which most of the air has been withdrawn. Thermionic tubes are devices in which electrons are produced by heating an electrode.

Vacuum tubes have been around for many years, and many of them are still available for equipment that is still in operating condition but occasionally needs a new tube.

There are vacuum tube diodes, amplifiers, and high-current rectifiers made of gas-filled tubes.

TUBE BASES

Since the first vacuum tubes were made, the internal complexity of the tubes has increased. A system of locating the many connections was needed. This was done by establishing a series of standard bases for connecting tubes into circuits. (See Figure 8-8.) Tubes are read from the keyway in a clockwise direction. There are three basic types of tube bases: miniature with seven pins, miniature with nine pins, and octal base with eight pins. Older tubes used four-pin bases, and some

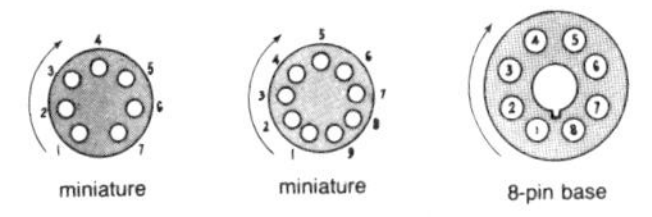

Figure 8-8. Three basic types of vacuum tube bases.

had different pin sizes to distinguish connections. Picture tubes for television sets and for computer monitors may have as many as twelve pins. They also use the clockwise-from-the-keyway system to identify their connections.

TYPES OF TUBES

COMMUNICATIONS AND INDUSTRIAL TUBES

Tubes are made for communications transmitters and for industrial control jobs. They have special numbering that differs somewhat from that used for domestic receiving and transmitting tubes. Some types of communications and industrial tubes are still available and in service, usually found in older equipment.

INCANDESCENT DIGITAL READOUT TUBES

These bright, low-voltage devices may be used in DC or multiplex mode. They are rugged, with single-plane unit construction for high reliability and 100,000 hours of life, so some of them are still around. (At one time, they were used in Burger King's cash registers.) Many industrial machines used them for readouts before they were replaced by LED readouts. They have either nine-pin or ten-pin bases. It is important to replace them with the right number of pins, or they won't fit in the sockets. (See Figure 8-9.)

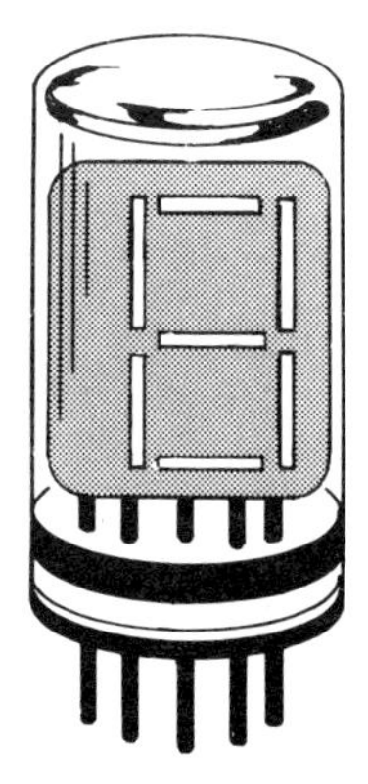

Figure 8-9. Incandescent digital readout tubes come with nine-pin or ten-pin bases.

LOW-LIGHT-LEVEL CAMERA TUBES

This type of camera tube is needed when a camera is used for closed-circuit monitoring of warehouses, apartment entrances, stores, and other places where there is little if any light.

Table 8-3 lists the types used with their numbering system. Some have the vacuum tube numbering of 20PE11, showing that the tube is a 20-volt filament with eleven pins. The four-number types are usually in-house numbers for particular manufacturers. Note the designation for a fiber-optic camera tube.

TABLE 8–3
Low-Light-Level Camera Tubes

Type
20PE11
20PE13A
20PE14
20PE19
20PE20
4848RCA
4848PANA
7262A
7735A
8134
8507
8541
20PE13A
8844
9677D2 (fiber optic)

MEDICAL AND BROADCAST MONITOR TUBES

These look like picture tubes for television sets. However, the tubes used in industrial equipment, medical monitors, and broadcast monitors usually are more rugged than home-entertainment types. They also cost more.

NUMERICAL NEON-GLOW READOUT TUBES

This type of tube was used before the advent of the incandescent digital readout tubes and the LED displays. (See Figure 8-10.) Some of them are still found in equipment that may need a single tube to be operational. Numerical neon-glow readout tubes vary in price from $26 to $40, and in some cases it may be less expensive to replace the whole device rather than a tube. These tubes operate with a high-voltage power supply. So when you encounter one, make sure you turn off the power supply before checking it out. The high voltage is very dangerous.

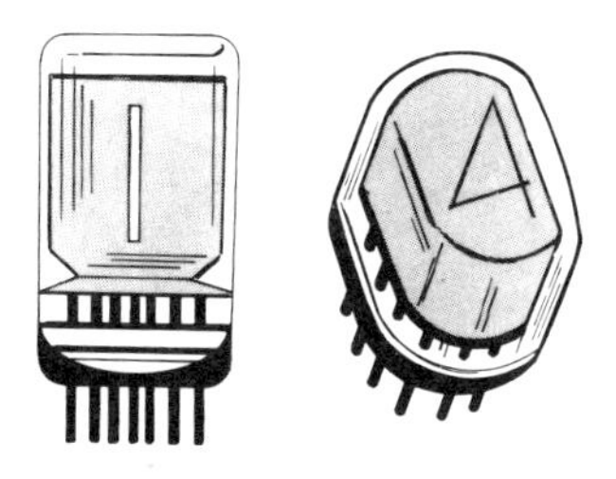

Figure 8-10. Numerical neon-glow readout tubes operate with a high-voltage power supply.

RECEIVING AND TRANSMITTING TUBES

There are still many receivers and transmitters that use older types of vacuum tubes. (See Figure 8-11.) Many of these tubes are still available for replacement purposes on all types of equipment.

VIDICONS, INDUSTRIAL GRADE

There are several television camera tubes, but the vidicon has become the most popular camera tube. It contains a target consisting of a thin layer of photoconductive material and a transparent conductive layer. (See Figure 8-12.) The resistance of the photoconductive layer diminishes with an increase in light intensity. An electron beam scans the target, producing a current that is proportional to the light intensity at the spot on the target being struck by the beam. A lens system focuses an image on the target.

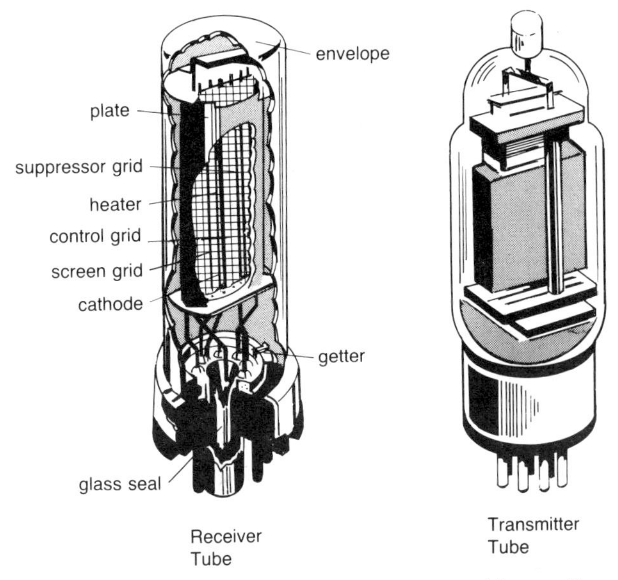

Figure 8-11. Many old-type receiver and transmitter vacuum tubes are still available for replacement purposes.

VACUUM TUBE IDENTIFICATION

There is no strict standard for assigning numbers to vacuum tubes, but usually the numbers have some meaning. As an example, a common tube is numbered 6SK7. The 6 indicates a 6.3-volt filament. The S stands for the word *sans* (French for "without"); this means the tube is without a grid cap. In other words, the grid connection is made to one of the tube base pins. The K means that it is an rf amplifier tube. The 7 means that seven of the eight pins are used for connections to the internal elements. These elements are the cathode, filaments, plate, grid, and the other two grids.

If the letter Z, X, Y, or U is used, it has a special meaning. These letters indicate that a tube is a rectifier type. The letter L, V, C, or B indicates that the tube is an audio power amplifier.

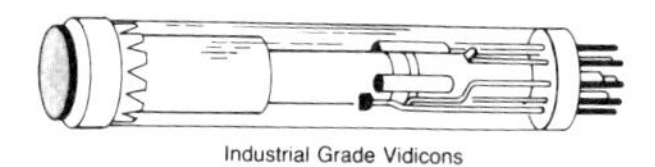

Figure 8-12. The vidicon is the most popular television camera tube.

There are many exceptions to these designations. The most accurate information on any tube can be found in a tube manual. RCA, General Electric, and Sylvania publish tube manuals.

VACUUM TUBE TROUBLES

The greatest problem with vacuum tubes is the filaments, specifically filament burnout. The red glow in the top of the tube indicates the filament is working. If there is no red glow and the tube is not warm, it should be checked and replaced. Tube testers are available to check tubes quickly and accurately. Most problems with filament burnout occur when the tube is first turned on.

The next greatest problem with vacuum tubes is shorts that may occur from rough handling. Shorts can cause resistors and capacitors to burn out or malfunction. Leaky or gaseous tubes sometimes have a purplish glow around them. This does not represent a problem with high-amplification-type tubes because they have a glow in normal operation. Gas-filled tubes used as high-current rectifiers have a purple glow in normal operation.

CHARGE-COUPLED DEVICE (CCD)

Charge-coupled devices have made completely solid-state television cameras possible. Charge-coupled light sensors are made on a single chip of semiconductor material covered with an array of very small electrodes. This device has a number of advantages over camera tubes. It does not require high voltage, a scanning beam, or a vacuum envelope. It does, however, require a rather elaborate electronic circuit to scan the scene. CCDs make possible handheld cameras that are capable of the same picture quality as large cameras.

9

How to Read Blueprints

The electrician must be able to read blueprints to determine what is to be done and where at a building site, to make estimates for bidding on a job, and to order supplies for a job that has been contracted. To understand blueprints, the electrician must be familiar with the conventions and drawing standards used in the preparation of blueprints throughout the United States.

A blueprint is a graphic language that allows a great deal of information to be presented in a condensed form that is easily read and readily stored when not in use. The graphic language is made up of pictorial and multiview drawings at various scales, and it uses standard symbols to specify materials, sizes, shapes, and other data.

This chapter presents a basic overview of information the electrician and apprentice should know to understand blueprints.

SYMBOLS

The symbols for electrical fixtures and devices are nothing more than the shorthand used in the electrical field. Symbols are used

Electrical Symbols Used on Blueprints

Object or Device	Symbol on Blueprint		
bell		range outlet	R
push button		refrigerator outlet	R
electric door opener		water heater outlet	WH
intercommunication		garbage-disposal outlet	GD
telephone outlet		dishwasher outlet	DW
telephone jack		iron outlet with pilot light	IP
dimmer switch	S_{DM}	washer outlet	W
heavy-duty outlet 220 voltage		dryer outlet	D
special-purpose outlet 110 voltage	XX	motor outlet	M

strip outlet	6"	lighting outlet—ceiling	
grounded outlet	GR	lighting outlet—recessed	
freezer outlet	FR	lighting outlet—wall	
wall bracket light with switch	S	lighting outlet—ceiling pull switch	PS
lighting fluorescent		flood light	
exit light	X	spot light	
illuminated house number	N	lighting outlet—vapor proof	VP
clock outlet	C	switch single-pole	S
buzzer		switch double-pole	S_2
chime	CH	switch three-way	S_3

switch four-way	S_4	radio aerial outlet	R
switch weatherproof	S_{WP}	duplex outlet	
switch automatic door	S_D	single outlet	1
switch pilot light	S_P	triple outlet	3
switch low-voltage system	$\underline{S}$	weatherproof outlet	WP
switch circuit breaker	S_{CB}	floor outlet	
switch low-voltage master	MS	split wire outlet	
two switches	SS	outlet with switch	S
three switches	SSS		
television antenna outlet	TV		

Figure 9-1. Electrical symbols used on blueprints.

to indicate where a receptacle or overhead light is to be located, where a switch is located, and what light or outlet that switch will control. In some cases, switches are used in three different locations to control one light. Seeing the symbols for this on the blueprint, the electrician will know from previous experience that two three-way and one four-way switches are needed to turn the light on and off from three locations. Figure 9-1 presents the electrical symbols used on blueprints. Anyone working in the field must be thoroughly familiar with these symbols.

Some electrical systems involve low-voltage wiring. The symbols used for low-voltage wiring are slightly different from those used for most wiring systems. (See Figure 9-2.)

Electricians also work with drawings that indicate how electric motors are wired into a circuit and how other electronic devices are used in the control circuitry for electric motors. Figure 9-3 shows the symbols used in drawings of devices with electric motors.

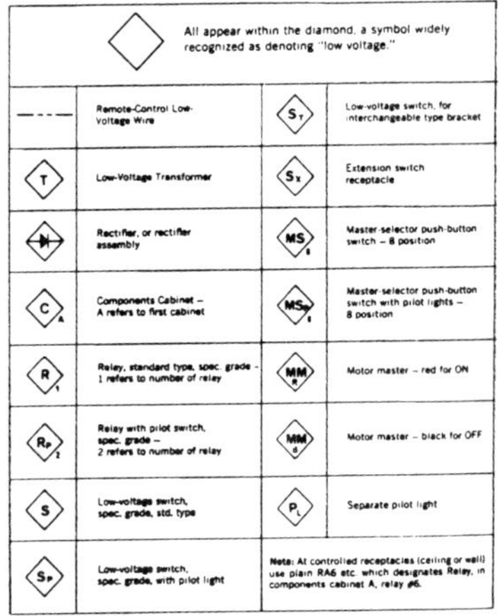

Figure 9-2. Symbols used for low-voltage systems.

ABBREVIATIONS

In some situations the symbol does not provide enough information. The architect or electrical contractor wants to provide the electrician with additional information. In these cases, an abbreviation is used alongside the symbol. Table 9-1 lists abbreviations frequently used alongside symbols on blueprints.

TABLE 9–1
Abbreviations Used With Electrical Symbols[1]

Abbreviation	Meaning
CSP	Central Switch Panel
DCP	Dimmer Control Panel
DT	Dust-Tight
ESP	Emergency Switch Panel
MT	Empty
EP	Explosion-Proof
G	Grounded
NL	Night Light
PC	Pull Chain
RT	Rain-Tight
R	Recessed
XFER	Transfer
XFRMR	Transformer
VT	Vapor-Tight
WT	Watertight
WP	Weatherproof

[1]These abbreviations are usually placed alongside the symbol on a blueprint.

LINES AND DIMENSIONS

All drawings are made with lines, the individual characteristics of which communicate specific information. What differentiates one line from another is its thickness and the way it is drawn. The set of conventional lines used in architectural and engineering blueprints is sometimes called the alphabet of lines. (See Figure 9-4.)

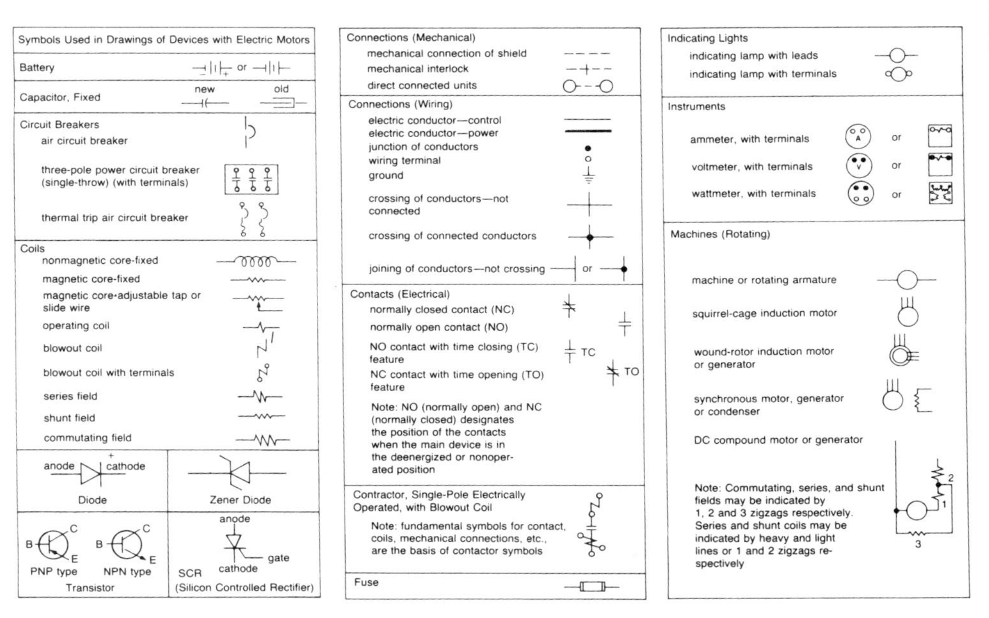

Symbols Used in Drawings of Devices with Electric Motors
Battery
or
Capacitor, Fixed
new
old
Circuit Breakers
air circuit breaker
three-pole power circuit breaker (single-throw) (with terminals)
thermal trip air circuit breaker
Coils
nonmagnetic core-fixed
magnetic core-fixed
magnetic core-adjustable tap or slide wire
operating coil
blowout coil
blowout coil with terminals
series field
shunt field
commutating field
anode
+
cathode
Diode
Zener Diode
B
C
E
PNP type
NPN type
Transistor
anode
gate
cathode
SCR
(Silicon Controlled Rectifier)
Connections (Mechanical)
mechanical connection of shield
mechanical interlock
direct connected units
Connections (Wiring)
electric conductor—control
electric conductor—power
junction of conductors
wiring terminal
ground
crossing of conductors—not connected
crossing of connected conductors
joining of conductors—not crossing
or
Contacts (Electrical)
normally closed contact (NC)
normally open contact (NO)
NO contact with time closing (TC) feature
TC
NC contact with time opening (TO) feature
TO
Note: NO (normally open) and NC (normally closed) designates the position of the contacts when the main device is in the deenergized or nonoperated position
Contractor, Single-Pole Electrically Operated, with Blowout Coil
Note: fundamental symbols for contact, coils, mechanical connections, etc., are the basis of contactor symbols
Fuse
Indicating Lights
indicating lamp with leads
indicating lamp with terminals
Instruments
ammeter, with terminals
A
or
voltmeter, with terminals
V
wattmeter, with terminals
Machines (Rotating)
machine or rotating armature
squirrel-cage induction motor
wound-rotor induction motor or generator
synchronous motor, generator or condenser
DC compound motor or generator
Note: Commutating, series, and shunt fields may be indicated by 1, 2 and 3 zigzags respectively. Series and shunt coils may be indicated by heavy and light lines or 1 and 2 zigzags respectively
1
2
3

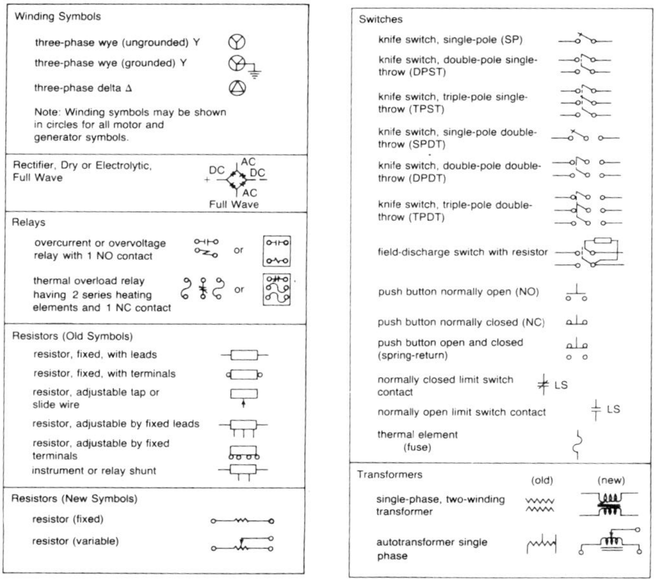

Winding Symbols

- three-phase wye (ungrounded) Y
- three-phase wye (grounded) Y
- three-phase delta Δ

Note: Winding symbols may be shown in circles for all motor and generator symbols.

Rectifier, Dry or Electrolytic, Full Wave

DC + AC DC – AC Full Wave

Relays

- overcurrent or overvoltage relay with 1 NO contact — or
- thermal overload relay having 2 series heating elements and 1 NC contact — or

Resistors (Old Symbols)

- resistor, fixed, with leads
- resistor, fixed, with terminals
- resistor, adjustable tap or slide wire
- resistor, adjustable by fixed leads
- resistor, adjustable by fixed terminals
- instrument or relay shunt

Resistors (New Symbols)

- resistor (fixed)
- resistor (variable)

Switches

- knife switch, single-pole (SP)
- knife switch, double-pole single-throw (DPST)
- knife switch, triple-pole single-throw (TPST)
- knife switch, single-pole double-throw (SPDT)
- knife switch, double-pole double-throw (DPDT)
- knife switch, triple-pole double-throw (TPDT)
- field-discharge switch with resistor
- push button normally open (NO)
- push button normally closed (NC)
- push button open and closed (spring-return)
- normally closed limit switch contact — LS
- normally open limit switch contact — LS
- thermal element (fuse)

Transformers (old) (new)

- single-phase, two-winding transformer
- autotransformer single phase

Figure 9-3. Symbols used in drawings for devices with electric motors.

ALPHABET OF LINES

The *object line,* sometimes known as the *visible line,* is a heavy continuous line. It represents the visible outlines and edges of objects.

The *extension line* is a thin continuous line. It is a continuation of an object line, extending from the visible lines of an object to indicate the limits of the object.

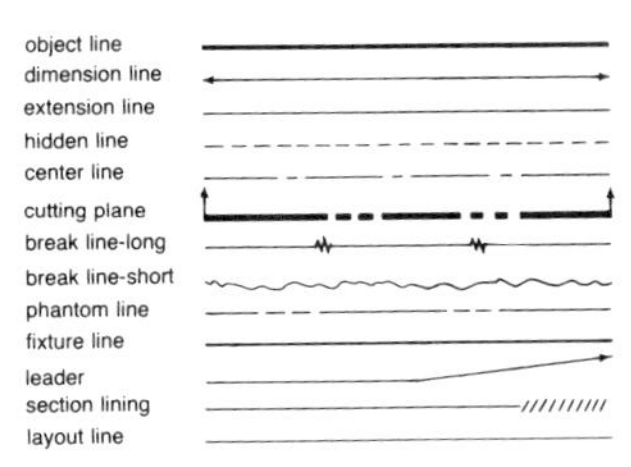

Figure 9-4. Alphabet of lines.

A *dimension line* is also a continuous line of the same weight as an extension line. Drawn between extension lines, a dimension line has arrowheads at its ends to indicate the start and end of a dimension. (The numerical value of the dimension is usually placed above the line.)

A *hidden line* is drawn as thin, short dashes. It is used to show hidden outlines or the location of something that may not be in sight or obvious from the vantage point of the viewer.

A *center line,* a thin line drawn with alternating long and short dashes, is used to indicate the center of symmetrical objects.

The *cutting plane* line is extra heavy, drawn with two short dashes between long dashed lines. It has arrowheads to show the direction of sighting the section view. Representing an imaginary cutting plane, it shows where an object would be sliced if a section of it were cut away.

Break lines indicate that a part is too long for a drawing and must be foreshortened.

A *phantom line* is a thin line consisting of a long line followed by two short dashes. It is used to show alternate positions of moving parts, adjacent positions of related parts, and repeated detail.

A *fixture line* is a continuous thin line that represents the outline of the shape of kitchen, laundry, or bathroom fixtures, or built-in furniture.

A *leader line* is used to note a specific feature or part. An arrowhead is placed at the feature end of the line, a note at the other end of the line. It is sometimes curved to prevent confusion with other lines.

A *section line* shows where a surface has been cut away. It is used to draw the section lining of sectional drawings. Different symbol patterns are used for each building material.

A *layout line* is very thin and is used to outline a material used in construction, such as lines to indicate siding or paneling.

DIMENSIONING

There are a number of ways in which distances from one point to another are indicated on blueprints. Arrowheads, triangles, slash lines, perpendicular lines, and circles may be used to indicate the start and end of a dimension. (See Figure 9-5.)

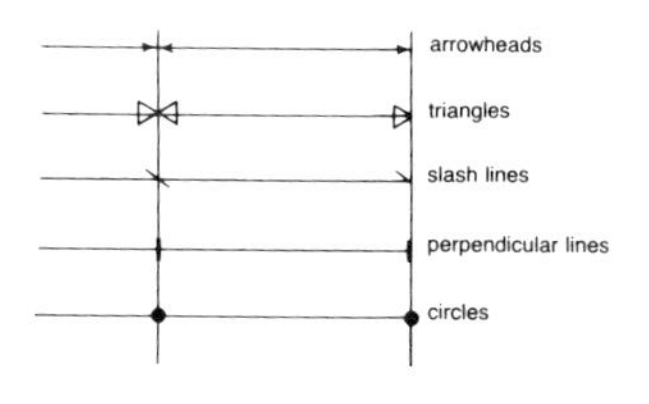

Figure 9-5. Methods to dimension blueprints.

The numerical value of the dimension is usually placed above the dimension line. Dimension lines show the width, height, and length of a building or any part of a building. They are used to show the location of windows, doors, fireplaces, stairs, and other objects. The number of specific dimensions used on an architectural plan reveals how much freedom of interpretation the architect wants to allow the builder.

SCALES

There are three types of scales used to make drawings for buildings: the architect's scale, the engineer's scale, and the metric scale. (See Figure 9-6.)

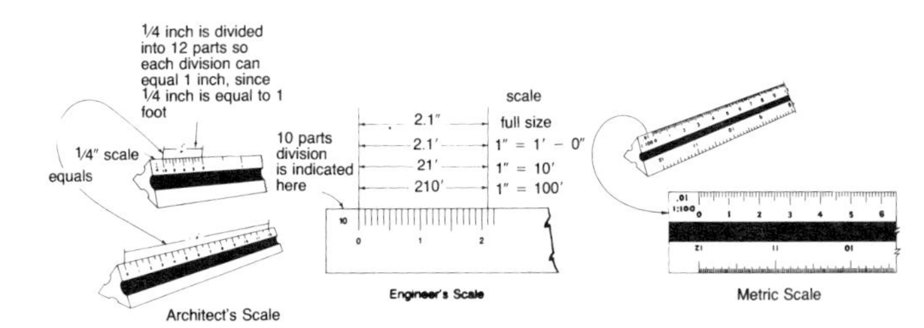

Figure 9-6. Ways of measuring: an architect's scale, an engineer's scale, and a metric scale.

ARCHITECT'S SCALE

The architect's scale is divided into eleven scales. One of these scales is a regular 1-through-12-inch scale. It is divided into sixteenths of an inch. The other ten scales available are open-divided scales—that is, scales that are divided into smaller parts only on the ends. These ten scales have ratios of 3/32, 1/8, 3/16, 1/4, 3/8, 1/2, 3/4, 1, 1 1/2, and 3.

There are two types of architect's scales: a bevel type and a triangular type. The triangular scale is made so that there are two scales on each face. One scale reads from left to right; the other, twice as large, from right to left.

The architect's scale is used most often to measure distance where the divisions of the scale equal 1 foot or 1 inch. Any of the scales—say, the 1/2—can be used to equal 1 foot or 1 inch depending on the drafter's needs in putting objects on the available blueprint or drawing.

ENGINEER'S SCALE

A civil engineer's scale differs from an architect's scale in the way an inch is divided. The architect's scale uses sixteenths of an inch; the civil engineer's scale divides an inch into tenths. Each unit can represent any linear value—an inch, a foot, a yard, a mile, or any other measurement unit. This means that the 10-ratio scale can be used so that its 10 units together equal 1 inch or 1 foot or 1 mile, or the scale can be used so that each subdivision is a tenth of the particular measuring unit being used.

The engineer's scale is most often used by civil engineers. These engineers frequently work in tenths of a foot up to hundredths of a foot, as well as in miles, so it is important that the scale they use accommodate their needs. The electrician encounters the engineer's scale designations when checking a plot plan for a house or the plan for the layout of transmission lines over a number of square miles.

METRIC SCALE

The metric method of measurement uses the meter (m) as the basic unit of measurement for length. In the building trades the meter and millimeter are the metric units used. However, the metric system has not been used much in the United States, and many of the sizes specified in architectural drawings are according to standards established for inches and feet. These measurements are not easily rounded off in metric units.

Some drawings have both customary and metric dimensioning. For example, the customary measurement will be given in feet and inches and the metric in millimeters often in parentheses alongside the customary unit or above or below the customary unit.

Table 9-2 lists the prefixes used with metric units to indicate multiples or subdivisions of the unit.

TABLE 9–2
Metric Prefixes

Prefix	Symbol	Prefix + Meter	Symbol	Powers of 10	Value
kilo	k	kilometer	km	1×10^3	1000
hecto	h	hectometer	hm	1×10^2	100
deka	da	dekameter	dam	1×10^1	10
—	m	meter	m	1×10^0	1
deci	d	decimeter	dm	1×10^{-1}	0.1
centi	c	centimeter	cm	1×10^{-2}	0.01
milli	m	millimeter	mm	1×10^{-3}	0.001

SCHEDULES

A schedule for an electrician lists the equipment to be installed in tabular form on a drawing or separate sheet attached to the drawing. Since the electrician usually has continual access to the working drawings, the presence of the schedule on the drawings is a timesaving device. Based on the detailed blueprints for the job and the specifications for the equipment, the schedule provides all of the necessary information in a clear, concise, accurate, and convenient manner. This allows all of those concerned with a project to carry out their particular assignments efficiently and quickly.

For residential situations, a schedule may be developed for the kitchen and its appliances, for the heating and air conditioning systems, for the lighting and the fixtures needed for the lighting, and for the panelboard and circuit breakers to be used. In large commercial and industrial buildings, there may be many more schedules—in fact, a whole book of schedules relating to the electrical equipment to be built into the building.

Schedules may be written in many forms, but all that is really necessary is a list of appliances and their specifications. Table 9-3 provides a typical schedule for a kitchen and its appliances; Table 9-4 shows a schedule for residential lighting fixtures.

SPECIFICATIONS

Specifications (specs) are written instructions that describe the basic requirements for constructing and equipping a building. Specifications describe the sizes, kinds, and quality of all building materials and the methods of construction, fabrication, and installation. Basically, specifications are orders to the contractor that show precisely what materials must be used and exactly how they

TABLE 9–3
Appliance Schedule

Room	Appliance	Type	Size	Color	Manufacturer	Model No.
Kitchen	Electric Stove	Cook Top	4-burner	White	Roper	2201-S
Kitchen	Electric Oven	Built-in	30″ × 24″ × 24″	White	Roper	2201-0
Kitchen	Dishwasher	Built-in	To fit under the counter	White	Roper	3210-DW
Kitchen	Garbage Disposer	Built-in	To fit under the sink	Stainless Steel	Magic Chef	0832-GD

TABLE 9–4
Fixture Schedule

Room	Fixture	Type	Material	Manufacturer	Model No.
Living	2 elec. lights	Hanging	Brass Reflec.	Lightolier	L-3221-a
Bedroom 1	2 spot lights	Wall Brac.	Flex. Neck-Alum.	Eagle	S-22220-B
Bedroom 2	2 elec. lights	Wall Brac.	Alum.-Water Res.	Marcal	M-3214-12

must be used. They also outline conditions under which the contractor undertakes the job. The specs guarantee the purchaser that the contractor will complete the building exactly as described in written form.

There are several different types of specifications. Of interest here are electrical fixture and wiring specifications.

The length and detail of an electrical specification set depends on the size and complexity of the job. However, nearly all electrical specifications contain the same basic sections within a division. For small jobs, each section may consist of only a few concise paragraphs; for large projects, each section may contain several pages of detailed instructions.

SECTIONS OF A TYPICAL ELECTRICAL SPECIFICATION

General Provisions. This section covers instructions to the electrical contractor and states that the drawings are diagrammatic. It indicates in general terms the location of material and equipment. This section also usually states that the drawings are to be followed as closely as possible and that the contractor shall coordinate the work with architectural, plumbing, heating, ventilation, and other drawings and shall cooperate with all other trades to make minor field adjustments to accommodate the work of others and coordinate the entire project.

This section also usually states that all fixtures and equipment are to be approved by the architect before installation, and, very important, that the contractor must follow the National Electrical Code® or local codes and certify that the building was wired to specifications within the Code.

Service and Distribution. This section of the specifications specifies the type and size of service, the manufacturer and size of panelboard, the wire and conduit size, the number and size of circuit breakers to be contained within the panelboard, the lighting fixtures and equipment, and other factors.

Other Sections. Other sections of typical electrical specifications are:

Basic Materials and Methods— spelled out.

Raceways— specified.

Conductors and Cables— identified.

Outlet boxes— described.

Panelboards and load centers— detailed.

Switches and receptacles— detailed.

Motors and motor controls— described in general terms.

In general, the specifications contain all of the supplemental data that it is not practical to include in the schedule.

ELECTRICAL AND ELECTRONIC DIAGRAMS

Electrical diagrams indicate the basic plan according to which electrical equipment is connected, specify the purpose of the equipment, and show the location of electrical wiring. Electrical diagrams are *not* drawn to scale. There are two basic types of electrical diagrams: elementary block diagrams and schematic diagrams.

A *block diagram* is useful for indicating the overall structure of an electrical system. A block diagram is made up of rectangular or square blocks. The blocks represent various pieces of electrical equipment with the name of the equipment usually spelled out in the block. Small blocks can be used for such things as a time clock, with the letters TC alongside the symbol. Large blocks are needed for panelboards so that there is sufficient room to allow for proper labeling of circuits and breakers. Lines are used to indicate electrical feeders or conduit raceways containing conductors. Notes on the drawing indicate such things as conduit and wire size and the catalog numbers of electrical equipment.

A *schematic electrical diagram* uses symbols to represent electrical devices and shows connections by both solid and broken lines. The number of lines used to connect the various devices may indicate the number of conductors. For example, a line with three hash marks indicates a wire with three conductors. (See Figure 9-7.)

Electronic schematics employ special symbols to indicate electronic devices and the connections necessary to make a complete

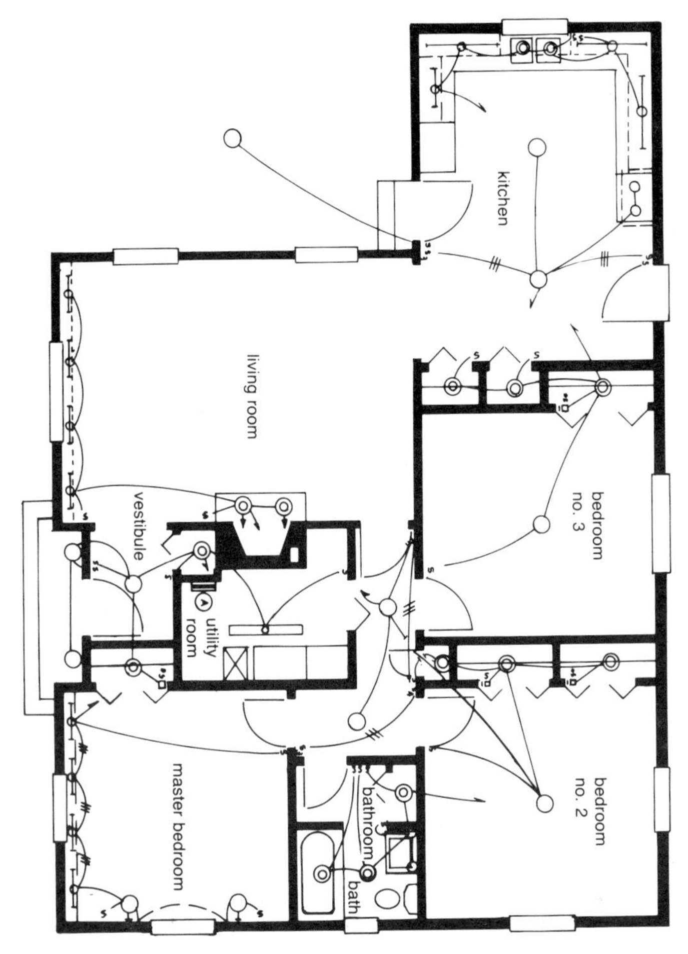

Figure 9-7. A typical schematic electrical diagram.

circuit. See Chapter 8 for a summary of basic electronics and hints on reading an electronic schematic.

CIRCUITS AND CODE REGULATIONS

The National Electrical Code® has placed restrictions on the layout of circuits for various residential rooms. For example, according to the NEC® code, a typical room must have receptacles spaced to permit a lamp or appliance equipped with a 6-foot cord to be located anywhere in the room. (See Figure 9-8.) A receptacle is required if any wall width is 2 feet or more, and a receptacle must be not more than 6 feet away from any point along a wall at the floor line. Refer to NEC® Article 210 for rules about branch circuits and safe practices mandated to make sure that electrical service is available without the use of extension cords. (See Figure 9-9.)

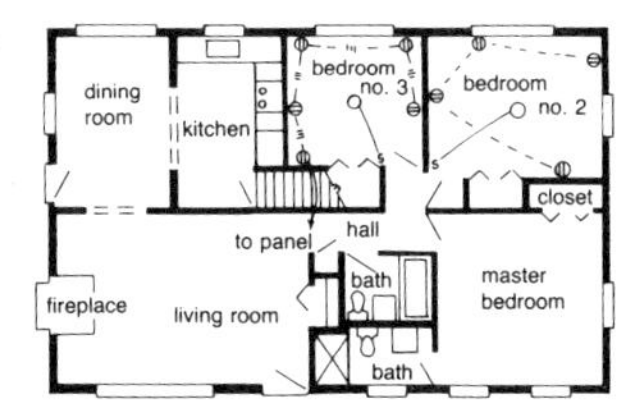

Figure 9-8. Code requirements govern the location of receptacles.

Although small branch circuits may serve certain outlets in specified areas, they are not permitted to serve other types of outlets, such as those that might be connected to exhaust hoods or fans, disposal units, or dishwashers.

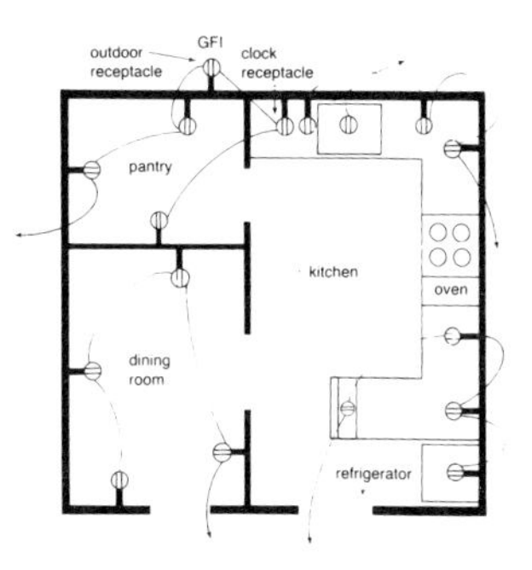

Figure 9-9. Certain outlets may not be served by a branch circuit, and some must be equipped with a GFI.

A most important suggestion: be sure to keep a current copy of the NEC handbook handy whenever circuit planning and layout is done.

10

How to Measure

The electrician frequently needs to measure when on the job. The measurements may be linear—for example, measuring lines on a blueprint and understanding what they represent or measuring lengths of wire or distances between switches or outlets. Measurements may also be of wire diameter or of electrical phenomena such as current, voltage, or resistance.

Devices and tools are available for all of these and for other types of measurement the electrician may be called on to do. Some of these were discussed in Part One—Tools of the Trade. Here we shall discuss how to use some of the most frequently employed measuring devices. However, before doing this, we should point out that there are many terms and abbreviations used in measuring electrical phenomena—terms not used in other fields and often not familiar to the layperson. The electrician and apprentice must be thoroughly familiar with all of these terms. Appendix 3 defines the terms used in electrical measurement and gives the abbreviations used for each unit.

One other thing to remember when measuring is the differences between the customary (U.S.) system of measurement and the SI, or metric, system used throughout much of the world. In many cases the electrician may need to convert measurement units from one system to another—for example, from feet to meters or from horsepower to watts. Chapter 6 provides guidelines on how to convert from one system to another, and Appendix 4 contains many tables that relate the two systems.

USING SCALES AND RULES

The engineer uses several types of scales and rules on the job—to read blueprints, to measure the length of wire, to locate switch boxes, and to do many other things.

ARCHITECT'S SCALE

The architect's scale, also discussed in Chapter 9, is most often used by the electrician when reading blueprints, enabling him or her to use the same scale as that used to make the drawing.

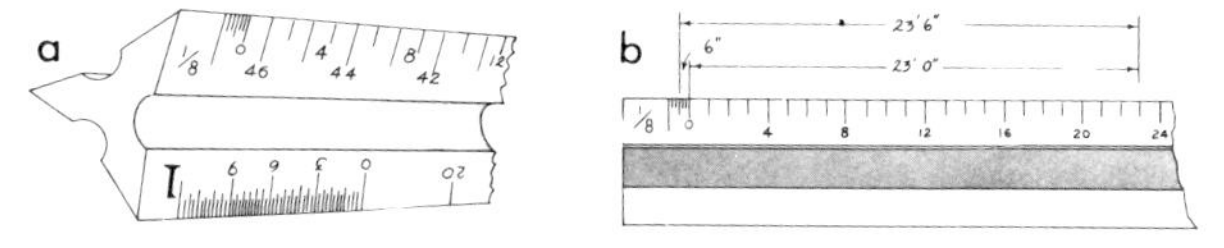

Figure 10-1. Architect's scale.

Figure 10-1 shows how the architect's scale is used to measure to scale. Figure 10-1a shows a scale that is 12 inches long (see top part of drawing). A 1/8-inch scale is visible at the top edge. This means that 1/8 inch is equal to 1 foot, unless otherwise noted on the drawing. Figure 10-1b shows the same scale, extended to 24 inches. Notice that units are numbered at intervals of 4. Remember, however, that each long line equals 1 foot. If, for example, you read between 0 and 23 on the scale, you have 23 feet. Then, if the line you are measuring is a little more than 23 feet but not 24 feet, read backward from 0 on the graduated scale. Say you find that the line you are measuring is halfway between 0 and the end of the scale. The halfway mark is 1/2 foot, or 6 inches. (Each of the small lines in the portion between the 1/8 scale marking and 0 represents 2 inches.) The line you have measured therefore represents 23 feet 6 inches.

FLAT RULE

The flat 6-inch rule is used the same as any scale if it is calibrated the same way. (See Figure 10-2.) The top rule is divided into

16ths of an inch and 32nds of an inch. This type of rule comes in handy in a machine shop and in reading a map scale where, for example, 1/32 inch equals 50 miles. An electrician may use this type of ruler when studying a large map showing where a building is to be erected.

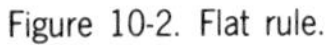

Figure 10-2. Flat rule.

The bottom rule in Figure 10-2 is a 6-inch ruler that is divided into 4 spaces per 1 inch and 8 spaces per 1 inch. This type of ruler can be used to make accurate measurements on blueprints.

FOLDING RULE

The 6-foot folding rule is easily carried in the pocket; it is rugged and can take the abuse given it on the job. Electricians need this type of measuring device to estimate wire lengths and to measure the height of boxes off the floor and the locations of switch boxes, among other uses. (See Figure 10-3.)

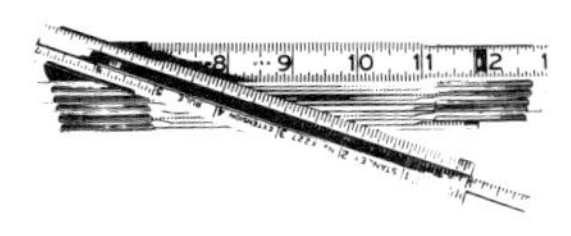

Figure 10-3. Folding rule.

USING WIRE GAGES

A wire gage is a tool used to measure the size of a wire. The wire being checked is pushed from the outside to the inside of the gage slot. The size of the slot is the actual size of the wire. However, if the wire is covered with insulation or other covering, you must allow for the thickness of the covering. For example, an insulated wire that reads No. 13 on the gage is really a No. 14 wire with a one-size coating.

One commonly used wire gage is the American Standard Wire Gage. It is especially useful for measuring sheets, plates, and wire made from nonferrous metals, such as aluminum, brass, and copper. This gage has a table on the back that gives the diameter of the various wires in decimal form. (See Figure 10-4.)

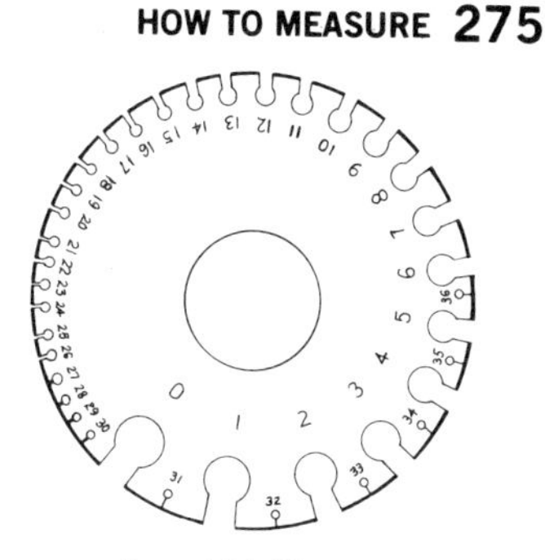

Figure 10-4. Wire gage.

Another commonly used wire gage is the Stub's Iron Wire Gage, commonly known as the Birmingham Gage. This gage is used to measure drawn steel wire or drill rod and gives the sizes in Stub's soft-wire sizes.

Table 10-1 shows dimensions of wire in decimal parts of an inch for different sizes of wire on several commonly used wire gages.

USING ELECTRICAL METERS

There are many types of meters used to measure electrical quantities. Here the clamp-on meter, the digital readout meter, and the VOM are discussed because they are the ones electricians use most frequently in the field.

AC CLAMP-ON METERS

The AC clamp-on meter is inserted over a wire carrying alternating current. The magnetic field around the wire induces a small amount of current in the meter. Because the wire is run through the large loop extending past the meter movement, it is possible to read the current without removing the insulation from the wire. These meters are very useful when working with AC motors and

TABLE 10–1

Standards for Wire Gages

(Dimensions of Sizes in Decimal Parts of an Inch.)

No. of Wire	American or Brown & Sharpe for Non Ferrous Metals	Birmingham or Stub's Iron Wire	American S. & W. Co.'s (Washburn & Moen) Std. Steel Wire	American S. & W. Co.'s Music Wire	Imperial Wire	Stub's Steel Wire	U.S. Std. Gage for Sheet & Plate Iron & Steel	No. of Wire
7–0's	0.651354	. . .	0.4900	. . .	0.500	. . .	0.500	7–0's
6–0's	0.580049	. . .	0.4615	0.004	0.464	. . .	0.46875	6–0's
5–0's	0.516549	0.500	0.4305	0.005	0.432	. . .	0.4375	5–0's
4–0's	0.460	0.454	0.3938	0.006	0.400	. . .	0.40625	4–0's
000	0.40964	0.425	0.3625	0.007	0.372	. . .	0.375	000
00	0.3648	0.380	0.3310	0.008	0.348	. . .	0.34375	00
0	0.32486	0.340	0.3065	0.009	0.324	. . .	0.3125	0
1	0.2893	0.300	0.2830	0.010	0.300	0.227	0.28125	1
2	0.25763	0.284	0.2625	0.011	0.276	0.219	0.265625	2
3	0.22942	0.259	0.2437	0.012	0.252	0.212	0.250	3
4	0.20431	0.238	0.2253	0.013	0.232	0.207	0.234375	4
5	0.18194	0.220	0.2070	0.014	0.212	0.204	0.21875	5
6	0.16202	0.203	0.1920	0.016	0.192	0.201	0.203125	6
7	0.14428	0.180	0.1770	0.018	0.176	0.199	0.1875	7
8	0.12849	0.165	0.1620	0.020	0.160	0.197	0.171875	8
9	0.11443	0.148	0.1483	0.022	0.144	0.194	0.15625	9
10	0.10189	0.134	0.1350	0.024	0.128	0.191	0.140625	10
11	0.090742	0.120	0.1205	0.026	0.116	0.188	0.125	11
12	0.080808	0.109	0.1055	0.029	0.104	0.185	0.109375	12
13	0.07196	0.095	0.0915	0.031	0.092	0.182	0.09375	13
14	0.064084	0.083	0.0800	0.033	0.080	0.180	0.078125	14

other high-current devices. Volts can be measured by using test leads and probes that fit into holes on the side of the meter. (See Figure 10-5.)

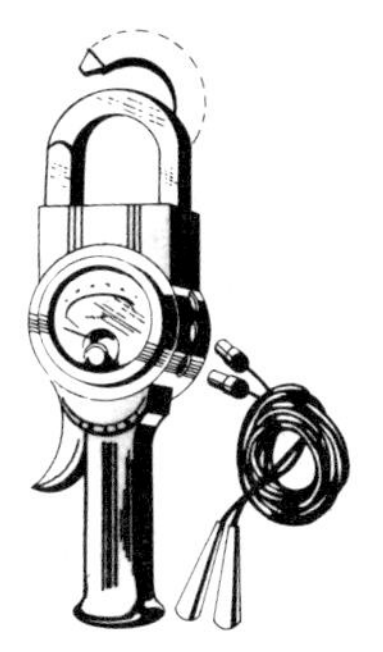

Figure 10-5. AC clamp-on meter.

DIGITAL METER

The digital meter is a portable, self-contained, battery-powered meter that is entirely electronic. It uses printed circuits and integrated circuit chips to measure and calculate voltage, resistance, or current. There are no coils or magnets. The digital reading is indicated on a liquid crystal display (LCD). (See Figure 10-6.)

Most digital meters have a number of voltage ranges, and you must select the range each time you measure it. Others are auto-ranging, selecting the proper range and measuring the voltage without the need for such preselections. The operator must, though, indicate if volts, ohms, or amps are to be measured.

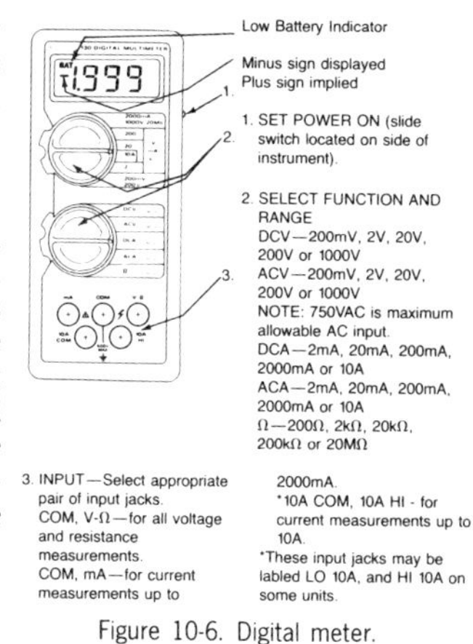

Figure 10-6. Digital meter.

VOM

The volt-ohm-milliammeter (VOM) is a multirange, multipurpose meter that has been around for a long time. It serves many useful functions for electricians and electronic technicians.

A VOM has its own batteries for measuring ohms on a number of ranges and shunts to increase the range for measuring amperes as well as milliamperes. The voltage scale is extended by using multipliers that allow measuring DC voltage and small diodes that allow the reading of AC volts. (See Figure 10-7.)

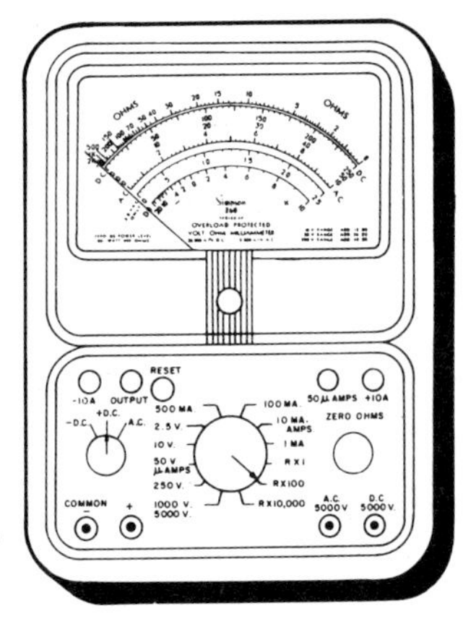

Figure 10-7. Volt-ohm-milliammeter (VOM).

The main thing to keep in mind when using a VOM is to be sure the meter is set to measure the factor you wish to measure. In other words, if you wish to measure ohms, be sure the meter is set to measure ohms, not voltage. If, for example, you try to measure voltage when the meter is set to measure resistance (ohms), the meter resistors may burn out and the meter may be rendered inoperative for other uses. Another caution: be sure the proper range is set for reading amperes so that the meter movement is not destroyed by too much current.

PART FOUR

APPENDICES

Appendix 1

Rules of Safety

Strange as it may seem, most fatal electrical shocks happen to people who should know better. Here are some electromedical facts that should make you think twice before taking chances.

It's not the voltage but the current that kills. People have been killed by 100 volts AC in the home and with as little as 42 volts DC. The real measure of a shock's intensity lies in the amount of current (in milliamperes) forced through the body. Any electrical device used on a house wiring circuit can, under certain conditions, transmit a fatal amount of current.

Currents between 100 and 200 milliamperes (0.1 ampere and 0.2 ampere) are fatal. Anything in the neighborhood of 10 milliamperes (0.01) is capable of producing painful to severe shock. Take a look at Table A1-1.

As the current rises, the shock becomes more severe. Below 20 milliamperes, breathing becomes labored; it ceases completely even at values below 75 milliamperes. As the current approaches 100 milliamperes ventricular fibrillation occurs. This is an uncoordinated twitching of the walls of the heart's ventricles.

Since you don't know how much current went through the body, it is necessary to perform artificial respiration to try to get the person breathing again; or if the heart is not beating, cardiopulmonary resuscitation (CPR) is necessary.

TABLE A1–1

Physiological Effects of Electric Currents*

Readings		Effects
Safe Current Values		Causes no sensation—not felt.
	1 mA or less 1 Ma to 8 mA	Sensation of shock, not painful; individual can let go at will since muscular control is not lost.
Unsafe Current Values	8 mA to 15 mA	Painful shock; individual can let go at will since muscular control is not lost.
	15 mA to 20 mA	Painful shock; control of adjacent muscles lost; victim cannot let go.
	50 mA to 100 mA	Ventricular fibrillation—a heart condition that can result in instant death—is possible.
	100 mA to 200 mA	Ventricular fibrillation occurs.
	200 mA and over	Severe burns, severe muscular contractions—so severe that chest muscles clamp the heart and stop it for the duration of the shock. (This prevents ventricular fibrillation.)

*Information provided by National Safety Council.

FIRST AID FOR ELECTRIC SHOCK

Shock is a common occupational hazard associated with working with electricity. A person who has stopped breathing is not necessarily dead but is in immediate danger. Life is dependent on oxygen, which is breathed into the lungs and then carried by the blood to every body cell. Since body cells cannot store oxygen and since the blood can hold only a limited amount (and only for a short time), death will surely result from continued lack of breathing.

However, the heart may continue to beat for some time after breathing has stopped, and the blood may still be circulated to the body cells. Since the blood will, for a short time, contain a small supply of oxygen, the body cells will not die immediately. For a very few minutes, there is some chance that the person's life may be saved.

The process by which a person who has stopped breathing can be saved is called artificial ventilation (respiration). The purpose of artificial respiration is to force air out of the lungs and into the lungs, in rhythmic alternation, until natural breathing is reestablished. Records show that seven out of ten victims of electric shock were revived when artificial respiration was started in less than three minutes. After three minutes, the chances of revival decrease rapidly.

Artificial ventilation should be given only when the breathing has stopped. *Do not give artificial ventilation to any person who is breathing naturally.* You should not assume that an individual who is unconscious due to electrical shock has stopped breathing. To tell if someone suffering from an electrical shock is breathing, place your hands on the person's sides at the level of the lowest ribs. If the victim is breathing, you will usually be able to feel movement.

Once it has been determined that breathing has stopped, the person nearest the victim should start the artificial ventilation without delay and send others for assistance and medical aid. The only logical, permissible delay is that required to free the victim from contact with the electricity in the quickest, safest way. This step, while it must be taken quickly, must be done with great care; otherwise, there may be two victims instead of one.

In the case of portable electric tools, lights, appliances, equipment, or portable outlet extensions, the victim should be freed from contact with the electricity by turning off the supply switch or by removing the plug from its receptacle. If the switch or receptacle cannot be quickly located, the suspected electrical device may be pulled free of the victim. Other persons arriving on the scene must be clearly warned not to touch the suspected equipment until it is deenergized.

The injured person should be pulled free of contact with stationary equipment (such as a bus bar) if the equipment cannot be quickly deenergized or if the survival of others relies on the electricity and prevents immediate shutdown of the circuits. This can be done quickly and easily by carefully applying the following procedures:

1. Protect yourself with dry insulating material.
2. Use a dry board, belt, clothing, or other available nonconductive material to free the victim from electrical contact. Do NOT touch the victim until the source of electricity has been removed.

Once the victim has been removed from the electrical source, it should be determined whether the person is breathing. If the person is not breathing, a method of artificial respiration is used.

CARDIOPULMONARY RESUSCITATION (CPR)

Sometimes victims of electrical shock suffer cardiac arrest or heart stoppage as well as loss of breathing. Artificial ventilation alone is not enough in cases where the heart has stopped. A technique known as CPR has been developed to provide aid to a person who has stopped breathing and suffered a cardiac arrest. Because you are working with electricity, the risk of electrical shock is higher than in other occupations. You should, at the earliest opportunity, take a course to learn the latest techniques used in CPR. The techniques are relatively easy to learn and are taught in courses available through the American Red Cross.

Note: A heart that is in fibrillation cannot be restricted by closed-chest cardiac massage. A special device called a defibrillator is available in some medical facilities and ambulance services.

Muscular contractions are so severe with 200 milliamperes and over that the heart is forcibly clamped during the shock. This clamping prevents the heart from going into ventricular fibrillation, making the victim's chances for survival better.

SAFETY MEASURES

Working with electricity can be dangerous. However, electricity can be safe if properly respected.

USING GROUND FAULT CIRCUIT INTERRUPTERS (GFCI)

Some dangerous situations have been minimized by using ground fault circuit interrupters (GFCI). (See Figure A1-1.) Since 1975, the National Electrical Code (NEC)® has required installation of GFCIs in outdoor and bathroom outlets in new construction, but most homes built before 1975 have no GFCI protection.

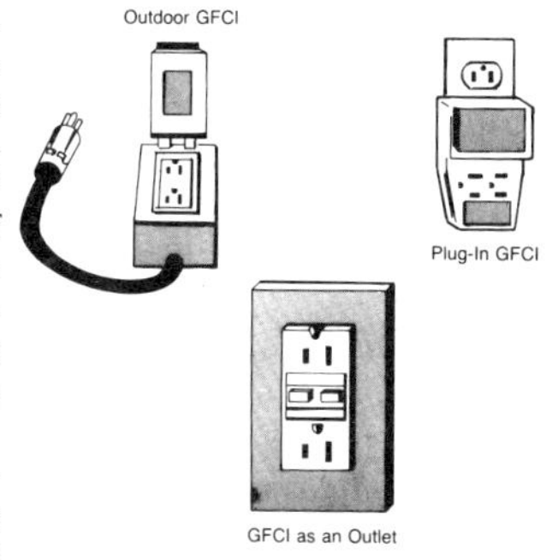

Figure A1-1. A ground fault circuit interrupter (GFCI) is now required by the NEC in all outdoor and bathroom outlets.

Retrofit GFCIs that can protect one outlet or an entire circuit with multiple outlets can be installed in older homes to reduce the danger. One of the simplest ways to achieve this protection in an outdoor outlet or other outlets in which shock dangers are high is to use a plug-in GFCI.

Two kinds of plug-ins are available. One has contact prongs attached to the housing, and it is simply plugged into a grounded outlet. The device to be used is plugged into a receptacle in the housing of the GFCI.

Another GFCI, more suitable for outdoor use, has a heavy-duty housing attached to a short extension cord. The extension cord type of GFCI is easily plugged into an outdoor outlet without its getting tangled with the outlet's lid or cover.

Keep in mind that GFCIs are not foolproof. They do, however, switch off the current in less than 0.025 second if a leakage is detected. They can detect as little as 2 milliamperes, although most operate at a threshold of 5 milliamperes—well below the level that would affect a person.

FOLLOWING CODE REQUIREMENTS FOR GROUNDING CONDUCTORS

The National Electrical Code® requires that a system grounding conductor be connected to any local metallic water-piping system available on the premises, provided that the length of the buried water piping is a minimum of 10 feet. If the system is less than 10 feet, or if the electrical continuity is broken by either disconnection or nonmetallic fittings, then it should be supplemented by the use of an additional electrode of a type specified by NEC® Section 250-81 or 250-83.

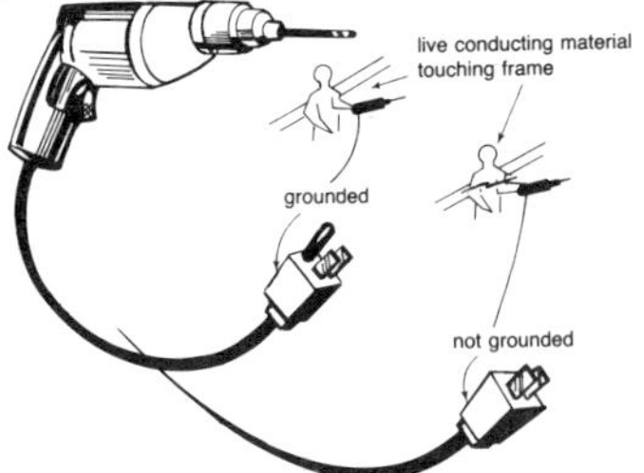

Figure A1-2. A person using a tool with an ungrounded plug may receive a shock, especially if he or she is touching a metal object.

AVOIDING HAZARDOUS SITUATIONS

Ground fault circuit interrupters are not always effective, especially under wet conditions. Fatal shocks are most likely under damp or wet conditions or if the user of an electrical device is touching a metal object such as a ladder or pipe. (See Figure A1-2.) In such a situation, the person completes the circuit to ground. The ground fault circuit interrupter detects the leakage of current through the circuit, which includes the person. (See Figure A1-3.)

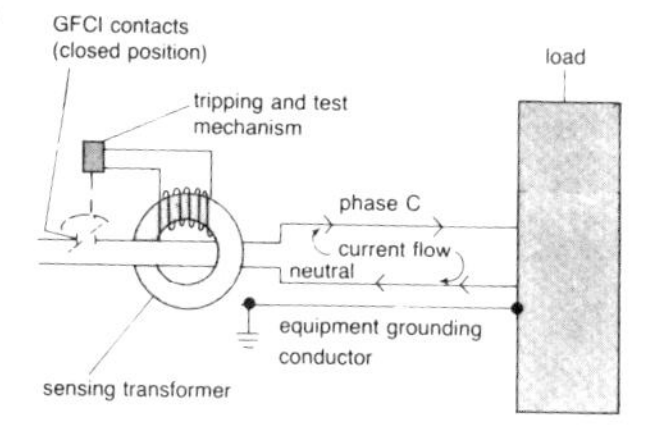

Figure A1-3. A GFCI detects any leakage of current in a circuit and thus prevents shock.

It is impractical to list all of the electrical shock hazards that might exist on the job, but those listed on Table A1-2 may be helpful in locating hazardous situations.

EXTENSION CORDS

Extension cords are used on the job for many purposes. However, if not carefully chosen for the job and properly cared for, they too can be hazardous.

The main concern is the insulation and the wire size needed to carry the current. Table A1-3 shows the current-carrying capacity of flexible cords.

Also, if the wrong length of cord is selected for a particular tool, then it is possible to reduce the voltage available at the piece of equipment, relying on the power source to supply the correct voltage. Table A1-4 shows the size of extension cords that should be used for portable electric tools.

PLUGS AND RECEPTACLES

Plugs and receptacles must match the job at hand. Each type of receptacle is designed to handle a specific amount of voltage and current. Table A1-5 shows how this is done by simply looking up the circuit requirements. The table shows which receptacle to install for the current and voltage requirements of the circuit or device. Some plug-in electrical devices are designed to reduce the danger of electrical shock and have plastic housings, double insulation, and other safety features.

TABLE A1–2[1]

Common Electrical Shock Hazards in Industry

Unsafe Physical Conditions	Control Measures
Worn insulation on extension and drop cords. Splices on cords.	Install a system of inspection and preventive maintenance to uncover dangerous conditions and to correct them. Use UL-approved materials only. Spliced cords should be removed from service.
Exposed conductors at rear of switchboard.	Enclose rear of switchboard to prevent exposure of unauthorized persons. Provide rubber mats for workers who must enter the enclosure.
Open switches and control apparatus on panel and switchboards. Location of machine switches.	Provide enclosed safety switches. Insulate with rubber mats in front of switch and control equipment. Locate machine switches so as not to create hazard to the operator.
Unsafe wiring practices, such as using wires too small for the current being carried; open wiring not in conduit; temporary wiring; wiring improperly located.	Comply with recognized electrical code. Remove temporary wiring as soon as it has served its purpose.
The accidental energizing of non-current-carrying parts of machines and tools by means of short circuits, breaks in insulation, etc.	Properly ground all non-current-carrying parts of machines, tools, and frames of control equipment.

Unsafe Actions	Control Measures
Working on "live" low-voltage circuits in the belief that they are not hazardous.	Educate and train workers in the hazards of low-voltage circuits.
Working on "live" circuits that are thought to be "dead."	Require that switches on all circuits being worked on be locked open and properly tagged. Use protective equipment such as rubber gloves, blankets, etc.
Replacing fuses by hand on "live" circuits.	Open switch before replacing fuses; use fuse pullers.
Using 120-volt lighting circuits for work in boiler or other similar enclosures.	Use low-voltage circuits: 6 volts for lighting, not over 30 volts for power.
Overloading circuits beyond their capacity.	Lock fuse boxes to prevent bridging or replacing with heavier fuse.
Abusing electrical equipment and poor housekeeping about electrical equipment.	Institute safe work practices, with inspection and preventive maintenance of equipment. Improve housekeeping practices.

[1]Courtesy of the National Safety Council.

TABLE A1–3

Flexible-Cord Ampacities*

AWG**	Type S, SJ SO, SJO	Type SJT
18	7	10
16	10	10
14	15	18
12	20	25
10	25	30
8	35	40
6	45	55
4	60	70
2	80	95

***Source: National Electrical Code,® Table 400–5.**
****American Wire Gage.**

HAZARDOUS LOCATIONS

Hazardous locations are classified into three main classes, which are further broken down into divisions. The classes and divisions of hazardous locations established by the National Electrical Code® are listed in Table A1-6.

SAFETY EQUIPMENT AND CLOTHING

Electricians who work on lines must wear the proper clothing and safety equipment. However, most of this work is done by persons trained by local utilities.

TABLE A1–4
Size of Extension Cords for Portable Electric Tools

For 115-Volt Tools						
Full-load ampere rating of tool	0 to 2 A	2.1 to 3.4 A	3.5 to 5 A	5.1 to 7 A	7.1 to 12 A	12.1 to 16 A
Length of Cord	Wire Size (AWG)					
25 feet	18	18	18	16	14	14
50 feet	18	18	18	16	14	12
75 feet	18	18	16	14	12	10
100 feet	18	16	14	12	10	8
200 feet	16	14	12	10	8	6
300 feet	14	12	10	8	6	4
400 feet	12	10	8	6	4	4
500 feet	12	10	8	4	6	2
600 feet	10	8	6	4	2	2
800 feet	10	8	6	4	2	1
1000 feet	8	6	4	2	1	0

Note: **If the voltage is already low at the source (outlet), increase to standard voltage or use a much larger cable than listed in order to prevent any further loss in voltage.**

TABLE A1–5

Voltage and Current Ratings for Plugs and Receptacles

	Rating	Receptacle Configuration	NEMA *ANSI*	Wiring Diagram
3P3W	30A 3ϕ 250V		11-30 *C73.56*	
3P3W	50A 3ϕ 250V		11-50 *C73.57*	
3 POLE 4 WIRE	15A 125/250V		14-15 *C73.49*	
3 POLE 4 WIRE	20A 125/250V		14-20 *C73.50*	
3 POLE 4 WIRE	30A 125/250V		14-30 *C73.16*	
3 POLE 4 WIRE	50A 125/250V		14-50 *C73.17*	
3 POLE 4 WIRE	60A 125/250V		14-60 *C73.18*	
3 POLE 4 WIRE	15A 3ϕ 250V		15-15 *C73.58*	
3 POLE 4 WIRE	20A 3ϕ 250V		15-20 *C73.59*	
3 POLE 4 WIRE	30A 3ϕ 250V		15-30 *C73.60*	
3 POLE 4 WIRE	50A 3ϕ 250V		15-50 *C73.61*	
3 POLE 4 WIRE	60A 3ϕ 250V		15-60 *C73.62*	
4 POLE 4 WIRE	15A 3ϕY 120/208V		18-15 *C73.15*	
4 POLE 4 WIRE	20A 3ϕY 120/208V		18-20 *C73.26*	
4 POLE 4 WIRE	30A 3ϕY 120/208V		18-30 *C73.47*	
4 POLE 4 WIRE	50A 3ϕY 120/208V		18-50 *C73.48*	
4 POLE 4 WIRE	60A 3ϕY 120/208V		18-60 *C73.27*	

Wiring Diagram	NEMA *ANSI*	Receptacle Configuration	Rating	
	5-15 *C73.11*		15A 125V	2 POLE 3 WIRE
	5-20 *C73.12*		20A 125V	2 POLE 3 WIRE
	5-30 *C73.45*		30A 125V	2 POLE 3 WIRE
	5-50 *C73.46*		50A 125V	2 POLE 3 WIRE
	6-15 *C37.20*		15A 250V	2 POLE 3 WIRE
	6-20 *C73.51*		20A 250V	2 POLE 3 WIRE
	6-30 *C73.52*		30A 250V	2 POLE 3 WIRE
	6-50 *C73.53*		50A 250V	2 POLE 3 WIRE
	7-15 *C73.28*		15A 277V	2 POLE 3 WIRE
	7-20 *C73.63*		20A 277V	2 POLE 3 WIRE
	7-30 *C73.64*		30A 277V	2 POLE 3 WIRE
	7-50 *C73.65*		50A 277V	2 POLE 3 WIRE
	10-20 *C73.23*		20A 125/250V	3 POLE 3 WIRE
	10-30 *C73.24*		30A 125/250V	3 POLE 3 WIRE
	10-50 *C73.25*		50A 125/250V	3 POLE 3 WIRE
	11-15 *C73.54*		15A 3ϕ 250V	3 POLE 3 WIRE
	11-20 *C73.55*		20A 3ϕ 250V	3 POLE 3 WIRE

Table A1–5 (*continued*)

Wiring Diagram	NEMA *ANSI*	Receptacle Configuration	Rating	
	ML2 *C73.44*		15A 125V	2 POLE 3 WIRE
	L5-15 *C73.42*		15A 125V	2 POLE 3 WIRE
	L5-20 *C73.72*		20A 125V	2 POLE 3 WIRE
	L5-30 *C73.73*		30A 125V	2 POLE 3 WIRE
	L6-15 *C73.74*		15A 250V	2 POLE 3 WIRE
	L6-20 *C73.75*		20A 250V	2 POLE 3 WIRE
	L6-30 *C73.76*		30A 250V	2 POLE 3 WIRE
	L7-15 *C73.43*		15A 277V	2 POLE 3 WIRE
	L7-20 *C73.77*		20A 277V	2 POLE 3 WIRE
	L7-30 *C73.78*		30A 277V	2 POLE 3 WIRE
	L8-20 *C73.79*		20A 480V	2 POLE 3 WIRE
	L8-30 *C73.80*		30A 480V	2 POLE 3 WIRE
	L9-20 *C73.81*		20A 600V	2 POLE 3 WIRE
	L9-30 *C73.82*		30A 600V	2 POLE 3 WIRE
	ML3 *C73.30*		15A 125/250V	3 POLE 3 WIRE
	L10-20 *C73.96*		20A 125/250V	3 POLE 3 WIRE
	L10-30 *C37.97*		30A 125/250V	3 POLE 3 WIRE
	L11-15 *C73.98*		15A 3ϕ 250V	3 POLE 3 WIRE
	L11-20 *C73.99*		20A 3ϕ 250V	3 POLE 3 WIRE
	L11-30 *C73.100*		30A 3ϕ 250V	3 POLE 3 WIRE
	L12-20 *C73.101*		20A 3ϕ 480V	3 POLE 3 WIRE

	Rating	Receptacle Configuration	NEMA *ANSI*	Wiring Diagram
3P3W	30A 3ϕ 480V		L12-30 *C73.102*	
3P3W	30A 3ϕ 600V		L13-30 *C73.103*	
3 POLE 4 WIRE	20A 125/250V		L14-20 *C73.83*	
3 POLE 4 WIRE	30A 125/250V		L14-30 *C73.84*	
3 POLE 4 WIRE	20A 3ϕ 250V		L15-20 *C73.85*	
3 POLE 4 WIRE	30A 3ϕ 250V		L15-30 *C73.86*	
3 POLE 4 WIRE	20A 3ϕ 480V		L16-20 *C73.87*	
3 POLE 4 WIRE	30A 3ϕ 480V		L16-30 *C73.88*	
3 POLE 4 WIRE	30A 3ϕ 600V		L17-30 *C73.89*	
4 POLE 4 WIRE	20A 3ϕY 120/208V		L18-20 *C73.104*	
4 POLE 4 WIRE	30A 3ϕY 120/208V		L18-30 *C73.105*	
4 POLE 4 WIRE	20A 3ϕY 277/480V		L19-20 *C73.106*	
4 POLE 4 WIRE	30A 3ϕY 277/480V		L19-30 *C73.107*	
4 POLE 4 WIRE	20A 3ϕY 347/600V		L20-20 *C73.108*	
4 POLE 4 WIRE	30A 3ϕY 347/600V		L20-30 *C73.109*	
4 POLE 5 WIRE	20A 3ϕY 120/208V		L21-20 *C73.90*	
4 POLE 5 WIRE	30A 3ϕY 120/208V		L21-30 *C73.91*	
4 POLE 5 WIRE	20A 3ϕY 277/480V		L22-20 *C73.92*	
4 POLE 5 WIRE	30A 3ϕY 277/480V		L22-30 *C73.93*	
4 POLE 5 WIRE	20A 3ϕY 347/600V		L23-20 *C73.94*	
4 POLE 5 WIRE	30A 3ϕY 347/600V		L23-30 *C73.95*	

TABLE A1–6

Hazardous Location Classifications

Class I— Highly Flammable Gases or Vapors		Class II— Combustible Dusts		Class III— Combustible Fibers or Flyings	
Division I	Division 2	Division 1	Division 2	Division 1	Division 2
Locations where hazardous concentrations are probable, or where accidental occurrence should be simultaneous with failure of electrical equipment	Locations where flammable concentrations are possible, but only in the event of process closures rupture, ventilation failure, etc.	Locations where hazardous concentrations are probable, where their existence would be simultaneous with electrical equipment failure, or where electrically conducting dusts are involved	Locations where hazardous concentrations are not likely, but where deposits of the dust might interfere with heat dissipation from electrical equipment, or be ignited by electrical equipment	Locations in which easily ignitable fibers or materials producing combustible flyings are handled, manufactured, or used	Locations in which such fibers or flyings are stored or handled, except in the process of manufacture

Groups:

A- Atmospheres containing acetylene
B- Atmospheres containing hydrogen or gases or vapors of equivalent hazard
C- Atmospheres containing ethyl ether vapors, ethylene, or cyclopropane
D- Atmospheres containing gasoline, hexane, naphtha, benzine, butane, propane, alcohol, acetone, benzol, or natural gas
E- Atmospheres containing metal dust, including aluminum, magnesium, and other metals of equally hazardous characteristics
F- Atmospheres containing carbon black, coke, or coal dust
G- Atmospheres containing flour, starch, or grain dusts

The lineman's equipment includes an electrician's safety belt and gloves. Proper headgear, eyeglasses, shoes, and outer clothing should be chosen for the job.

At increased voltages, the line worker has a special equipment checklist to follow and routines to use in making sure the equipment is in good condition and that the gloves have not become infected with bacterial growths that could cause electrical conduction.

Appendix 2
Codes, Standards, and Regulations

Since the inception of the electric utility industry, research has made it possible for electricity to transform the way we live at home, on the farm, and in the factory. Research has also transformed the industry.

Thomas Edison's first electric station could transmit electricity only 5,000 feet, while modern power pools enable a customer to use electricity produced in a power plant many states away. Early generators had a power-producing capacity of only 100 kilowatts, but generators with the ability to produce millions of kilowatts are now in operation. Since 1900, transmission voltages have increased from 30,000 volts to 500,000 volts, and lines of up to 765,000 volts are now in operation.

In the early 1800s, the New York Board of Fire Underwriters became concerned with the new method of electric lighting proposed by Thomas Edison. Although Edison did not realize the danger of the giant force that he was helping to harness, the New

York Board recognized that unless proper precautions were followed the new method of lighting could prove to be as hazardous as the open flame that it was replacing. In 1881, one man was appointed to inspect every electrical installation before power was turned on. This was the beginning of the Electrical Department of the New York Board. It was necessary for the inspector to check not only the installation within the building but also to carry his investigation back to the power station, which was then only one or two blocks distant. The Board at that time investigated the safety of the entire power system and required the power companies to make weekly tests for grounds and open circuits and to report to the Board the results of their tests. In 1882, the Committee of Surveys drew up a set of safeguards, which was the forerunner of the present *National Electrical Code®* for arc and incandescent lighting.

NATIONAL ELECTRICAL CODE®

The National Electrical Code (NEC)® is the most widely adopted code in the world. Over 1 million copies of the code are sold each time it is published, which is once every three years. The NEC is a nationally accepted guide for the safe installation of electrical conductors and equipment and is, in fact, the basis for all electrical codes used in the United States. It is also used extensively outside the United States, particularly where American-made equipment is installed. No electrician is without a copy in the toolbox.

The code is purely advisory as far as the National Fire Protection Association (NFPA) and American National Standards Institute (ANSI) are concerned, but it is offered for use in law and for regulatory purposes in the interest of life and property protection. Anyone noticing any errors is asked to notify the NFPA Executive Office and the Chairman and the Secretary of the Committee.

CODE BOOK

The *National Electrical Code® Handbook* is published by the National Fire Protection Association to assist those concerned with electrical safety in understanding the intent of the code. A verbatim reproduction of the 1987 NEC® is included, and comments, diagrams, and illustrations are added where necessary to clarify some of the intricate requirements of the code.

The NEC® may be bought in paperback or as a hardcover book. It contains much of the information needed by an electrician on the job. It has definitions of electrical terms; chapters on wiring design and protection, wiring methods and materials, equipment for general use, and special occupancies, equipment, and conditions; sections on communications and tables; an index; and an appendix.

A copy of the book can be obtained from a local bookstore or by writing directly to:

National Fire Protection Association (NFPA)
Batterymarch Park
Quincy, MA 02269

For an illustration of how extensive the code is, take a look at Table A2-1. It shows the articles and sections of the code that are concerned with wiring devices alone. There are a number of regulatory agencies (listed later in this chapter) that develop standards for wiring devices. Each agency's standards are incorporated into the code as it is updated.

UNDERWRITERS LABORATORIES

William Henry Merrill started the Underwriters Laboratories, Inc. (UL) in 1894. He was called to Chicago to test the installation of Thomas Edison's new incandescent electric lights at the Colum-

TABLE A2–1
Sections of the NEC® Dealing with Wiring Devices

Article 90 — Introduction
90-1. Purpose

Article 100 — Definitions

Article 110 — Requirements for Electrical Installations
110-18. Arcing Parts.
110-21. Marking.

Article 200 — Use and Identification of Grounded Conductors
200-9. Means of Identification of Terminals.
200-10. Identification of Terminals.

Article 210 — Branch Circuits
210-4. Multiwire Branch Circuits.
210-6. Maximum Voltage.
210-7. Receptacles and Cord Connectors.
210-8. Ground Fault Protection for Personnel.
210-21. Outlet Devices.
210-24. Branch-Circuit Requirements.
210-50. Required Outlets General.
210-52. Dwelling Unit Receptacle Outlets.
210-60. Guest Rooms.
210-62. Show Windows.
210-70. Lighting Outlets Required.

Article 220 — Branch Circuit and Feeder Calculations
220-3. Branch Circuits Required.

Articles 422 — Appliances
422-22. Disconnection of Cord- and Plug-Connected Appliances.

Article 517 — Health Care Facilities
517-2. Definitions.
517-10. Wiring Methods.
517-11. Grounding.
517-13. Receptacles with Insulated Grounds.
517-60. Essential Electrical Systems, General.
517-61. Emergency Systems.
517-83. General Care Areas.
517-84. Critical Care Areas.
517-90. Additional Protective Techniques.
517-101. Wiring and Equipment.
(a) Within Hazardous Anesthetizing Locations.
(b) Located Above Hazardous Anesthetizing Locations.
(c) Other-Than-Hazardous Anesthetizing Locations.
517-103. Grounding.
517-105. Low-Voltage Equipment and Instruments.

Article 550 — Mobile Homes and Mobile Home Parks
550-3(f). Attachment Plug Cap.
550-6. Receptacle Outlets.

Article 551 — Recreational Vehicles and Recreational Vehicle Parks
551-4. Combination Electrical Systems.
551-42. Type Receptacles Provided.

Article 250 — Grounding
250-45. Equipment Connected by Cord and Plug.
250-50. Equipment Grounding Conductor Connections.
250-51. Effective Grounding Path.
250-59. Cord- and Plug-Connected Equipment.
250-74. Connecting Receptacle Grounding Terminal to Box.

Article 380 — Switches
380-1. Scope.
380-2. Switch Connections.
380-8. Accessibility and Grouping.
380-9. Faceplates for Flush-Mounted Snap Switches.
380-14. Rating and Use of Snap Switches.
380-15. Marking.

Article 410 — Lighting Fixtures, Lampholders, Lamps, Receptacles and Rosettes
410-56. Rating and Type.
410-57. Receptacles in Damp or Wet Locations.
410-58. Grounding-Type Receptacles, Adapters, Cord Connectors and Attachment Plugs.

Article 555 — Marinas and Boatyards
555-3. Receptacles.
555-7. Grounding.

Article 680 — Swimming Pools, Fountains and Similar Installations
680-4. Definitions.
680-5. Transformers and Ground-Fault Circuit-Interrupters.
680-6. Receptacles, Lighting Fixtures and Lighting Outlets.
(a) Receptacles.
680-7. Cord- and Plug-Connected Equipment.
680-40. Outdoor Installations.
680-41. Indoor Installations.
680-51. Lighting Fixtures, Submersible Pumps and Other Submersible Equipment.
680-56. Cord- and Plug-Connected Equipment.
680-62(a). Ground-Fault Circuit-Interrupter.
680-62(c). Methods of Bonding.

bian Exposition. The display had a nasty habit of setting itself on fire. Merrill started the UL to test products for electric and fire hazards for insurance companies. It continued as a testing laboratory for insurance underwriters until 1917. Then it became an independent, self-supporting safety-testing laboratory. The National Board of Fire Underwriters continued as sponsors of the UL until 1968. At that time, sponsorship and membership were broadened to include representatives of consumer interests, governmental bodies or agencies, education, public-safety bodies, public utilities, and the insurance industry, in addition to safety and standardization experts.

The UL has offices in Chicago and Northbrook, Illinois; Melville, New York; and Santa Clara, California.

UL LABEL

The UL label, Figure A2-1, tells you that the product on which the label appears is reasonably free from fire, electric shock, and related accident hazards.

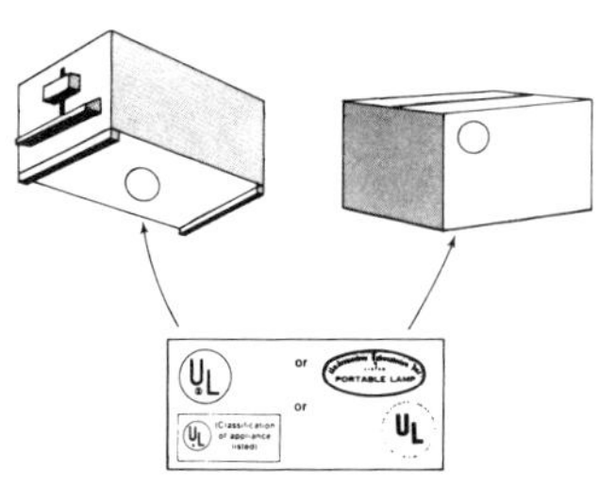

Figure A2-1. UL symbol.

The UL's engineers test products that are voluntarily submitted by manufacturers to see whether the products meet the UL's requirements for safety. A product is tested and analyzed for all reasonably foreseeable hazards. The UL label is placed on products as they come off the assembly line. The issuance of the label is firmly controlled by the UL and can be obtained only through them. The label should be on the body of the product or on the carton.

The electrician is especially interested in the label on metal boxes for electrical wiring systems. The UL label will be stamped on the side of the metal box so that it cannot be removed. It is indented into the metal. The UL label is placed on cords for lamps, drills, and other electrical equipment if the cord (not necessarily the product it serves) meets the standards of the UL. In some instances, the cord may be approved and not the product it serves.

Make sure the electrical device also has the UL label attached for safe operation.

UL ELECTRICAL DEPARTMENT

The Electrical Department of the UL is the largest of the organization's six engineering departments. Safety evaluations are made on hundreds of different types of appliances for use in the home, commercial buildings, schools, and factories. The scope of the work in this department also includes electrical construction materials used within buildings to distribute electrical power from the meter location to the electrical outlet.

The UL publishes more than 500 *Standards for Safety* for materials, devices, constructions, and methods. Copies of these are available to interested persons, and a free catalog is available.

CANADIAN STANDARDS ASSOCIATION

The Canadian Standards Association (CSA) has more authority as an agency than the UL does. The UL program is strictly voluntary. If an electrical product (or in some cases other type of product) used in Canada is connected in any way with the consumption of power from the electrical power sources owned by the provinces, the product must have CSA approval.

The basic objectives of the CSA are:

- to develop voluntary national standards,
- to provide certification services for national standards, and
- to represent Canada in international standards activities.

CSA headquarters are located in Rexdale, Ontario. Professional engineers and skilled lab technicians work with a management team to keep the standards current. Regional offices and test facilities are located in Montreal and Vancouver. A branch office is located in Edmonton, and executive offices are located in the capital city of Ottawa.

STANDARDS DEVELOPMENT

An early role of the CSA was directed toward safety in the operation of electrical appliances and equipment. The *Canadian Electrical Code* was first published by the CSA in 1927. Its recommendations soon became mandatory requirements in every province, and the CSA established the testing and inspection facilities that enable manufacturers to obtain certification of products that conform to the code. This early example has been followed in many fields: fuel-burning equipment, plumbing products, and plastics, to name a few.

CERTIFICATION PROGRAM

A manufacturer, who may be from any part of the world, files an application with the CSA. The CSA engineers and technicians inspect and test the product for compliance with an applicable standard. If the product meets the standard (modifications may be required), the manufacturer may apply a CSA mark to the product indicating certification. (See Figure A2-2.) The CSA is the safety testing authority that is accepted by inspection authorities in Canada. Inspectors may drop in unannounced at factories to inspect the quality of the product to see if it can keep its certification and the CSA label.

Figure A2-2. CSA symbol.

The CSA symbol on electrical equipment you purchase is an assurance that such equipment has passed rigid inspection. Many electrical wiring devices made in the United States will have the CSA emblem stamped or tagged on them next to the UL label.

STATE AND LOCAL CODES

Some states have prepared special codes and ordinances of their own. These usually apply to electrical systems in public buildings. In most cases, the local rules and regulations are taken from the

NEC®Many municipalities adopt the NEC® verbatim. This provides some degree of authority for the code, and, in most instances, the municipality will also appoint electrical inspectors to check all new construction to make sure it conforms to the code.

In some localities, the soil may be a little different, and special standards have to be established for grounding conductors. There may also be differences in weather conditions that call for some special rules and regulations. The rules have to be adopted by the city, town, county, or state in order for inspection to have any meaning.

Before any building is begun in a community, it is best to check with local regulations for electrical installations.

UTILITY COMPANIES

In most instances, the local utility will have information available that deals with its suggestions for hookup to the utility's lines. Most electric companies have an Adequate Wiring Bureau that you will want to consult before trying to wire a building. If the building does not meet the Bureau's standards, the electric company can refuse to connect the building to the company's lines.

WIRING-DEVICE STANDARDS AND REGULATORY AGENCIES

Standards for the manufacture of electrical equipment have to be established in order for the equipment to be used in the wiring of buildings. The important part is to standardize the devices so that they will still be able to fit regardless of who manufactures the devices. There are agencies that develop standards, rules, and regulations that govern the manufacture of electrical boxes, switches, and wire. Most of these standards are incorporated into the NEC and made law by federal, state, and local governments.

The agencies that have the most to do with establishing these regulations and standards appear on the next page.

National Electrical Manufacturers Association (NEMA)
155 East 44th Street
New York, NY 10017

American National Standards Institute, Inc. (ANSI)
1430 Broadway
New York, NY 10018

Underwriter's Laboratories, Inc. (UL)
Chicago and Northbrook, IL; Melville, NY; Santa Clara, CA

Canadian Standards Association (CSA)
178 Rexdale Boulevard
Rexdale, Ontario, Canada
M9W 1R3

General Services Administration (GSA)
Federal Supply Service
Crystal Mall, Bldg. 4
Washington, DC 20406

Occupational Safety and Health Administration (OSHA)
U.S. Department of Labor
200 Constitution Avenue, NW
Washington, DC 20210

National Fire Protection Association (NFPA)
Batterymarch Park
Quincy, MA 02269

Standards can take many forms, dictated by the particular subject or object under consideration. For instance, the National Electrical Manufacturers Association (NEMA) has a standard, WD-1, that deals with the definition of terms used in the wiring-device industry (See Table A2-2.)

TABLE A2-2

Wiring Device Industry Terms/Definitions
In Accordance with NEMA* Standard WD-1

WD 1-1.01 Cord Connector. A cord connector is a portable receptacle which is provided with means for attachment to a flexible cord and which is not intended for permanent mounting.

NEMA Standard 7-13-1967.

WD 1-1.02 Grounded Conductor (System Ground). A grounded conductor is a circuit conductor (normally current carrying) which is intentionally connected to earth ground. (It is identified as the white conductor.)

NEMA Standard 7-13-1967.

WD 1-1.03 Grounding Conductor (Equipment Ground). A grounding conductor is a conductor which connects non-current-carrying metal parts of equipment to earth ground to provide an intentional path for fault current to ground. (It is bare or, when covered, is identified as the green or green with yellow stripes conductor.)

NEMA Standard 7-13-1967.

WD 1-1.04 Lampholder. A lampholder is a device which is intended to support an electric lamp mechanically and to connect it electrically to a circuit.

NEMA Standard 7-13-1967.

WD 1-1.05 Male Base (Inlet). A male base is a plug which is intended for flush or surface mounting on an appliance or equipment and which serves to connect utilization equipment to a connector.

NEMA Standard 7-13-1967.

WD 1-1.06 Outlet. An outlet is a point on the wiring system at which current is taken to supply utilization equipment.

NEMA Standard 7-13-1967.

WD 1-1.07 Plug. A plug is a device with male blades which, when inserted into a receptacle, establishes connection between the conductors of the attached flexible cord and the conductors connected to the receptacle.

NEMA Standard 7-13-1967.

WD 1-1.08 Polarization (Plugs and Receptacles). Polarization is a means of assuring the mating of plugs and receptacles of the same rating in only the correct position.

NEMA Standard 7-13-1967.

WD 1-1.09 Pole. The term "pole" as used in designating plugs and receptacles refers to a terminal to which a circuit conductor (normally current-carrying) is connected.

In switches, the number of poles indicates the number of conductors being controlled.

NEMA Standard 7-13-1967.

WD 1-1.10 Receptacle. A receptacle is a device with female contacts which is primarily installed at an outlet or on equipment and which is intended to establish electrical connection with an inserted plug.

NEMA Standard 7-13-1967

WD 1-1.11 Slant Symbol (/). The "slant" line (/) as used in wiring device ratings indicates that two or more voltage potentials are present simultaneously between different terminals of a wiring device.

NEMA Standard 7-13-1967.

WD 1-1.12 Switch. A switch is a device for making, breaking, or changing the connections in an electric circuit.

A. **Single-pole Switch (Single-pole Single-throw)** A switch which makes or breaks the connection of one conductor.

B. **Double-pole Switch (Double-pole Single-throw)** A switch which makes or breaks the connection of two conductors of a single branch circuit.

C. **Three-way Switch (Single-pole Double-throw)** A switch which changes the connection of one conductor and which is normally used in pairs to control one utilization equipment from two locations.

D. **Four-way Switch (Double-pole Double-throw Reversing)** A form of double-pole switch which is used in conjunction with two three-way switches to control one utilization equipment from three or more locations.

NEMA Standard 7-13-1967.

WD 1-1.13 Terminal (On a Wiring Device). A terminal is a fixed location on a wiring device where a conductor is intended to be connected.

NEMA Standard 7-13-1967.

WD 1-1.14 Wire (Plugs and Receptacles). The term "wire" as used in designating plugs and receptacles indicates the number of either normally current-carrying or equipment grounding connected conductors.

NEMA Standard 7-13-1967.

*Courtesy National Electrical Manufacturers' Association.

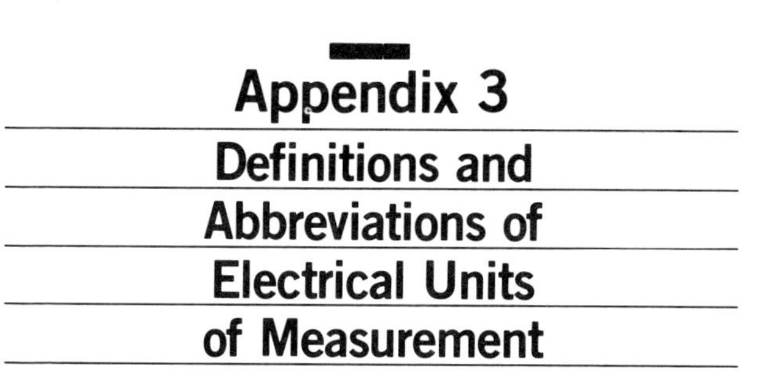

Appendix 3

Definitions and Abbreviations of Electrical Units of Measurement

Ampere (A). Unit of current strength or current flow; the current that 1 volt can send through a resistance of 1 ohm.

Candela. The basic unit of light intensity, defined as the intensity of light given off by 5 square millimeters of platinum at its solidification temperature of 1773.5°C.

Coulomb. Unit of measure of static electricity, 6.25×10^{18} electrons on a surface.

Cycle (~). Now called hertz (Hz)—the period of a complete alternation of an alternating current circuit, comprising a positive and a negative alternation.

Frequency (Hz). Measure of how frequently alternating current takes place. In this case it is the number of cycles (Hz) per second. Hertz has replaced cycles as a measurement of frequency. The "per second" has been dropped and it is just hertz abbreviated Hz.

Horsepower (hp). Mechanical unit for the measurement of power or rate of doing work. It is the work required to raise 33,000 pounds 1 foot in 1 minute and equals 746 watts, or 0.746 kilowatts.

Hertz (Hz). Unit of frequency that tells how many times per second alternating current changes direction and makes complete trips through a circuit to the generation source.

Impedance (Z). Opposition to alternating-current flow presented by resistance, capacitance, and inductance. It is measured in ohms. The impedance is always the vector sum of resistance and reactance and has to be calculated rather than measured.

Joule (j). Unit of energy equal to 1 watt-second.

Kilovolt-amperes (kVA). 1000 volt-amperes.

Kilowatt (kW). 1000 watts.

Kilowatt-hour (kWh). 1000 watts for 1 hour.

Lumen. A unit of measurement of luminous flux, approximately equal to the light output of one candle. It is defined more precisely as a flux on a unit surface such as 1 square foot or square centimeter. All points on the surface are at unit distance 1 foot or 1 centimeter from a uniform point source of 1 candela intensity.

Ohm (Ω). The unit of measurement of resistance. One ohm is the amount of opposition put up by a resistor to 1 ampere of current with a voltage of 1 volt.

Polyphase. A general term applied to an alternating-current system with more than one phase (e.g., two-phase, three-phase.)

Power Factor (PF). The ratio of true power to apparent power of an alternating-current system. The power factor may be expressed as a percentage but is usually written as a decimal.

Single-phase (1ϕ). The term applied to an alternating-current circuit energized by a single alternating emf (electromotive force) source. (Such a circuit is usually supplied through two wires.)

Three Phase (3ϕ). The term applied to an alternating system comprising three electric circuits energized by alternating emf's that differ in phase by one-third of a hertz.

Two Phase (2ϕ). The term applied to an alternating-current system comprising two electrical circuits energized by alternating emf's that differ in phase by a quarter of a hertz. Frequently referred to as quarter-phase, it is an old term seldom used today.

Volt (V). Unit of electrical pressure, electromotive force (emf), or difference of potential; force required to send 1 ampere of current through 1 ohm of resistance.

Volt-Ampere (VA). Unit of measurement of volts and amperes used for measuring alternating current—1 volt times 1 ampere.

Watt (W). Electrical unit of energy. The product of volts times amperes in a DC circuit or AC circuit with resistance only. It is 1 ampere times 1 volt.

Watt-Hour (Wh). Electrical unit of work or energy. Watt-hours equals watts times hours.

Appendix 4

Tables and Charts

There are many tables and charts that provide quick access to information the electrician or apprentice may need to read blueprints and understand instructions or do mathematical calculations. This appendix includes many of the tables the electrician will find most useful.

GREEK ALPHABET

The Greek alphabet (Table A4-1) is used to designate any number of electrical quantities and mathematical equivalents.

TABLE A4–1
Greek Alphabet

Letter			
Lower Case	Upper Case	Name	Designates
α	A	alpha	Angles, coefficients, attenuation, absorption factor, area.
β	B	beta	Angles, coefficients, phase constant.
γ	Γ	gamma	Specific quantity, angles, electrical conductivity, propagation constant (capacitance).
δ	Δ	delta	Density, angles, increment or decrement (upper or lower case), determinant (upper case), permittivity.
ϵ	E	epsilon	Dielectric constant, permittivity, base of natural logarithms, electric intensity.
ζ	Z	zeta	Coordinate, coefficient.
η	H	eta	Intrinsic impedance, efficiency, surface charge density, hysteresis coordinates.
θ	Θ	theta	Angular phase displacement, time constant, reluctance, angles.
ι	I	iota	Unit vector.
κ	K	kappa	Susceptibility, coupling coefficient.

Table A4–1 (*continued*)

Greek Alphabet			
Letter			
Lower Case	Upper Case	Name	Designates
λ	Λ	lambda	Wavelength (lower case), attenuation constant, permeance (upper case).
μ	M	mu	Prefix *micro*, permeability, amplification factor (lower case).
ν	N	nu	Reluctivity, frequency.
ξ	Ξ	xi	Coordinates.
o	O	omicron	—
π	Π	pi	3.141592654 (circumference divided by the diameter of a circle [lower case]).
ρ	P	rho	Resistivity, volume charge density, coordinates (lower case).
σ, s	Σ	sigma	Surface charge density, complex propagation constant, electrical conductivity, leakage coefficient, sign of summation (upper case).
τ	T	tau	Time constant, volume resistivity time-phase displacement, transmission factor, density.
υ	Υ	upsilon	—
ϕ	Φ	phi	Magnetic flux, angles, scalar potential (upper case).
χ	X	chi	Electrical susceptibility, angles.
ψ	Ψ	psi	Dielectric flux, phase differences, coordinates, angles (lower case).
ω	Ω	omega	Angular velocity (2f), resistance in ohms (upper case), solid angles (upper case).

PROPERTIES OF MATERIALS

The electrician works with different types of materials. Familiarity with the properties of these materials is important if the electrician is to choose which materials to use in a particular situation, know how certain materials should be handled, and know what to expect in terms of the performance of the material. Table A4-2 lists some important properties of materials electricians often come into contact with.

SQUARE ROOTS AND SQUARES

Many electrical formulas (see Chapter 7) require that a number be squared or that the square root of a number be taken. Table A4-3 will help the electrician in computation.

TRIGONOMETRIC TABLES

The electrician sometimes needs to use formulas that involve trigonometry, especially in calculating the power factor. Although many calculators are now programmed with trigonometric functions, the electrician may at some time need to refer to a table of trigonometric values and should be familiar with how to use it. Two commonly used tables are presented here. (See Tables A4-4 and A4-5.)

FIGURING POWER FACTOR USING THE SINE-COSINE TABLE.

The power factor (PF) is equal to the cosine of the phase angle. Look up the phase angle and then refer to the column labeled *cos* to find the power factor. If you want to have the power factor in a percentage form, just move the decimal point two places to the

TABLE A4–2

Properties of Materials

Liquid	Lb./Gal.	Element	Symbol	Melting Point °F	Coef. of Exp. Per °F	Electrical Conductivity % Pure Copper	Lb./Cu. In.
Acetone	6.6	Aluminum	Al	1215	.0000133	64.9	.098
Alcohol (100%)	6.8	Antimony	Sb	1167	.00000627	4.42	.239
Ammonia	7.4	Beryllium	Be	2345	.0000068	9.32	.066
Benzine	6.4	Bismuth	Bi	520	.00000747	1.50	.354
Benzol	7.4	Cadmium	Cd	610	.00000166	22.7	.313
Carbon Tetrachloride	13.3	Chromium	Cr	2822	.0000045	13.2	.258
Castor Oil	8.1	Cobalt	Co	2714	.00000671	17.8	.322
Gasoline	6.1	Copper	Cu	1981	.0000091	100.	.323
Glue Liquid	10.7	Gold	Au	1945	.0000080	71.2	.697
Hydrochloric Acid	9.4	Iron	Fe	2795	.0000066	17.6	.284
Inerteen	12.9	Lead	Pb	621	.0000164	8.35	.409
Kerosene	6.7	Magnesium	Mg	1204	.0000143	38.7	.063
Lard Oil	7.7	Mercury	Hg	-38	—	1.80	.489
Linseed Oil	7.8	Molybdenum	Mo	4748	.00000305	36.1	.368
Machine Oil	7.5	Nickel	Ni	2646	.0000076	25.0	.322
Paints	10.3–13.5	Platinum	Pt	3224	.0000043	17.5	.774
Shellac	7.5	Selenium	Se	428	.0000206	14.4	.174
Sodium Silicate	12.0	Silver	Ag	1761	.0000105	106.	.380
Sulphuric Acid	15.3	Tellurium	Te	846	.0000093	—	.224
Tung Oil	7.8	Tin	Sn	450	.0000124	15.0	.264
Turpentine	7.3	Tungsten	W	6098	.0000022	31.5	.698
Varnish-Ins.	7.0	Vanadium	V	3110	—	6.63	.205
Water	8.34	Zinc	Zn	787	.0000219	29.1	.258
Wemco Oil	7.5						

TABLE A4–3

Table of Square Roots and Squares

n	n^2	$\sqrt{n}$	n	n^2	$\sqrt{n}$
			50	2500	7.0711
1	1	1.0000	51	2601	7.1414
2	4	1.4142	52	2704	7.2111
3	9	1.7321	53	2809	7.2801
4	16	2.0000	54	2916	7.3485
5	25	2.2361	55	3025	7.4162
6	36	2.4495	56	3136	7.4833
7	49	2.6458	57	3249	7.5498
8	64	2.8284	58	3364	7.6158
9	81	3.0000	59	3481	7.6811
10	100	3.1623	60	3600	7.7460
11	121	3.3166	61	3721	7.8102
12	144	3.4641	62	3844	7.8740
13	169	3.6056	63	3969	7.9373
14	196	3.7417	64	4096	8.0000
15	225	3.8730	65	4225	8.0623
16	256	4.0000	66	4356	8.1240
17	289	4.1231	67	4489	8.1854
18	324	4.2426	68	4624	8.2462
19	361	4.3589	69	4761	8.3066
20	400	4.4721	70	4900	8.3666
21	441	4.5826	71	5041	8.4261
22	484	4.6904	72	5184	8.4853
23	529	4.7958	73	5329	8.5440
24	576	4.8990	74	5476	8.6023
25	625	5.0000	75	5625	8.6603
26	676	5.0990	76	5776	8.7178
27	729	5.1962	77	5929	8.7750
28	784	5.2915	78	6084	8.8318
29	841	5.3852	79	6241	8.8882
30	900	5.4772	80	6400	8.9443
31	961	5.5678	81	6561	9.0000
32	1024	5.6569	82	6724	9.0554
33	1089	5.7446	83	6889	9.1104

TABLE A4–3

Table of Square Roots and Squares

n	n^2	$\sqrt{n}$	n	n^2	$\sqrt{n}$
34	1156	5.8310	84	7056	9.1652
35	1225	5.9161	85	7225	9.2195
36	1296	6.0000	86	7396	9.2736
37	1369	6.0828	87	7569	9.3274
38	1444	6.1644	88	7744	9.3808
39	1521	6.2450	89	7921	9.4340
40	1600	6.3246	90	8100	9.4868
41	1681	6.4031	91	8281	9.5394
42	1764	6.4807	92	8464	9.5917
43	1849	6.5574	93	8649	9.6437
44	1936	6.6332	94	8836	9.6954
45	2025	6.7082	95	9025	9.7468
46	2116	6.7823	96	9216	9.7980
47	2209	6.8557	97	9409	9.8489
48	2304	6.9282	98	9604	9.8995
49	2401	7.0000	99	9801	9.9499
50	2500	7.0711	100	10,000	10.0000

right. Conversely, if you know the power factor and need to find the phase angle, check the *cos* column to find the number closest to the power factor (remember it is in decimal form in the table) and then look at the adjacent column, labeled x, to find the angle. For example, if you know the phase angle is 45°, you find that the cosine listing is 0.7071, and then you know that the power factor is 70.71%.

FIGURING POWER FACTOR USING THE ARC TAN TABLE

If the tangent is known, use the Arc Tan x table to determine the angle. Then turn to the Sine-Cosine table, find the angle, and look at what is listed under the *cos* column. This will tell you the power factor, or, in other words, the cosine of the angle produces the power factor.

TABLE A4–4

TABLE OF VALUES: Sin X and Cos X. X = 00.00° to 14.90°

X	Sin X	Cos X	X	Sin X	Cos X	X	Sin X	Cos X
.00	.0000	1.000	5.00	.0872	.9962	10.00	.1736	.9848
.10	.0017	1.000	5.10	.0889	.9960	10.10	.1754	.9845
.20	.0035	1.000	5.20	.0906	.9959	10.20	.1771	.9842
.30	.0052	1.000	5.30	.0924	.9957	10.30	.1788	.9839
.40	.0070	1.000	5.40	.0941	.9956	10.40	.1805	.9836
.50	.0087	1.000	5.50	.0958	.9954	10.50	.1822	.9833
.60	.0105	.9999	5.60	.0976	.9952	10.60	.1840	.9829
.70	.0122	.9999	5.70	.0993	.9951	10.70	.1857	.9826
.80	.0140	.9999	5.80	.1011	.9949	10.80	.1874	.9823
.90	.0157	.9999	5.90	.1028	.9947	10.90	.1891	.9820
1.00	.0175	.9998	6.00	.1045	.9945	11.00	.1908	.9816
1.10	.0192	.9998	6.10	.1063	.9943	11.10	.1925	.9813
1.20	.0209	.9998	6.20	.1080	.9942	11.20	.1942	.9810
1.30	.0227	.9997	6.30	.1097	.9940	11.30	.1959	.9806
1.40	.0244	.9997	6.40	.1115	.9938	11.40	.1977	.9803
1.50	.0262	.9997	6.50	.1132	.9936	11.50	.1994	.9799
1.60	.0279	.9996	6.60	.1149	.9934	11.60	.2011	.9796
1.70	.0297	.9996	6.70	.1167	.9932	11.70	.2028	.9792
1.80	.0314	.9995	6.80	.1184	.9930	11.80	.2045	.9789
1.90	.0332	.9995	6.90	.1201	.9928	11.90	.2062	.9785
2.00	.0349	.9994	7.00	.1219	.9925	12.00	.2079	.9781
2.10	.0366	.9993	7.10	.1236	.9923	12.10	.2096	.9778
2.20	.0384	.9993	7.20	.1253	.9921	12.20	.2113	.9774
2.30	.0401	.9992	7.30	.1271	.9919	12.30	.2130	.9770
2.40	.0419	.9991	7.40	.1288	.9917	12.40	.2147	.9767
2.50	.0436	.9990	7.50	.1305	.9914	12.50	.2164	.9763
2.60	.0454	.9990	7.60	.1323	.9912	12.60	.2181	.9759
2.70	.0471	.9989	7.70	.1340	.9910	12.70	.2198	.9755
2.80	.0488	.9988	7.80	.1357	.9907	12.80	.2215	.9751
2.90	.0506	.9987	7.90	.1374	.9905	12.90	.2233	.9748
3.00	.0523	.9986	8.00	.1392	.9903	13.00	.2250	.9744
3.10	.0541	.9985	8.10	.1409	.9900	13.10	.2267	.9740
3.20	.0558	.9984	8.20	.1426	.9898	13.20	.2284	.9736
3.30	.0576	.9983	8.30	.1444	.9895	13.30	.2300	.9732
3.40	.0593	.9982	8.40	.1461	.9893	13.40	.2317	.9728
3.50	.0610	.9981	8.50	.1478	.9890	13.50	.2334	.9724
3.60	.0628	.9980	8.60	.1495	.9888	13.60	.2351	.9720
3.70	.0645	.9979	8.70	.1513	.9885	13.70	.2368	.9715
3.80	.0663	.9978	8.80	.1530	.9882	13.80	.2385	.9711
3.90	.0680	.9977	8.90	.1547	.9880	13.90	.2402	.9707
4.00	.0698	.9976	9.00	.1564	.9877	14.00	.2419	.9703

Table A4–4 (*continued*)

X	Sin X	Cos X	X	Sin X	Cos X	X	Sin X	Cos X
4.10	.0715	.9974	9.10	.1582	.9874	14.10	.2436	.9699
4.20	.0732	.9973	9.20	.1599	.9871	14.20	.2453	.9694
4.30	.0750	.9972	9.30	.1616	.9869	14.30	.2470	.9690
4.40	.0767	.9971	9.40	.1633	.9866	14.40	.2487	.9686
4.50	.0785	.9969	9.50	.1650	.9863	14.50	.2504	.9681
4.60	.0802	.9968	9.60	.1668	.9860	14.60	.2521	.9677
4.70	.0819	.9966	9.70	.1685	.9857	14.70	.2538	.9673
4.80	.0837	.9965	9.80	.1702	.9854	14.80	.2554	.9668
4.90	.0854	.9963	9.90	.1719	.9851	14.90	.2571	.9664

TABLE OF VALUES: Sin X and Cos X. X = 15.00° to 29.90°

X	Sin X	Cos X	X	Sin X	Cos X	X	Sin X	Cos X
15.00	.2588	.9659	20.00	.3420	.9397	25.00	.4226	.9063
15.10	.2605	.9655	20.10	.3437	.9391	25.10	.4242	.9056
15.20	.2622	.9650	20.20	.3453	.9385	25.20	.4268	.9048
15.30	.2639	.9646	20.30	.3469	.9379	25.30	.4274	.9041
15.40	.2656	.9641	20.40	.3486	.9373	25.40	.4289	.9033
15.50	.2672	.9636	20.50	.3502	.9367	25.50	.4305	.9026
15.60	.2689	.9632	20.60	.3518	.9361	25.60	.4321	.9018
15.70	.2706	.9627	20.70	.3535	.9354	25.70	.4337	.9011
15.80	.2723	.9622	20.80	.3551	.9348	25.80	.4352	.9003
15.90	.2740	.9617	20.90	.3567	.9342	25.90	.4368	.8996
16.00	.2756	.9613	21.00	.3584	.9336	26.00	.4384	.8988
16.10	.2773	.9608	21.10	.3600	.9330	26.10	.4399	.8980
16.20	.2790	.9603	21.20	.3616	.9323	26.20	.4415	.8973
16.30	.2807	.9598	21.30	.3633	.9317	26.30	.4431	.8965
16.40	.2823	.9593	21.40	.3649	.9311	26.40	.4446	.8957
16.50	.2840	.9588	21.50	.3665	.9304	26.50	.4462	.8949
16.60	.2857	.9583	21.60	.3681	.9298	26.60	.4478	.8942
16.70	.2874	.9578	21.70	.3697	.9291	26.70	.4493	.8934
16.80	.2890	.9573	21.80	.3714	.9285	26.80	.4509	.8926
16.90	.2907	.9568	21.90	.3730	.9278	26.90	.4524	.8918
17.00	.2924	.9563	22.00	.3746	.9272	27.00	.4540	.8910
17.10	.2940	.9558	22.10	.3762	.9265	27.10	.4555	.8902
17.20	.2957	.9553	22.20	.3778	.9259	27.20	.4571	.8894
17.30	.2974	.9548	22.30	.3795	.9252	27.30	.4586	.8886
17.40	.2990	.9542	22.40	.3811	.9245	27.40	.4602	.8878
17.50	.3007	.9537	22.50	.3827	.9239	27.50	.4617	.8870
17.60	.3024	.9532	22.60	.3843	.9232	27.60	.4633	.8862

Table A4–4 (*continued*)

X	Sin X	Cos X	X	Sin X	Cos X	X	Sin X	Cos X
17.70	.3040	.9527	22.70	.3859	.9225	27.70	.4648	.8854
17.80	.3057	.9527	22.80	.3875	.9219	27.80	.4664	.8846
17.90	.3074	.9516	22.90	.3891	.9212	27.90	.4679	.8838
18.00	.3090	.9511	23.00	.3907	.9205	28.00	.4695	.8829
18.10	.3107	.9505	23.10	.3923	.9198	28.10	.4710	.8821
18.20	.3123	.9500	23.20	.3939	.9191	28.20	.4726	.8813
18.30	.3140	.9494	23.30	.3955	.9184	28.30	.4741	.8805
18.40	.3156	.9489	23.40	.3971	.9178	28.40	.4756	.8796
18.50	.3173	.9483	23.50	.3987	.9171	28.50	.4772	.8788
18.60	.3190	.9478	23.60	.4003	.9164	28.60	.4787	.8780
18.70	.3206	.9472	23.70	.4019	.9157	28.70	.4802	.8771
18.80	.3223	.9466	23.80	.4035	.9150	28.80	.4818	.8763
18.90	.3239	.9461	23.90	.4051	.9143	28.90	.4833	.8755
19.00	.3256	.9455	24.00	.4067	.9135	29.00	.4848	.8746
19.10	.3272	.9449	24.10	.4083	.9128	29.10	.4863	.8738
19.20	.3289	.9444	24.20	.4099	.9121	29.20	.4879	.8729
19.30	.3305	.9438	24.30	.4115	.9114	29.30	.4894	.8721
19.40	.3322	.9432	24.40	.4131	.9107	29.40	.4909	.8712
19.50	.3338	.9426	24.50	.4147	.9100	29.50	.4924	.8704
19.60	.3355	.9421	24.60	.4163	.9092	29.60	.4939	.8695
19.70	.3371	.9415	24.70	.4179	.9085	29.70	.4955	.8686
19.80	.3387	.9409	24.80	.4195	.9078	29.80	.4970	.8678
19.90	.3404	.9403	24.90	.4210	.9070	29.90	.4985	.8669

TABLE OF VALUES: Sin X and Cos X. X = 30.00° to 44.90°

X	Sin X	Cos X	X	Sin X	Cos X	X	Sin X	Cos X
30.00	.5000	.8660	35.00	.5736	.8192	40.00	.6428	.7660
30.10	.5015	.8652	35.10	.5750	.8181	40.10	.6441	.7649
30.20	.5030	.8643	35.20	.5764	.8171	40.20	.6455	.7638
30.30	.5045	.8634	35.30	.5779	.8161	40.30	.6468	.7627
30.40	.5060	.8625	35.40	.5793	.8151	40.40	.6481	.7615
30.50	.5075	.8616	35.50	.5807	.8141	40.50	.6494	.7604
30.60	.5090	.8607	35.60	.5821	.8131	40.60	.6508	.7593
30.70	.5105	.8599	35.70	.5835	.8121	40.70	.6521	.7581
30.80	.5120	.8590	35.80	.5850	.8111	40.80	.6534	.7570
30.90	.5135	.8581	35.90	.5864	.8100	40.90	.6547	.7559

Table A4–4 (*continued*)

X	Sin X	Cos X	X	Sin X	Cos X	X	Sin X	Cos X
31.00	.5150	.8572	36.00	.5878	.8090	41.00	.6561	.7547
31.10	.5165	.8563	36.10	.5892	.8080	41.10	.6574	.7536
31.20	.5180	.8554	36.20	.5906	.8070	41.20	.6587	.7524
31.30	.5195	.8545	36.30	.5920	.8059	41.30	.6600	.7513
31.40	.5210	.8536	36.40	.5934	.8049	41.40	.6613	.7501
31.50	.5225	.8526	36.50	.5948	.8039	41.50	.6626	.7490
31.60	.5240	.8517	36.60	.5962	.8028	41.60	.6639	.7478
31.70	.5255	.8508	36.70	.5976	.8018	41.70	.6652	.7466
31.80	.5270	.8499	36.80	.5990	.8007	41.80	.6665	.7455
31.90	.5284	.8490	36.90	.6004	.7997	41.90	.6678	.7443
32.00	.5299	.8480	37.00	.6018	.7986	42.00	.6691	.7431
32.10	.5314	.8471	37.10	.6032	.7976	42.10	.6704	.7420
32.20	.5329	.8462	37.20	.6046	.7965	42.20	.6717	.7408
32.30	.5344	.8453	37.30	.6060	.7955	42.30	.6730	.7396
32.40	.5358	.8443	37.40	.6074	.7944	42.40	.6743	.7385
32.50	.5373	.8434	37.50	.6088	.7934	42.50	.6756	.7373
32.60	.5388	.8425	37.60	.6101	.7923	42.60	.6769	.7361
32.70	.5402	.8415	37.70	.6115	.7912	42.70	.6782	.7349
32.80	.5417	.8406	37.80	.6129	.7902	42.80	.6794	.7337
32.90	.5432	.8396	37.90	.6143	.7891	42.90	.6807	.7325
33.00	.5446	.8387	38.00	.6157	.7880	43.00	.6820	.7314
33.10	.5461	.8377	38.10	.6170	.7869	43.10	.6833	.7302
33.20	.5476	.8368	38.20	.6184	.7859	43.20	.6845	.7290
33.30	.5490	.8358	38.30	.6198	.7848	43.30	.6858	.7278
33.40	.5505	.8348	38.40	.6211	.7837	43.40	.6871	.7266
33.50	.5519	.8339	38.50	.6225	.7826	43.50	.6884	.7254
33.60	.5534	.8329	38.60	.6239	.7815	43.60	.6896	.7242
33.70	.5548	.8320	38.70	.6252	.7804	43.70	.6909	.7230
33.80	.5563	.8310	38.80	.6266	.7793	43.80	.6921	.7218
33.90	.5577	.8300	38.90	.6280	.7782	43.90	.6934	.7206
34.00	.5592	.8290	39.00	.6293	.7771	44.00	.6947	.7193
34.10	.5606	.8281	39.10	.6307	.7760	44.10	.6959	.7181
34.20	.5621	.8271	39.20	.6320	.7749	44.20	.6972	.7169
34.30	.5635	.8261	39.30	.6334	.7738	44.30	.6984	.7157
34.40	.5650	.8251	39.40	.6347	.7727	44.40	.6997	.7145
34.50	.5664	.8241	39.50	.6361	.7716	44.50	.7009	.7133
34.60	.5678	.8231	39.60	.6374	.7705	44.60	.7022	.7120
34.70	.5693	.8221	39.70	.6388	.7694	44.70	.7034	.7108
34.80	.5707	.8211	39.80	.6401	.7683	44.80	.7046	.7096
34.90	.5721	.8202	39.90	.6414	.7672	44.90	.7059	.7083

Table A4–4 (*continued*)

TABLE OF VALUES: Sin X and Cos X.						X = 45.00° to 59.90°		
X	Sin X	Cos X	X	Sin X	Cos X	X	Sin X	Cos X
45.00	.7071	.7071	50.00	.7660	.6428	55.00	.8192	.5736
45.10	.7083	.7059	50.10	.7672	.6414	55.10	.8202	.5721
45.20	.7096	.7046	50.20	.7683	.6401	55.20	.8211	.5707
45.30	.7108	.7034	50.30	.7694	.6388	55.30	.8221	.5693
45.40	.7120	.7022	50.40	.7705	.6374	55.40	.8231	.5678
45.50	.7133	.7009	50.50	.7716	.6361	55.50	.8241	.5664
45.60	.7145	.6997	50.60	.7727	.6347	55.60	.8251	.5650
45.70	.7157	.6984	50.70	.7738	.6334	55.70	.8261	.5635
45.80	.7169	.6972	50.80	.7749	.6320	55.80	.8271	.5621
45.90	.7181	.6959	50.90	.7760	.6307	55.90	.8281	.5606
46.00	.7193	.6947	51.00	.7771	.6293	56.00	.8290	.5592
46.10	.7206	.6934	51.10	.7782	.6280	56.10	.8300	.5577
46.20	.7218	.6921	51.20	.7793	.6266	56.20	.8310	.5563
46.30	.7230	.6909	51.30	.7804	.6252	56.30	.8320	.5548
46.40	.7242	.6896	51.40	.7815	.6239	56.40	.8329	.5534
46.50	.7254	.6884	51.50	.7826	.6225	56.50	.8339	.5519
46.60	.7266	.6871	51.60	.7837	.6211	56.60	.8348	.5505
46.70	.7278	.6858	51.70	.7848	.6198	56.70	.8358	.5490
46.80	.7290	.6845	51.80	.7859	.6184	56.80	.8368	.5476
46.90	.7302	.6833	51.90	.7869	.6170	56.90	.8377	.5461
47.00	.7314	.6820	52.00	.7880	.6157	57.00	.8387	.5446
47.10	.7325	.6807	52.10	.7891	.6143	57.10	.8396	.5432
47.20	.7337	.6794	52.20	.7902	.6129	57.20	.8406	.5417
47.30	.7349	.6782	52.30	.7912	.6115	57.30	.8415	.5402
47.40	.7361	.6769	52.40	.7923	.6101	57.40	.8425	.5388
47.50	.7373	.6756	52.50	.7934	.6088	57.50	.8434	.5373
47.60	.7385	.6743	52.60	.7944	.6074	57.60	.8443	.5358
47.70	.7396	.6730	52.70	.7955	.6060	57.70	.8453	.5344
47.80	.7408	.6717	52.80	.7965	.6046	57.80	.8462	.5329
47.90	.7420	.6704	52.90	.7976	.6032	57.90	.8471	.5314
48.00	.7431	.6691	53.00	.7986	.6018	58.00	.8480	.5299
48.10	.7443	.6678	53.10	.7997	.6004	58.10	.8490	.5284
48.20	.7455	.6665	53.20	.8007	.5990	58.20	.8499	.5270
48.30	.7466	.6652	53.30	.8018	.5976	58.30	.8508	.5255
48.40	.7478	.6639	53.40	.8028	.5962	58.40	.8517	.5240
48.50	.7490	.6626	53.50	.8039	.5948	58.50	.8526	.5225
48.60	.7501	.6613	53.60	.8049	.5934	58.60	.8536	.5210
48.70	.7513	.6600	53.70	.8059	.5920	58.70	.8545	.5195
48.80	.7524	.6587	53.80	.8070	.5906	58.80	.8554	.5180
48.90	.7536	.6574	53.90	.8080	.5892	58.90	.8563	.5165
49.00	.7547	.6561	54.00	.8090	.5878	59.00	.8572	.5150

Table A4–4 (*continued*)

X	Sin X	Cos X	X	Sin X	Cos X	X	Sin X	Cos X
49.10	.7559	.6547	54.10	.8100	.5864	59.10	.8581	.5135
49.20	.7570	.6534	54.20	.8111	.5850	59.20	.8590	.5120
49.30	.7581	.6521	54.30	.8121	.5835	59.30	.8599	.5105
49.40	.7593	.6508	54.40	.8131	.5821	59.40	.8607	.5090
49.50	.7604	.6494	54.50	.8141	.5807	59.50	.8616	.5075
49.60	.7615	.6481	54.60	.8151	.5793	59.60	.8625	.5060
49.70	.7627	.6468	54.70	.8161	.5779	59.70	.8634	.5045
49.80	.7638	.6455	54.80	.8171	.5764	59.80	.8643	.5030
49.90	.7649	.6441	54.90	.8181	.5750	59.90	.8652	.5015

TABLE OF VALUES: Sin X and Cos X. X = 60.00° to 74.90°

X	Sin X	Cos X	X	Sin X	Cos X	X	Sin X	Cos X
60.00	.8660	.5000	65.00	.9063	.4226	70.00	.9397	.3420
60.10	.8669	.4985	65.10	.9070	.4210	70.10	.9403	.3404
60.20	.8678	.4970	65.20	.9078	.4195	70.20	.9409	.3387
60.30	.8686	.4955	65.30	.9085	.4179	70.30	.9415	.3371
60.40	.8695	.4939	65.40	.9092	.4163	70.40	.9421	.3355
60.50	.8704	.4924	65.50	.9100	.4147	70.50	.9426	.3338
60.60	.8712	.4909	65.60	.9107	.4131	70.60	.9432	.3322
60.70	.8721	.4894	65.70	.9114	.4115	70.70	.9438	.3305
60.80	.8729	.4879	65.80	.9121	.4099	70.80	.9444	.3289
60.90	.8738	.4863	65.90	.9128	.4083	70.90	.9449	.3272
61.00	.8746	.4848	66.00	.9135	.4067	71.00	.9455	.3256
61.10	.8755	.4833	66.10	.9143	.4051	71.10	.9461	.3239
61.20	.8763	.4818	66.20	.9150	.4035	71.20	.9466	.3223
61.30	.8771	.4802	66.30	.9157	.4019	71.30	.9472	.3206
61.40	.8780	.4787	66.40	.9164	.4003	71.40	.9478	.3190
61.50	.8788	.4772	66.50	.9171	.3987	71.50	.9483	.3173
61.60	.8796	.4756	66.60	.9178	.3971	71.60	.9489	.3156
61.70	.8805	.4741	66.70	.9184	.3955	71.70	.9494	.3140
61.80	.8813	.4726	66.80	.9191	.3939	71.80	.9500	.3123
61.90	.8821	.4710	66.90	.9198	.3923	71.90	.9505	.3107
62.00	.8829	.4695	67.00	.9205	.3907	72.00	.9511	.3090
62.10	.8838	.4679	67.10	.9212	.3891	72.10	.9516	.3074
62.20	.8846	.4664	67.20	.9219	.3875	72.20	.9521	.3057
62.30	.8854	.4648	67.30	.9225	.3859	72.30	.9527	.3040
62.40	.8862	.4633	67.40	.9232	.3843	72.40	.9532	.3024
62.50	.8870	.4617	67.50	.9239	.3827	72.50	.9537	.3007
62.60	.8878	.4602	67.60	.9245	.3811	72.60	.9542	.2990
62.70	.8886	.4586	67.70	.9252	.3795	72.70	.9548	.2974

Table A4–4 (*continued*)

X	Sin X	Cos X	X	Sin X	Cos X	X	Sin X	Cos X
62.80	.8894	.4571	67.80	.9259	.3778	72.80	.9553	.2957
62.90	.8902	.4555	67.90	.9265	.3762	72.90	.9558	.2940
63.00	.8910	.4540	68.00	.9272	.3746	73.00	.9563	.2924
63.10	.8918	.4524	68.10	.9278	.3730	73.10	.9568	.2907
63.20	.8926	.4509	68.20	.9285	.3714	73.20	.9573	.2890
63.30	.8934	.4493	68.30	.9291	.3697	73.30	.9578	.2874
63.40	.8942	.4478	68.40	.9298	.3681	73.40	.9583	.2857
63.50	.8949	.4462	68.50	.9304	.3665	73.50	.9588	.2840
63.60	.8957	.4446	68.60	.9311	.3649	73.60	.9593	.2823
63.70	.8965	.4431	68.70	.9317	.3633	73.70	.9598	.2807
63.80	.8973	.4415	68.80	.9323	.3616	73.80	.9603	.2790
63.90	.8980	.4399	68.90	.9330	.3600	73.90	.9608	.2773
64.00	.8988	.4384	69.00	.9336	.3584	74.00	.9613	.2756
64.10	.8996	.4368	69.10	.9342	.3567	74.10	.9617	.2740
64.20	.9003	.4352	69.20	.9348	.3551	74.20	.9622	.2723
64.30	.9011	.4337	69.30	.9354	.3535	74.30	.9627	.2706
64.40	.9018	.4321	69.40	.9361	.3518	74.40	.9632	.2689
64.50	.9026	.4305	69.50	.9367	.3502	74.50	.9636	.2672
64.60	.9033	.4289	69.60	.9373	.3486	74.60	.9641	.2656
64.70	.9041	.4274	69.70	.9379	.3469	74.70	.9646	.2639
64.80	.9048	.4258	69.80	.9385	.3453	74.80	.9650	.2622
64.90	.9056	.4242	69.90	.9391	.3437	74.90	.9655	.2605

TABLE OF VALUES: Sin X and Cos X. X = 75.00° to 90.00°

X	Sin X	Cos X	X	Sin X	Cos X	X	Sin X	Cos X
75.00	.9659	.2588	80.00	.9848	.1736	85.00	.9962	.0872
75.10	.9664	.2571	80.10	.9851	.1719	85.10	.9963	.0854
75.20	.9668	.2554	80.20	.9854	.1702	85.20	.9965	.0837
75.30	.9673	.2538	80.30	.9857	.1685	85.30	.9966	.0819
75.40	.9677	.2521	80.40	.9860	.1668	85.40	.9968	.0802
75.50	.9681	.2504	80.50	.9863	.1650	85.50	.9969	.0785
75.60	.9686	.2487	80.60	.9866	.1633	85.60	.9971	.0767
75.70	.9690	.2470	80.70	.9869	.1616	85.70	.9972	.0750
75.80	.9694	.2453	80.80	.9871	.1599	85.80	.9973	.0732
75.90	.9693	.2436	80.90	.9874	.1582	85.90	.9974	.0715
76.00	.9703	.2419	81.00	.9877	.1564	86.00	.9976	.0698

Table A4–4 (*continued*)

X	Sin X	Cos X	X	Sin X	Cos X	X	Sin X	Cos X
76.10	.9707	.2402	81.10	.9880	.1547	86.10	.9977	.0680
76.20	.9711	.2385	81.20	.9882	.1530	86.20	.9978	.0663
76.30	.9715	.2368	81.30	.9885	.1513	86.30	.9979	.0645
76.40	.9720	.2351	81.40	.9888	.1495	86.40	.9980	.0628
76.50	.9724	.2334	81.50	.9890	.1478	86.50	.9981	.0610
76.60	.9728	.2317	81.60	.9893	.1461	86.60	.9982	.0593
76.70	.9732	.2300	81.70	.9895	.1444	86.70	.9983	.0576
76.80	.9736	.2284	81.80	.9898	.1426	86.80	.9984	.0558
76.90	.9740	.2267	81.90	.9900	.1409	86.90	.9985	.0541
77.00	.9744	.2250	82.00	.9903	.1392	87.00	.9986	.0523
77.10	.9743	.2233	82.10	.9905	.1374	87.10	.9987	.0506
77.20	.9751	.2215	82.20	.9907	.1357	87.20	.9988	.0488
77.30	.9755	.2198	82.30	.9910	.1340	87.30	.9989	.0471
77.40	.9759	.2181	82.40	.9912	.1323	87.40	.9990	.0454
77.50	.9763	.2164	82.50	.9914	.1305	87.50	.9990	.0436
77.60	.9767	.2147	82.60	.9917	.1288	87.60	.9991	.0419
77.70	.9770	.2130	82.70	.9919	.1271	87.70	.9992	.0401
77.80	.9774	.2113	82.80	.9921	.1253	87.80	.9993	.0384
77.90	.9778	.2096	82.90	.9923	.1236	87.90	.9993	.0366
78.00	.9781	.2079	83.00	.9925	.1219	88.00	.9994	.0349
78.10	.9785	.2062	83.10	.9928	.1201	88.10	.9995	.0332
78.20	.9789	.2045	83.20	.9930	.1184	88.20	.9995	.0314
78.30	.9792	.2028	83.30	.9932	.1167	88.30	.9996	.0297
78.40	.9796	.2011	83.40	.9934	.1149	88.40	.9996	.0279
78.50	.9799	.1994	83.50	.9936	.1132	88.50	.9997	.0262
78.60	.9803	.1977	83.60	.9938	.1115	88.60	.9997	.0244
78.70	.9806	.1959	83.70	.9940	.1097	88.70	.9997	.0227
78.80	.9810	.1942	83.80	.9942	.1080	88.80	.9998	.0209
78.90	.9813	.1925	83.90	.9943	.1063	88.90	.9998	.0192
79.00	.9816	.1908	84.00	.9945	.1045	89.00	.9998	.0175
79.10	.9820	.1891	84.10	.9947	.1028	89.10	.9999	.0157
79.20	.9823	.1874	84.20	.9949	.1011	89.20	.9999	.0140
79.30	.9826	.1857	84.30	.9951	.0993	89.30	.9999	.0122
79.40	.9829	.1840	84.40	.9952	.0976	89.40	.9999	.0105
79.50	.9833	.1822	84.50	.9954	.0958	89.50	1.000	.0087
79.60	.9836	.1805	84.60	.9956	.0941	89.60	1.000	.0070
79.70	.9839	.1788	84.70	.9957	.0924	89.70	1.000	.0052
79.80	.9842	.1771	84.80	.9959	.0906	89.80	1.000	.0035
79.90	.9845	.1754	84.90	.9960	.0889	89.90	1.000	.0017
						90.00	1.000	.0000

TABLE A4–5

TABLE OF VALUES: Arc Tan X X = 0.000 to 0.499

X	.000	.001	.002	.003	.004	.005	.006	.007	.008	.009
.00	.0000	.0573	.1146	.1719	.2292	.2865	.3438	.4011	.4584	.5156
.01	.5729	.6302	.6875	.7448	.8021	.8594	.9167	.9739	1.031	1.088
.02	1.146	1.203	1.260	1.318	1.375	1.432	1.489	1.547	1.604	1.661
.03	1.718	1.776	1.833	1.890	1.947	2.005	2.062	2.119	2.176	2.233
.04	2.291	2.348	2.405	2.462	2.519	2.577	2.634	2.691	2.748	2.805
.05	2.862	2.920	2.977	3.034	3.091	3.148	3.205	3.262	3.319	3.377
.06	3.434	3.491	3.548	3.605	3.662	3.719	3.776	3.833	3.890	3.947
.07	4.004	4.061	4.118	4.175	4.232	4.289	4.346	4.403	4.460	4.517
.08	4.574	4.631	4.688	4.745	4.802	4.858	4.915	4.972	5.029	5.086
.09	5.143	5.200	5.256	5.313	5.370	5.427	5.484	5.540	5.597	5.654
.10	5.711	5.767	5.824	5.881	5.937	5.994	6.051	6.107	6.164	6.221
.11	6.277	6.334	6.390	6.447	6.504	6.560	6.617	6.673	6.730	6.786
.12	6.843	6.899	6.956	7.012	7.069	7.125	7.181	7.238	7.294	7.351
.13	7.407	7.463	7.520	7.576	7.632	7.688	7.745	7.801	7.857	7.913
.14	7.970	8.026	8.082	8.138	8.194	8.250	8.306	8.363	8.419	8.475
.15	8.531	8.587	8.643	8.699	8.755	8.811	8.867	8.923	8.979	9.034
.16	9.090	9.146	9.202	9.258	9.314	9.369	9.425	9.481	9.537	9.592
.17	9.648	9.704	9.759	9.815	9.871	9.926	9.982	10.04	10.09	10.15
.18	10.20	10.26	10.31	10.37	10.43	10.48	10.54	10.59	10.65	10.70
.19	10.76	10.81	10.87	10.92	10.98	11.03	11.09	11.14	11.20	11.25
.20	11.31	11.37	11.42	11.48	11.53	11.59	11.64	11.70	11.75	11.80
.21	11.86	11.91	11.97	12.02	12.08	12.13	12.19	12.24	12.30	12.35
.22	12.41	12.46	12.52	12.57	12.63	12.68	12.73	12.79	12.84	12.90
.23	12.95	13.01	13.06	13.12	13.17	13.22	13.28	13.33	13.39	13.44
.24	13.50	13.55	13.60	13.66	13.71	13.77	13.82	13.87	13.93	13.98
.25	14.04	14.09	14.14	14.20	14.25	14.31	14.36	14.41	14.47	14.52
.26	14.57	14.63	14.68	14.74	14.79	14.84	14.90	14.95	15.00	15.06
.27	15.11	15.16	15.22	15.27	15.32	15.38	15.43	15.48	15.54	15.59
.28	15.64	15.70	15.75	15.80	15.85	15.91	15.96	16.01	16.07	16.12
.29	16.17	16.22	16.28	16.33	16.38	16.44	16.49	16.54	16.59	16.65
.30	16.70	16.75	16.80	16.86	16.91	16.96	17.01	17.07	17.12	17.17
.31	17.22	17.28	17.33	17.38	17.43	17.48	17.54	17.59	17.64	17.69
.32	17.74	17.80	17.85	17.90	17.95	18.00	18.06	18.11	18.16	18.21
.33	18.26	18.31	18.37	18.42	18.47	18.52	18.57	18.62	18.68	18.73
.34	18.78	18.83	18.88	18.93	18.98	19.03	19.09	19.14	19.19	19.24
.35	19.29	19.34	19.39	19.44	19.49	19.54	19.60	19.65	19.70	19.75
.36	19.80	19.85	19.90	19.95	20.00	20.05	20.10	20.15	20.20	20.25
.37	20.30	20.35	20.41	20.46	20.51	20.56	20.61	20.66	20.71	20.76
.38	20.81	20.86	20.91	20.96	21.01	21.06	21.11	21.16	21.21	21.26
.39	21.31	21.36	21.41	21.45	21.50	21.55	21.60	21.65	21.70	21.75
.40	21.80	21.85	21.90	21.95	22.00	22.05	22.10	22.15	22.20	22.24
.41	22.29	22.34	22.39	22.44	22.49	22.54	22.59	22.64	22.68	22.73
.42	22.78	22.83	22.88	22.93	22.98	23.03	23.07	23.12	23.17	23.22
.43	23.27	23.32	23.36	23.41	23.46	23.51	23.56	23.61	23.65	23.70
.44	23.75	23.80	23.85	23.89	23.94	23.99	24.04	24.08	24.13	24.18
.45	24.23	24.28	24.32	24.37	24.42	24.47	24.51	24.56	24.61	24.66
.46	24.70	24.75	24.80	24.84	24.89	24.94	24.99	25.03	25.08	25.13

Table A4–5 (*continued*)

X	.000	.001	.002	.003	.004	.005	.006	.007	.008	.009
.47	25.17	25.22	25.27	25.31	25.36	25.41	25.45	25.50	25.55	25.59
.48	25.64	25.69	25.73	25.78	25.83	25.87	25.92	25.97	26.01	26.06
.49	26.10	26.15	26.20	26.24	26.29	26.34	26.38	26.43	26.47	26.52

TABLE OF VALUES: Arc Tan X — X = 0.500 to 0.999

X	.000	.001	.002	.003	.004	.005	.006	.007	.008	.009
.50	26.57	26.61	26.66	26.70	26.75	26.79	26.84	26.89	26.93	26.98
.51	27.02	27.07	27.11	27.16	27.20	27.25	27.29	27.34	27.36	27.43
.52	27.47	27.52	27.56	27.61	27.65	27.70	27.74	27.79	27.85	27.88
.53	27.92	27.97	28.01	28.06	28.10	28.15	28.19	28.84	28.28	28.39
.54	28.37	28.41	28.46	28.60	28.55	28.59	28.65	28.66	28.72	28.77
.55	28.81	28.85	28.90	28.94	28.99	29.03	29.07	29.12	29.16	29.21
.56	29.25	29.29	29.34	29.38	29.42	29.47	29.51	29.55	29.60	29.64
.57	29.68	29.73	29.77	29.81	29.86	29.90	29.94	29.98	30.03	30.07
.58	30.11	30.16	30.20	30.24	30.28	30.33	30.37	30.41	30.46	30.50
.59	30.54	30.58	30.63	30.67	30.71	30.75	30.79	30.84	30.88	30.92
.60	30.96	31.01	31.05	31.09	31.13	31.17	31.22	31.26	31.30	31.34
.61	31.38	31.42	31.47	31.51	31.55	31.59	31.63	31.67	31.72	31.76
.62	31.80	31.84	31.88	31.92	31.96	32.01	32.05	32.09	32.13	32.17
.63	32.21	32.25	32.29	32.33	32.37	32.42	32.46	32.50	32.54	32.58
.64	32.62	32.66	32.70	32.74	32.78	32.82	32.86	32.90	32.94	32.98
.65	33.02	33.06	33.10	33.14	33.18	33.22	33.26	33.30	33.34	33.38
.66	33.42	33.46	33.50	33.54	33.58	33.62	33.66	33.70	33.74	33.78
.67	33.82	33.86	33.90	33.94	33.98	34.02	34.06	34.10	34.14	34.18
.68	34.22	34.25	34.29	34.33	34.37	34.41	34.45	34.49	34.53	34.57
.69	34.61	34.64	34.68	34.72	34.76	34.80	34.84	34.88	34.92	34.95
.70	34.99	35.03	35.07	35.11	35.15	35.18	35.22	35.26	35.30	35.34
.71	35.37	35.41	35.45	35.49	35.53	35.56	35.60	35.64	35.68	35.72
.72	35.75	35.79	35.83	35.87	35.90	35.94	35.98	36.02	36.05	36.09
.73	36.13	36.17	36.20	36.24	36.28	36.32	36.35	36.39	36.43	36.46
.74	36.50	36.54	36.58	36.61	36.65	36.69	36.72	36.76	36.80	36.83
.75	36.87	36.91	36.94	36.98	37.02	37.05	37.09	37.13	37.16	37.20
.76	37.23	37.27	37.31	37.34	37.38	37.42	37.45	37.49	37.52	37.56
.77	37.60	37.63	37.67	37.70	37.74	37.78	37.81	37.85	37.88	37.92
.78	37.95	37.99	38.03	38.06	38.10	38.13	38.17	38.20	38.24	38.27
.79	38.31	38.34	38.38	38.41	38.45	38.48	38.52	38.55	38.59	38.62
.80	38.66	38.69	38.73	38.76	38.80	38.83	38.87	38.90	38.94	38.97
.81	39.01	39.04	39.08	39.11	39.15	39.18	39.21	39.25	39.28	39.32
.82	39.35	39.39	39.42	39.45	39.49	39.52	39.56	39.59	39.62	39.66
.83	39.69	39.73	39.76	39.79	39.83	39.86	39.90	39.93	39.96	40.00
.84	40.03	40.06	40.10	40.13	40.16	40.20	40.23	40.26	40.30	40.33
.85	40.36	40.40	40.43	40.46	40.50	40.53	40.56	40.60	40.63	40.66
.86	40.70	40.73	40.76	40.79	40.83	40.89	40.89	40.93	40.96	40.99
.87	41.02	41.06	41.09	41.12	41.15	41.19	41.22	41.25	41.28	41.32
.88	41.35	41.38	41.41	41.44	41.48	41.51	41.54	41.57	41.61	41.64

Table A4–5 (*continued*)

X	.000	.001	.002	.003	.004	.005	.006	.007	.008	.009
.89	41.67	41.70	41.73	41.76	41.80	41.83	41.86	41.89	41.92	41.96
.90	41.99	42.02	42.05	42.08	42.11	42.15	42.18	42.21	42.24	42.27
.91	42.30	42.33	42.36	42.40	42.43	42.46	42.49	42.52	42.55	42.58
.92	42.61	42.65	42.68	42.71	42.74	42.77	42.80	42.83	42.86	42.89
.93	42.92	42.95	42.98	43.01	43.05	43.08	43.11	43.14	43.17	43.20
.94	43.23	43.26	43.29	43.32	43.35	43.38	43.41	43.44	43.47	43.50
.95	43.53	43.56	43.59	43.62	43.65	43.68	43.71	43.74	43.77	43.80
.96	43.83	43.86	43.89	43.92	43.95	43.98	44.01	44.04	44.07	44.10
.97	44.13	44.16	44.19	44.22	44.25	44.27	44.30	44.33	44.36	44.39
.98	44.42	44.45	44.48	44.51	44.54	44.57	44.60	44.63	44.65	44.68
.99	44.71	44.74	44.77	44.80	44.83	44.86	44.89	44.91	44.94	44.97

TABLE OF VALUES: Arc Tan X — X = 5.00 to 9.99

X	.00	.01	.02	.03	.04	.05	.06	.07	.08	.09
5.50	79.70	79.71	79.73	79.75	79.77	79.79	79.80	79.82	79.84	79.86
5.60	79.88	79.89	79.91	79.93	79.95	79.96	79.96	80.00	80.02	80.03
5.70	80.05	80.07	80.08	80.10	80.12	80.13	80.15	80.17	80.18	80.20
5.80	80.22	80.23	80.25	80.27	80.28	80.30	80.32	80.33	80.35	80.38
5.90	80.38	80.40	80.41	80.43	80.44	80.46	80.48	80.49	80.51	80.52
6.00	80.54	80.55	80.57	80.58	80.60	80.61	80.63	80.64	80.66	80.68
6.10	80.69	80.71	80.72	80.73	80.75	80.76	80.78	80.79	80.81	80.82
6.20	80.84	80.85	80.87	80.88	80.90	80.91	80.92	80.94	80.95	80.97
6.30	80.98	80.99	81.01	81.02	81.04	81.05	81.06	81.08	81.09	81.11
6.40	81.12	81.13	81.15	81.16	81.17	81.19	81.20	81.21	81.23	81.24
6.50	81.25	81.27	81.28	81.29	81.31	81.32	81.33	81.35	81.36	81.37
6.60	81.38	81.40	81.41	81.42	81.44	81.45	81.46	81.47	81.49	81.50
6.70	81.51	81.52	81.54	81.55	81.56	81.57	81.59	81.60	81.61	81.62
6.80	81.63	81.65	81.66	81.67	81.68	81.69	81.71	81.72	81.73	81.74
6.90	81.75	81.77	81.78	81.79	81.80	81.81	81.82	81.84	81.85	81.86
7.00	81.87	81.88	81.89	81.90	81.92	81.93	81.94	81.95	81.96	81.97
7.10	81.98	81.99	82.01	82.02	82.03	82.04	82.05	82.06	82.07	82.08
7.20	82.09	82.10	82.11	82.13	82.14	82.15	82.16	82.17	82.18	82.19
7.30	82.20	82.21	82.22	82.23	82.24	82.25	82.26	82.27	82.28	82.29
7.40	82.30	82.31	82.32	82.33	82.34	82.35	82.37	82.38	82.39	82.40
7.50	82.41	82.42	82.43	82.44	82.45	82.46	82.46	82.47	82.48	82.49
7.60	82.50	82.51	82.52	82.53	82.54	82.55	82.56	82.57	82.58	82.59
7.70	82.60	82.61	82.62	82.63	82.64	82.65	82.66	82.67	82.68	82.68
7.80	82.69	82.70	82.71	82.72	82.73	82.74	82.75	82.76	82.77	82.78
7.90	82.79	82.79	82.80	82.81	82.82	82.83	82.84	82.85	82.86	82.87
8.00	82.87	82.88	82.89	82.90	82.91	82.92	82.93	82.94	82.94	82.95
8.10	82.96	82.97	82.98	82.99	83.00	83.00	83.01	83.02	83.03	83.04
8.20	83.05	83.06	83.06	83.07	83.08	83.09	83.10	83.11	83.11	83.12
8.30	83.13	83.14	83.15	83.15	83.16	83.17	83.18	83.19	83.19	83.20
8.40	83.21	83.22	83.23	83.23	83.24	83.25	83.26	83.27	83.27	83.28

Table A4–5 (*continued*)

X	.00	.10	.20	.30	.40	.50	.60	.70	.80	.90
8.50	83.29	83.30	83.31	83.31	83.32	83.33	83.34	83.34	83.35	83.36
8.60	83.37	83.38	83.38	83.39	83.40	83.41	83.41	83.42	83.43	83.44
8.70	83.44	83.45	83.46	83.47	83.47	83.48	83.49	83.99	83.50	83.51
8.80	83.52	83.52	83.53	83.54	83.55	83.55	83.56	83.57	83.57	83.58
8.90	83.59	83.60	83.60	83.61	83.62	83.62	83.63	83.64	83.65	83.65
9.00	83.66	83.67	83.67	83.68	83.69	83.69	83.70	83.71	83.72	83.72
9.10	83.73	83.74	83.74	83.75	83.76	83.76	83.77	83.78	83.78	83.79
9.20	83.80	83.80	83.81	83.82	83.82	83.83	83.84	83.84	83.85	83.86
9.30	83.86	83.87	83.88	83.88	83.89	83.90	83.90	83.91	83.91	83.92
9.40	83.93	83.93	83.94	83.95	83.95	83.96	83.97	83.97	83.98	83.98
9.50	83.99	84.00	84.00	84.01	84.02	84.02	84.03	84.03	84.04	84.05
9.60	84.05	84.06	84.07	84.07	84.08	84.08	84.09	84.10	84.10	84.11
9.70	84.11	84.12	84.13	84.13	84.14	84.14	84.15	84.16	84.16	84.17
9.80	84.17	84.18	84.19	84.19	84.20	84.20	84.21	84.21	84.22	84.23
9.90	84.23	84.24	84.24	84.25	84.26	84.26	84.27	84.27	84.28	84.28

TABLE OF VALUES: Arc Tan X X = 10.00 to 54.90

X	.00	.10	.20	.30	.40	.50	.60	.70	.80	.90
10.00	84.29	84.35	84.40	84.45	84.51	84.56	84.61	84.66	84.71	84.76
11.00	84.81	84.85	84.90	84.94	84.99	85.03	85.07	85.11	85.16	85.20
12.00	85.24	85.28	85.31	85.35	85.39	85.43	85.46	85.50	85.53	85.57
13.00	85.60	85.63	85.67	85.70	85.73	85.76	85.79	85.83	85.86	85.89
14.00	85.91	85.94	85.97	86.00	86.03	86.05	86.08	86.11	86.13	86.16
15.00	86.19	86.21	86.24	86.26	86.28	86.31	86.33	86.36	86.38	86.40
16.00	86.42	86.45	86.47	86.49	86.51	86.53	86.55	86.57	86.59	86.61
17.00	86.63	86.65	86.67	86.69	86.71	86.73	86.75	86.77	86.78	86.80
18.00	86.82	86.84	86.86	86.87	86.89	86.91	86.92	86.94	86.96	86.97
19.00	86.99	87.00	87.02	87.03	87.05	87.06	87.08	87.09	87.11	87.12
20.00	87.14	87.15	87.17	87.18	87.19	87.21	87.22	87.23	87.25	87.26
21.00	87.27	87.29	87.30	87.31	87.32	87.34	87.35	87.36	87.37	87.39
22.00	87.40	87.41	87.42	87.43	87.44	87.46	87.47	87.48	87.49	87.50
23.00	87.51	87.52	87.53	87.54	87.55	87.56	87.57	87.58	87.59	87.60
24.00	87.61	87.62	87.63	87.64	87.65	87.66	87.67	87.68	87.69	87.70
25.00	87.71	87.72	87.73	87.74	87.75	87.75	87.76	87.77	87.78	87.79
26.00	87.80	87.81	87.81	87.82	87.83	87.84	87.85	87.86	87.86	87.87
27.00	87.88	87.89	87.89	87.90	87.91	87.92	87.92	87.93	87.94	87.95
28.00	87.95	87.96	87.97	87.98	87.98	87.99	88.00	88.00	88.01	88.02
29.00	88.03	88.03	88.04	88.05	88.05	88.06	88.07	88.07	88.08	88.08
30.00	88.09	88.10	88.10	88.11	88.12	88.12	88.13	88.13	88.14	88.15
31.00	88.15	88.16	88.16	88.17	88.18	88.18	88.19	88.19	88.20	88.20
32.00	88.21	88.22	88.22	88.23	88.23	88.24	88.24	88.25	88.25	88.26
33.00	88.26	88.27	88.27	88.28	88.29	88.29	88.30	88.30	88.31	88.31
34.00	88.32	88.32	88.33	88.33	88.33	88.34	88.34	88.35	88.35	88.36
35.00	88.36	88.37	88.37	88.38	88.38	88.39	88.39	88.40	88.40	88.40

Table A4–5 (*continued*)

X	.000	.001	.002	.003	.004	.005	.006	.007	.008	.009
36.00	88.41	88.41	88.42	88.42	88.43	88.43	88.43	88.44	88.44	88.45
37.00	88.45	88.46	88.46	88.46	88.47	88.47	88.48	88.48	88.48	88.49
38.00	88.49	88.50	88.50	88.50	88.51	88.51	88.52	88.52	88.52	88.53
39.00	88.53	88.53	88.54	88.54	88.55	88.55	88.55	88.56	88.56	88.56
40.00	88.57	88.57	88.58	88.58	88.58	88.59	88.59	88.59	88.60	88.60
41.00	88.60	88.61	88.61	88.61	88.62	88.62	88.62	88.63	88.63	88.63
42.00	88.64	88.64	88.64	88.65	88.65	88.65	88.66	88.66	88.66	88.66
43.00	88.67	88.67	88.67	88.68	88.68	88.68	88.69	88.69	88.69	88.70
44.00	88.70	88.70	88.70	88.71	88.71	88.71	88.72	88.72	88.72	88.72
45.00	88.73	88.73	88.73	88.74	88.74	88.74	88.74	88.75	88.75	88.75
46.00	88.75	88.76	88.76	88.76	88.77	88.77	88.77	88.77	88.78	88.78
47.00	88.78	88.78	88.79	88.79	88.79	88.79	88.80	88.80	88.80	88.80
48.00	88.81	88.81	88.81	88.81	88.82	88.82	88.82	88.82	88.83	88.83
49.00	88.83	88.83	88.84	88.84	88.84	88.84	88.84	88.85	88.85	88.85
50.00	88.85	88.86	88.86	88.86	88.86	88.87	88.87	88.87	88.87	88.87
51.00	88.88	88.88	88.88	88.88	88.89	88.89	88.89	88.89	88.89	88.90
52.00	88.90	88.90	88.90	88.90	88.91	88.91	88.91	88.91	88.91	88.92
53.00	88.92	88.92	88.92	88.93	88.93	88.93	88.93	88.93	88.94	88.94
54.00	88.94	88.94	88.94	88.94	88.95	88.95	88.95	88.95	88.95	88.96

TABLE OF VALUES: Arc Tan X — X = 55.00 to 99.90

X	.00	.10	.20	.30	.40	.50	.60	.70	.80	.90
55.00	88.96	88.96	88.96	88.96	88.97	88.97	88.97	88.97	88.97	88.98
56.00	88.98	88.98	88.98	88.98	88.98	88.99	88.99	88.99	88.99	88.99
57.00	88.99	89.00	89.00	89.00	89.00	89.00	89.01	89.01	89.01	89.01
58.00	89.01	89.01	89.02	89.02	89.02	89.02	89.02	89.02	89.03	89.03
59.00	89.03	89.03	89.03	89.03	89.04	89.04	89.04	89.04	89.04	89.04
60.00	89.05	89.05	89.05	89.05	89.05	89.05	89.06	89.06	89.06	89.06
61.00	89.06	89.06	89.06	89.07	89.07	89.07	89.07	89.07	89.07	89.07
62.00	89.08	89.08	89.08	89.08	89.08	89.08	89.09	89.09	89.09	89.09
63.00	89.09	89.09	89.09	89.09	89.10	89.10	89.10	89.10	89.10	89.10
64.00	89.10	89.11	89.11	89.11	89.11	89.11	89.11	89.11	89.12	89.12
65.00	89.12	89.12	89.12	89.12	89.12	89.13	89.13	89.13	89.13	89.13
66.00	89.13	89.13	89.13	89.14	89.14	89.14	89.14	89.14	89.14	89.14
67.00	89.14	89.15	89.15	89.15	89.15	89.15	89.15	89.15	89.15	89.16
68.00	89.16	89.16	89.16	89.16	89.16	89.16	89.16	89.17	89.17	89.17
69.00	89.17	89.17	89.17	89.17	89.17	89.18	89.18	89.18	89.18	89.18
70.00	89.18	89.18	89.18	89.19	89.19	89.19	89.19	89.19	89.19	89.19
71.00	89.19	89.19	89.20	89.20	89.20	89.20	89.20	89.20	89.20	89.20
72.00	89.20	89.21	89.21	89.21	89.21	89.21	89.21	89.21	89.21	89.21
73.00	89.22	89.22	89.22	89.22	89.22	89.22	89.22	89.22	89.22	89.22
74.00	89.23	89.23	89.23	89.23	89.23	89.23	89.23	89.23	89.23	89.24
75.00	89.24	89.24	89.24	89.24	89.24	89.24	89.24	89.24	89.24	89.25
76.00	89.25	89.25	89.25	89.25	89.25	89.25	89.25	89.25	89.25	89.25
77.00	89.26	89.26	89.26	89.26	89.26	89.26	89.26	89.26	89.26	89.26

Table A4–5 (*continued*)

X	.00	.10	.20	.30	.40	.50	.60	.70	.80	.90
78.00	89.27	89.27	89.27	89.27	89.27	89.27	89.27	89.27	89.27	89.27
79.00	89.27	89.28	89.28	89.28	89.28	89.28	89.28	89.28	89.28	89.28
80.00	89.28	89.28	89.29	89.29	89.29	89.29	89.29	89.29	89.29	89.29
81.00	89.29	89.29	89.29	89.30	89.30	89.30	89.30	89.30	89.30	89.30
82.00	89.30	89.30	89.30	89.30	89.30	89.31	89.31	89.31	89.31	89.31
83.00	89.31	89.31	89.31	89.31	89.31	89.31	89.31	89.32	89.32	89.32
84.00	89.32	89.32	89.32	89.32	89.32	89.32	89.32	89.32	89.32	89.33
85.00	89.33	89.33	89.33	89.33	89.33	89.33	89.33	89.33	89.33	89.33
86.00	89.33	89.33	89.34	89.34	89.34	89.34	89.34	89.34	89.34	89.34
87.00	89.34	89.34	89.34	89.34	89.34	89.35	89.35	89.35	89.35	89.35
88.00	89.35	89.35	89.35	89.35	89.35	89.35	89.35	89.35	89.35	89.36
89.00	89.36	89.36	89.36	89.36	89.36	89.36	89.36	89.36	89.36	89.36
90.00	89.36	89.36	89.36	89.37	89.37	89.37	89.37	89.37	89.37	89.37
91.00	89.37	89.37	89.37	89.37	89.37	89.37	89.37	89.38	89.38	89.38
92.00	89.38	89.38	89.38	89.38	89.38	89.38	89.38	89.38	89.38	89.38
93.00	89.38	89.38	89.39	89.39	89.39	89.39	89.39	89.39	89.39	89.39
94.00	89.39	89.39	89.39	89.39	89.39	89.39	89.39	89.39	89.40	89.40
95.00	89.40	89.40	89.40	89.40	89.40	89.40	89.40	89.40	89.40	89.40
96.00	89.40	89.40	89.40	89.41	89.41	89.41	89.41	89.41	89.41	89.41
97.00	89.41	89.41	89.41	89.41	89.41	89.41	89.41	89.41	89.41	89.41
98.00	89.42	89.42	89.42	89.42	89.42	89.42	89.42	89.42	89.42	89.42
99.00	89.42	89.42	89.42	89.42	89.42	89.42	89.42	89.43	89.43	89.43

TABLE OF VALUES: Arc Tan X X = 100 to 11,459

(min.)	(max.)	angle	(min.)	(max.)	angle
99.64	101.4	89.43°	201.0	208.3	89.72°
101.4	103.2	89.44°	208.3	216.2	89.73°
103.2	105.1	89.45°	216.2	224.7	89.74°
105.1	107.1	89.46°	224.7	233.9	89.75°
107.1	109.1	89.47°	233.9	243.8	89.76°
109.1	111.3	89.48°	243.8	254.6	89.77°
111.3	113.5	89.49°	254.6	266.5	89.78°
113.5	115.7	89.50°	266.5	279.5	89.79°
115.7	118.1	89.51°	279.5	293.8	89.80°
118.1	120.6	89.52°	293.8	309.7	89.81°
120.6	123.2	89.53°	309.7	327.4	89.82°
123.2	125.9	89.54°	327.4	347.2	89.83°
125.9	128.8	89.55°	347.2	369.6	89.84°
128.8	131.7	89.56°	369.6	395.1	89.85°
131.7	134.8	89.57°	395.1	424.4	89.86°
134.8	138.1	89.58°	424.4	458.4	89.87°
138.1	141.5	89.59°	458.4	498.2	89.88°
141.5	145.1	89.60°	498.2	545.7	89.89°
145.1	148.8	89.61°	545.7	603.1	89.90°
148.8	152.8	89.62°	603.1	674.1	89.91°

Table A4–5 (*continued*)

(min.)	(max.)	angle	(min.)	(max.)	angle
152.8	157.0	89.63°	674.1	763.9	89.92°
157.0	161.4	89.64°	763.9	881.5	89.93°
161.4	166.1	89.65°	881.5	1042.0	89.94°
166.1	171.0	89.66°	1042.0	1273.0	89.95°
171.0	176.3	89.67°	1273.0	1637.0	89.96°
176.3	181.9	89.68°	1637.0	2292.0	89.97°
181.9	187.9	89.69°	2292.0	3820.0	89.98°
187.9	194.2	89.70°	3820.0	11459.0	89.99°
194.2	201.0	89.71°	(values above 11,459)		90.00°

NATIONAL ELECTRICAL CODE® TABLES

The National Electrical Code provides many tables that present information useful to the electrician in a concise and convenient way. All of the tables are found in the Code, a copy of which every electrician should have in the toolbox at all times. Remember, the Code is updated every three years, and you can obtain new copies from

The National Fire Protection Association
Quincy, Massachusetts 02269.

For your convenience, several of the most frequently used National Electric Code® tables are reprinted here. The table numbers appearing in the actual Code have been retained.

Table 210-21(b)(2)
Maximum Cord- and Plug-Connected Load to Receptacle

Circuit Rating Amperes	Receptacle Rating Amperes	Maximum Load Amperes
15 or 20	15	12
20	20	16
30	30	24

Table 210-21(b)(3)

Circuit Rating Amperes	Receptacle Rating Amperes
15	Not over 15
20	15 or 20
30	30
40	40 or 50
50	50

Table 210-24
Summary of Branch-Circuit Requirements

(Type FEP, FEPB, RUW, SA, T, TW, RH, RUH, RHW, RHH, THHN, THW, THWN, and XHHW conductors in raceway or cable.)

Circuit Rating	15 Amp	20 Amp	30 Amp	40 Amp	50 Amp
Conductors (Min. Size)					
Circuit Wires*	14	12	10	8	6
Taps	14	14	14	12	12
Fixture Wires and Cords	Refer to Section 240-4				
Overcurrent Protection	15 Amp	20 Amp	30 Amp	40 Amp	50 Amp
Outlet Devices:					
Lampholders Permitted	Any Type	Any Type	Heavy Duty	Heavy Duty	Heavy Duty
Receptacle Rating**	15 Max. Amp	15 or 20 Amp	30 Amp	40 or 50 Amp	50 Amp
Maximum Load	15 Amp	20 Amp	30 Amp	40 Amp	50 Amp
Permissible Load	Refer to Section 210-23(a)	Refer to Section 210-23(a)	Refer to Section 210-23(b)	Refer to Section 210-23(c)	Refer to Section 210-23(c)

***These gages are for copper conductors.**
****For receptacle rating of cord-connected electric-discharge lighting fixtures, see Section 410-30(c).**

Table 220-18
Demand Factors for Household Electric Clothes Dryers

Number of Dryers	Demand Factor Percent
1	100
2	100
3	100
4	100
5	80
6	70
7	65
8	60
9	55
10	50
11-13	45
14-19	40
20-24	35
25-29	32.5
30-34	30
35-39	27.5
40 & over	25

Table 220-19
Demand Loads for Household Electric Ranges, Wall-Mounted Ovens, Counter-Mounted Cooking Units, and Other Household Cooking Appliances over 1¾ kW Rating. Column A to be used in all cases except as otherwise permitted in Note 3 below.

Number of Appliances	Maximum Demand (See Notes)	Demand Factors Percent (See Note 3)	
	Column A (Not over 12 kW Rating)	Column B (Less than 3 1/2 kW Rating)	Column C (3 1/2 kW to 8 3/4 kW Rating)
1	8 kW	80%	80%
2	11 kW	75%	65%
3	14 kW	70%	55%
4	17 kW	66%	50%
5	20 kW	62%	45%
6	21 kW	59%	43%
7	22 kW	56%	40%

Table 220–19 (*continued*)

8	23 kW	53%	36%
9	24 kW	51%	35%
10	25 kW	49%	34%
11	26 kW	47%	32%
12	27 kW	45%	32%
13	28 kW	43%	32%
14	29 kW	41%	32%
15	30 kW	40%	32%
16	31 kW	39%	28%
17	32 kW	38%	28%
18	33 kW	37%	28%
19	34 kW	36%	28%
20	35 kW	35%	28%
21	36 kW	34%	26%
22	37 kW	33%	26%
23	38 kW	32%	26%
24	39 kW	31%	26%
25	40 kW	30%	26%
26-30	15 kW plus 1 kW for each range	30%	24%
31-40		30%	22%
41-50	25 kW plus 3/4 kW for each range	30%	20%
51-60		30%	18%
61 & over		30%	16%

Note 1. Over 12 kW through 27 kW ranges all of same rating. For ranges individually rated more than 12 kW but not more than 27 kW, the maximum demand in Column A shall be increased 5 percent for each additional kW of rating or major fraction thereof by which the rating of individual ranges exceeds 12 kW.

The size of the conductors is to be determined by the rating of the range. By referring to Table 220-19, it can be seen that for a range not over 12 kW the demand load is 8 kW (8 kVA per Section 220-19) and a No. 8 AWG copper conductor, with 60°C insulation, would suffice.

Note 2. Over 12 kW through 27 kW ranges of unequal ratings. For ranges individually rated more than 12 kW and of different ratings but none exceeding 27 kW, an average value of rating shall be computed by adding together the ratings of all ranges to obtain the total connected load (using 12 kW for any range rated less than 12 kW) and dividing by the total number of ranges; and then the maximum demand in Column A shall be increased 5 percent for each KW or major fraction thereof by which this average value exceeds 12 kW.

Note 2 provides for ranges larger than 12 kW, and Note 4 covers situations in which the range consists of several components.

Note 3. Over 1 3/4 kW through 8 3/4 kW. In lieu of the method provided in Column A, it shall be permissible to add the nameplate ratings of all ranges rated more than 1 3/4 kW but not more than 8 3/4 kW, and multiply the sum by the demand factors specified in Column B or C for the given number of appliances.

Table 220-32
Optional Calculation—Demand Factors for Three or More Multifamily Dwelling Units

Number of Dwelling Units	Demand Factor Percent
3 – 5	45
6 – 7	44
8 – 10	43
11	42
12 – 13	41
14 – 15	40
16 – 17	39
18 – 20	38
21	37
22 – 23	36
24 – 25	35
26 – 27	34
28 – 30	33
31	32
32 – 33	31
34 – 36	30
37 – 38	29
39 – 42	28
43 – 45	27
46 – 50	26
51 – 55	25
56 – 61	24
62 & over	23

Table 250-94
Grounding Electrode Conductor for AC Systems

Size of Largest Service-Entrance Conductor or Equivalent Area for Parallel Conductors		Size of Grounding Electrode Conductor	
Copper	Aluminum or Copper-Clad Aluminum	Copper	*Aluminum or Copper-Clad Aluminum
2 or smaller	0 or smaller	8	6
1 or 0	2/0 or 3/0	6	4
2/0 or 3/0	4/0 or 250 MCM	4	2
Over 3/0 thru 350 MCM	Over 250 MCM thru 500 MCM	2	0

Table 250–94 (*continued*)

Over 350 MCM thru 600 MCM	Over 500 MCM thru 900 MCM	0	3/0
Over 600 MCM thru 1100 MCM	Over 900 MCM thru 1750 MCM	2/0	4/0
Over 1100 MCM	Over 1750 MCM	3/0	250 MCM

Where there are no service-entrance conductors, the grounding electrode conductor size shall be determined by the equivalent size of the largest service-entrance conductor required for the load to be served.

***See installation restrictions in Section 250-92(a).**

See Section 250-23(b).

Table 250-95
Minimum Size Equipment Grounding Conductors for Grounding Raceway and Equipment

Rating or Setting of Automatic Overcurrent Device in Circuit Ahead of Equipment, Conduit, etc., Not Exceeding (Amperes)	Size	
	Copper Wire No.	Aluminum or Copper-Clad Aluminum Wire No.*
15	14	12
20	12	10
30	10	8
40	10	8
60	10	8
100	8	6
200	6	4
300	4	2
400	3	1
500	2	1/0
600	1	2/0
800	0	3/0
1000	2/0	4/0
1200	3/0	250 MCM
1600	4/0	350 "
2000	250 MCM	400 "
2500	350 "	600 "
3000	400 "	600 "
4000	500 "	800 "
5000	700 "	1200 "
6000	800 "	1200 "

***See installation restrictions in Section 250-92(a).**

Table 300-19(a)
Spacings for Conductor Supports

Conductors		Aluminum or Copper-Clad Aluminum	Copper
No. 18 thru No. 8	Not greater than	100 feet	100 feet
No. 6 thru No. 0	" " "	200 feet	100 feet
No. 00 thru No. 0000	" " "	180 feet	80 feet
211,601 CM thru 350,000 CM	" " "	135 feet	60 feet
350,001 CM thru 500,000 CM	" " "	120 feet	50 feet
500,001 CM thru 750,000 CM	" " "	95 feet	40 feet
Above 750,000 CM	" " "	85 feet	35 feet

For SI units: one foot = 0.3048 meter.

In dwelling units, conductors, as listed below, shall be permitted to be utilized as three-wire, single-phase, service-entrance conductors and the three-wire, single-phase feeder that carries the total current supplied by that service. (This is Note 3.)

Conductor Types and Sizes
RH-RHH-RHW-THW-THWN-THHN-XHHW

Copper	Aluminum and Copper-Clad AL	Service Rating in Amps
AWG	AWG	
4	2	100
3	1	110
2	1/0	125
1	2/0	150
1/0	3/0	175
2/0	4/0	200

If a single set of three-wire, single-phase, service-entrance conductors supplies a one-family, two-family, or multifamily dwelling, the reduced conductor size permitted by Note 3 is applicable to the service-entrance conductors only. If there are panelboards on the load side of the main service-entrance equipment supplied by feeders, Note 3 does not permit a reduction in the conductor size for these feeders because they do carry the total current supplied by the service.

See also Section 550-3(a) for a description of a feeder assembly that carries the total current supplied by the service and Section 215-2.

It is the intent that all conductors, including the neutral, be the same size when applying Note 3.

Table 346-10
Radius of Conduit Bends (Inches)

Size of Conduit (In.)	Conductors Without Lead Sheath (In.)	Conductors With Lead Sheath (In.)
1/2	4	6
3/4	5	8
1	6	11

Table 346–10 (*continued*)

Size of Conduit (In.)	Conductors Without Lead Sheath (In.)	Conductors With Lead Sheath (In.)
1 1/4	8	14
1 1/2	10	16
2	12	21
2 1/2	15	25
3	18	31
3 1/2	21	36
4	24	40
5	30	50
6	36	61

For SI units: (Radius) one inch = 25.4 millimeters.

Table 346-10 Exception
Radius of Conduit Bends (Inches)

Size of Conduit (In.)	Radius to Center of Conduit (In.)
1/2	4
3/4	4 1/2
1	5 3/4
1 1/4	7 1/4
1 1/2	8 1/4
2	9 1/2
2 1/2	10 1/2
3	13
3 1/2	15
4	16
5	24
6	30

For SI units: (Radius) one inch = 25.4 millimeters.

Table 346-12
Supports for Rigid Metal Conduit

Conduit Size (Inches)	Maximum Distance Between Rigid Metal Conduit Supports (Feet)
1/2–3/4	10
1	12

Table 346–12 (*continued*)

1 1/4–1 1/2	14
2–2 1/2	16
3 and larger	20

For SI units: (Supports) one foot = 0.3048 meter.

Table 349-20(a)
Minimum Radii for Flexing Use

Trade Size	Minimum Radii
1/8 inch	10 inches
1/2 inch	12 1/2 inches
3/4 inch	17 1/2 inches

For SI units: (Radii) one inch = 25.4 millimeters.

Table 347-8
Support of Rigid Nonmetallic Conduit

Conduit Size (Inches)	Maximum Spacing Between Supports (Feet)
1/2–1	3
1 1/4–2	5
2 1/2–3	6
3 1/2–5	7
6	8

For SI units: (Supports) one foot = 0.3048 meter.

Table 349-20(b)
Minimum Radii for Fixed Bends

Trade Size	Minimum Radii
3/8 inch	3 1/2 inches
1/2 inch	4 inches
3/4 inch	5 inches

For SI units: (Radii) one inch = 25.4 millimeters.

Table 350-3
Maximum Number of Insulated Conductors in 3/8-Inch Flexible Metal Conduit*

Col. A = With fitting inside conduit.
Col. B = With fitting outside conduit.

Size AWG	Types RFH-2, SF-2		Types TF, T, XHHW, AF, TW, RUH, RUW		Types TFN, THHN, THWN		Types FEP, FEPB, PF, PGF	
	A	B	A	B	A	B	A	B
18	..	3	3	7	4	8	5	8
16	..	2	2	4	3	7	4	8
14	..	..	..	4	3	7	3	7
12	..	..	..	3	..	4	..	4
10	..	..	..	..	..	2	..	3

***In addition, one uninsulated grounding conductor of the same AWG size shall be permitted.**

Types of *Wiremold*® Surface Raceways

Type of Raceway	Wire Size Gage No.	Number of Wires: Types RHH, RHW	Type THW	Type TW	Types THHN, THWN
No. 200 	12		2	3	3
	14		2	3	5
No. 500 	8			2	2
	10	2	2	3	4
	12	2	3	4	7
	14	2	4	6	9
No. 700 	6				2
	8		2	2	3
	10	2	3	4	5
	12	2	4	6	8
	14	3	5	7	11
No. 1500 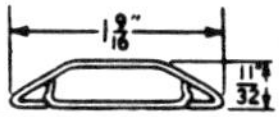	6				2
	8			2	3
	10	2	3	4	5
	12	2	3	5	7
	14	2	4	6	10
No. 2000† 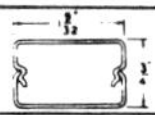	12			7	7
	14			7	7
No. 2100† 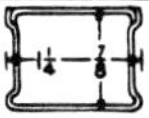	6	2	4	4	6
	8	4	6	8	10
	10	7	10	14	17
	12	8	13	19	28
	14	10	15	24	37
No. 2200† 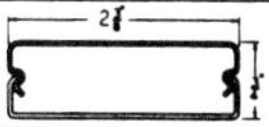	6	5	7	3• 7	11
	8	8	11	7• 14	19
	10	13	19	10• 26	32
	12	15	23	10• 34	51
	14	18	29	10• 44	69
No. 2600 	6	2	3	3	5
	8	4	5	7	9
	10	6	9	12	15
	12	7	11	16	24
	14	9	14	21	33
G-3000 	6	4• 11	6• 19	6• 17	6• 27
	8	6• 18	8• 26	8• 34	8• 44
	10	10• 30	10• 45	10• 62	10• 76
	12	14• 36	18• 55	18• 81	18• 119
	14	16• 42	26• 67	26• 103	26• 160
G-4000 — With Divider 	2	- 7	- 10	- 10	- 12
	3	- 8	- 11	- 11	- 15
	4	- 9	- 13	- 13	- 17
	6	4• 12	4• 18	4• 18	7• 28
	8	7• 19	7• 28	7• 36	8• 47
	10	11• 32	11• 48	11• 66	15• 81
	12	15• 39	15• 59	15• 86	24• 128
	14	17• 45	17• 72	17• 110	32• 171
Without Divider 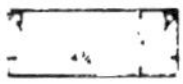	2	- 14	- 20	- 20	- 25
	3	- 16	- 23	- 23	- 30
	4	- 18	- 27	- 27	- 35
	6	8• 24	8• 36	8• 36	10• 57
	8	10• 39	10• 57	10• 78	15• 94
	10	15• 65	15• 96	12• 133	18• 163
	12	21• 78	21• 119	16• 174	34• 256
	14	21• 91	21• 145	17• 222	34• 344
G-6000 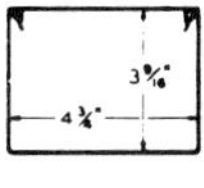	2/0	10• 17	12• 22	12• 22	15• 27
	1/0	11• 20	14• 26	14• 26	18• 33
	1	12• 23	17• 31	17• 31	21• 39
	2	16• 30	23• 43	23• 43	29• 53
	3	19• 34	27• 50	27• 50	34• 63
	4	21• 39	32• 58	32• 58	40• 74
	6	27• 51	42• 77	42• 77	66• 122
	8	40• 74	57• 106	73• 134	92• 169
	10	75• 137	111• 203	154• 282	187• 343
	12	90• 164	137• 252	200• 368	295• 540
	14	105• 193	167• 307	255• 469	396• 726

† Figures for Nos. 2000, 2100, 2200, G-3000, G-4000, and G-6000 are *without receptacles*, except where noted.
•With receptacles.

Types of Metal Surface Raceways[1]

Type of Raceway	Wire Size Gage No.	Number of Wires in One Raceway: Types RHH, RHW	Type RH	Type THW	Types T, TW	Types THHN, THWN
No. 111 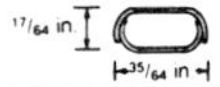 (17/64 in. × 35/64 in.)	12				2	3
	14				2	3
No. 222 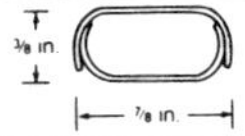 (3/8 in. × 7/8 in.)	8				2	2
	10	2		2	4	4
	12	2		3	5	7
	14	3		4	6	10
No. 333 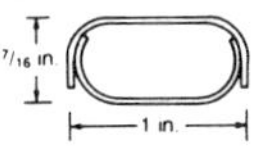 (7/16 in. × 1 in.)	6					2
	8			2	2	3
	10	2	2	3	5	6
	12	2	4	4	6	9
	14	3	4	5	8	12
No. 888 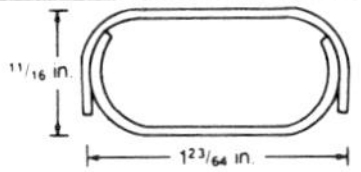 (11/16 in. × 1 23/64 in.)	6	2	2	3	3	5
	8	3	3	5	7	9
	10	6	6	9	13	16
	12	7	10	11	17	25
	14	9	12	14	22	34
No. 711 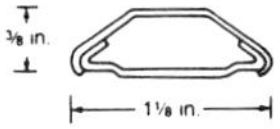 (3/8 in. × 1 1/8 in.)	8				2	2
	10			2	3	4
	12	2	2	3	4	7
	14	2	3	4	6	9
No. 733 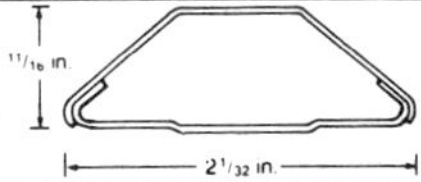 (11/16 in. × 2 1/32 in.)	6	3	3	4	4	7
	8	4	4	7	9	11
	10	8	8	11	16	20
	12	9	13	14	21	31
	14	11	16	17	27	42
No. 1700† 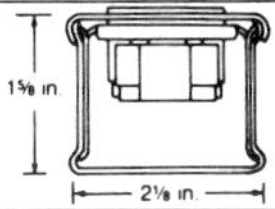 (1 5/8 in. × 2 1/8 in.)	6	3* / 8	3* / 8	5* / 13	5* / 13	8* / 21
	8	5* / 14	5* / 14	8* / 21	10* / 27	13* / 34
	10	9* / 24	9* / 24	13* / 35	19* / 49	23* / 60
	12	11* / 28	15* / 39	17* / 43	24* / 64	36* / 94
	14	13* / 33	18* / 48	20* / 53	31* / 81	49* / 126
No. 3400	Catalog No. 3400 is a raceway consisting of two No. 1700 housings in a common cover. Each channel has the same wire fill as 1700.					
No. 5100	Catalog No. 5100 is a raceway consisting of three No. 1700 housings in a common cover. Each channel has the same wire fill as 1700.					

[1] **Walker Div. of Butler Mfg. Co.**

†Figures for No. 1700 are *without devices*, except where noted.
*With devices.

Table 370-6(a)
Metal Boxes

Box Dimension, Inches Trade Size or Type	Min. Cu. In. Cap.	Maximum Number of Conductors				
		No. 14	No. 12	No. 10	No. 8	No. 6
4 × 1 1/4 Round or Octagonal	12.5	6	5	5	4	0
4 × 1 1/2 Round or Octagonal	15.5	7	6	6	5	0
4 × 2 1/8 Round or Octagonal	21.5	10	9	8	7	0
4 × 1 1/4 Square	18.0	9	8	7	6	0
4 × 1 1/2 Square	21.0	10	9	8	7	0
4 × 2 1/8 Square	30.3	15	13	12	10	6*
4 11/16 × 1 1/4 Square	25.5	12	11	10	8	0
4 11/16 × 1 1/2 Square	29.5	14	13	11	9	0
4 11/16 × 2 1/8 Square	42.0	21	18	16	14	6
3 × 2 × 1 1/2 Device	7.5	3	3	3	2	0
3 × 2 × 2 Device	10.0	5	4	4	3	0
3 × 2 × 2 1/4 Device	10.5	5	4	4	3	0
3 × 2 × 2 1/2 Device	12.5	6	5	5	4	0
3 × 2 × 2 1/4 Device	14.0	7	6	5	4	0
3 × 2 × 3 1/2 Device	18.0	9	8	7	6	0
4 × 2 1/8 × 1 1/2 Device	10.3	5	4	4	3	0
4 × 2 1/8 × 1 7/8 Device	13.0	6	5	5	4	0
4 × 2 1/8 × 2 1/8 Device	14.5	7	6	5	4	0
3 3/4 × 2 × 2 1/2 Masonry Box/Gang	14.0	7	6	5	4	0
3 3/4 × 2 × 3 1/2 Masonry Box/Gang	21.0	10	9	8	7	0
FS—Minimum Internal Depth 1 1/4 Single Cover/Gang	13.5	6	6	5	4	0
FD—Minimum Internal Depth 2 3/8 Single Cover/Gang	18.0	9	8	7	6	3
FS—Minimum Internal Depth 1 3/4 Multiple Cover/Gang	18.0	9	8	7	6	0
FD—Minimum Internal Depth 2 3/8 Multiple Cover/Gang	24.0	12	10	9	8	4

*Not to be used as a pull box. For termination only.

Table 370-6(b)
Volume Required per Conductor

Size of Conductor	Free Space Within Box for Each Conductor
No. 14	2. cubic inches
No. 12	2.25 cubic inches
No. 10	2.5 cubic inches
No. 8	3. cubic inches
No. 6	5. cubic inches

Where No. 6 conductors are installed the minimum wire bending space required in Table 373-6(a) shall be provided.

ARTICLE 373—CABINETS AND CUTOUT BOXES

Table 373-6(b)
Minimum Wire Bending Space at Terminals for Section 373-6(b)(2) in Inches

	Wires per Terminal							
Wire Size	1		2		3		4 or More	
14–10	Not Specified		—		—		—	
8	1 1/2		—		—		—	
6	2		—		—		—	
4	3		—		—		—	
3	3		—		—		—	
2	3 1/2		—		—		—	
1	4 1/2		—		—		—	
0	5 1/2		5 1/2		7		—	
2/0	6		6		7 1/2		—	
3/0	6 1/2	(1/2)	6 1/2	(1/2)	8			
4/0	7	(1)	7 1/2	(1 1/2)	8 1/2	(1/2)	—	
250	8 1/2	(2)	8 1/2	(2)	9	(1)	10	
300	10	(3)	10	(2)	11	(1)	12	
350	12	(3)	12	(3)	13	(3)	14	(2)
400	13	(3)	13	(3)	14	(3)	15	(3)
500	14	(3)	14	(3)	15	(3)	16	(3)
600	15	(3)	16	(3)	18	(3)	19	(3)
700	16	(3)	18	(3)	20	(3)	22	(3)
750	17	(3)	19	(3)	22	(3)	24	(3)
800	18		20		22		24	
900	19		22		24		24	
1000	20		—		—		—	
1250	22		—		—		—	
1500	24		—		—		—	
1750	24		—		—		—	
2000	24		—		—		—	

For SI units: one inch = 25.4 millimeters.

Bending space at terminals shall be measured in a straight line from the end of the lug or wire connector in a direction perpendicular to the enclosure wall.

For removable wire terminals intended for only one wire, bending space shall be permitted to be reduced by the number of inches shown in parentheses.

is represented only by the standard in its entirety.

Table 400-5
Ampacity of Flexible Cords and Cables
[Based on Ambient Temperature of 30°C (86°F). See Section 400-13 and Table 400-4.]

Size AWG	Thermoset Types TP, TS / Thermoplastic Types TPT, TST	Thermoset Types C, PD, E, EO, EN, S, SO, SRD, SJ, SJO, SV, SVO, SP / Thermoplastic Types ET, ETT, ETLB, ETP, ST, STO, SRDT, SJT, SJTO, SVT, SVTO, SPT		Types AFS, AFSJ, HPD, HSJ, HSJO, HS, HSO, HPN	Cotton Types CFPD* / Asbestos Types AFC* AFPD*
		A†	B†		
27**	0.5	..	..	..	..
20	..	5***	7***	..	..
18	..	7	10	10	6
17	..	..	12	..	..
16	..	10	13	15	8
15	..	..	..	17	..
14	..	15	18	20	17
12	..	20	25	30	23
10	..	25	30	35	28

Table 400-5 (*continued*)

Size AWG	Thermoset Types TP, TS ——— Thermoplastic Types TPT, TST	Thermoset Types C, PD, E, EO, EN, S SO, SRD, SJ, SJO, SV, SVO, SP ——— Thermoplastic Types ET, ETT, ETLB, ETP, ST, STO, SRDT, SJT, SJTO, SVT, SVTO, SPT		Types AFS, AFSJ, HPD, HSJ, HSJO, HS, HSO, HPN	Cotton Types CFPD* ——— Asbestos Types AFC* AFPD*
8	..	35	40	..	..
6	..	45	55	..	..
4	..	60	70	..	..
2	..	80	95	..	..

*These types are used almost exclusively in fixtures where they are exposed to high temperatures and ampere ratings are assigned accordingly.

**Tinsel cord.

***Elevator cables only.

†The ampacities under sub-heading A apply to 3-conductor cords and other multiconductor cords connected to utilization equipment so that only 3 conductors are current-carrying. The ampacities under sub-heading B apply to 2-conductor cords and other multiconductor cords connected to utilization equipment so that only 2 conductors are current carrying.

NOTE: Ultimate Insulation Temperature. In no case shall conductors be associated together in such a way with respect to the kind of circuit, the wiring method used, or the number of conductors that the limiting temperature of the conductors will be exceeded.

Table 430-10(b)
Minimum Wire Bending Space
at the Terminals of Enclosed Motor Controllers (In Inches)

AWG or Circular-Mil Size of Wire	*Wires per Terminal 1	2
14–10	Not specified	—
8–6	1 1/2	—
4–3	2	—
2	2 1/2	—
1	3	—
1/0	5	5
2/0	6	6
3/0–4/0	7	7
250	8	8
300	10	10
350–500	12	12
600–700	14	16
750–900	18	19

***Where provision for 3 or more wires per terminal exists the minimum wire bending space shall be in accordance with the requirements of Article 373.**

Glossary

A

AC generator. A generator that produces alternating current.

Alternator. Another name for an AC generator.

Ammeter. A meter that measures amperes.

Ampacity. The capacity in amperes of a wire, fuse, or circuit.

Ampere. The current produced when 6.25×10^{18} electrons flow past a given point in 1 second.

Armature. Rotating part of an electric motor or generator; may also be the moving part of a relay, buzzer, or speaker.

Assemblies. An electronic or electrical piece of equipment made up of a number of parts.

Atom. The smallest particle of an element that retains all of the properties of that element.

Automatic Load Transfer. A method of starting and switching for an emergency power plant when the utility power is interrupted.

Auxiliary Equipment. Equipment in addition to that needed as standard equipment, usually designed to make the piece of equipment do a specific job.

B

Ballast. A choke or inductor used in a gaseous-discharge (fluorescent) lamp circuit.

Battery. Two or more cells connected in series or parallel.

Bonding. Ensured electrical continuity of the grounding circuit by proper connections between service raceways, service cable armor, all service equipment enclosures containing service conductors, and any conduit or conductor that forms part of the grounding conductor to the service raceway.

Bonding Jumpers. Devices used to ensure continuity around concentric or eccentric knockouts that are punched or otherwise formed in such a manner that they impair electrical current flow.

Branch Circuit. A circuit that is used to feed small electrical loads such as lamps, small appliances, and kitchen equipment.

Brushes. Devices, usually made of carbon or metal, that make contact between rotating connections and the armature of a motor or generator.

C

Cable Tray. A metal try used to hold and protect large cables, usually found in commercial or industrial locations.

Canadian Standards Association. An organization, parallel to the UL in the United States, set up to develop voluntary national standards, provide certification services for national standards, and represent Canada in international standards activities.

Capacitor. A device, consisting of two plates and a dielectric, used to store an electrical charge.

Cell. A device used to generate an electrical current, or emf.

Circuit. A path for electrons to flow or electricity to move.

Circular Mil. A measurement unit for the cross-sectional area of a round wire, equal to one-thousandth (0.001) of an inch in diameter.

Class 1, 2, 3 conductors. Classes of wiring used in remote control circuits such as door bells, garage door openers, furnace thermostat circuits, and burglar alarm circuits.

Coil. A device, also called an inductor, made by turns of insulated wire wound around a core; sometimes with a hollow center portion.

Color Code. Colors used to indicate the resistance of carbon composition resistors. Bands are colored according to an established code:

Black	0	Yellow	4	Gray	8
Brown	1	Green	5	White	9
Red	2	Blue	6		
Orange	3	Violet	7		

Commutator. A device, made of segments of copper insulated by mica or some other material, used to reverse the direction of current flow from a generator or to a motor. Brushes are usually placed to make contact with the surface of the commutator. Segments of a commutator are connected to ends of the armature coils in a motor or generator.

Complete Circuit. A circuit made up of a source of electricity, a conductor, and a consuming device.

Conductor. A material through which electrons move.

Configuration. An arrangement of wires or components that may or may not be in a circuit.

Construction Electrician. One who assembles, installs, and wires systems for heat, light, power, air conditioning, and refrigeration and who may also install electrical machinery and electronic equipment as well as control and signal communications systems.

Controller. A device consisting of circuitry that will aid in the control of an electric motor. In robotics it is the entire controlling circuitry usually located in a large box or cabinet.

Convenience Outlet. Outlets placed for the convenience of the home owner in the everyday use of electricity.

Converter. An electromechanical device used to convert alternating current to direct current.

Coulomb. The unit of measurement of 6.25×10^{18} electrons.

Current. The movement of electrons in a negative-to-positive direction along a conductor.

Current Flow. The movement of free electrons in a given direction.

D

Dielectric. An insulating material used in a capacitor or other electrical device.

Dielectric Heating. The heating of a nominally insulating material due to its own dielectric losses when the material is placed in a varying electric field. (The term may be used in the NEC to include any equipment used for heating purposes whose heat is generated by induction or dielectric methods.)

Diode. A semiconductor or vacuum tube device that allows current to flow easily in one direction but retards or stops its flow in the other direction. Diodes are used in rectifier circuits and switching circuits.

Direct Current. Current that flows in one direction only.

Drafter. One who draws plans and electrical schematics with the aid of mechanical devices.

Dry Niche. A lighting fixture intended for installation in the wall of a pool or fountain in a niche that is sealed against the entry of pool water.

E

Electrical Engineer One who designs, develops, and supervises the manufacture of electrical and electronic equipment.

Electrician. One who works with electrical equipment and wiring.

Electrode. An element in a cell; a part of a vacuum tube; a terminal or connector of an apparatus used in the treatment or diagnosis of disease. (In arc welding the electrode is usually melted by the arc and becomes part of the weld. It is the actual contacting point between the electric welder and the work.)

Electrodynamometer. A type of meter that uses no permanent magnet but two fixed coils to produce a magnetic field. The meter also uses two moving coils. This meter can be used as a voltmeter or an ammeter.

Electrolyte. A solution capable of conducting electric current; the liquid part of a battery.

Electrolytic. A capacitor that has parts separated by an electrolyte. Thin film formed on one plate provides the dielectric. The electrolytic capacitor has polarity (− and +).

Electrolytic Cell. A receptacle or vessel in which electrochemical reactions are caused by applying electrical energy for the purpose of refining or producing usable materials.

Electromagnet. A magnet produced by current flow through a coil of wire. The core is usually used to concentrate the magnetic lines of force.

Electromagnetism. Magnetism produced by current flowing through a coil of wire or other conductor.

Electron. The smallest part of an atom with a negative charge.

Electroplating. The process of depositing a metal coating on an object by the use of electrical charges.

Elements. The most basic substances of the universe. Ninety-four elements, such as iron, copper, and nitrogen, have been found in nature. Others have been artificially produced. Every known material is composed of elements.

Encapsulation. The embodiment of a component or assembly in a solid or semisolid medium such as tar, wax, or epoxy.

Enclosed. Surrounded by a case, housing, fence, or walls that will prevent accidental contact with energized parts.

Enclosure. The case or housing of apparatus, or the fence or walls surrounding an installation to prevent people from accidentally contacting energized parts or to protect equipment from physical damage.

Energy. The ability to do work.

Entrance Signals. A device, usually consisting of a door bell or chime, to alert the home occupant to someone outside wishing to enter.

Equipment. A general term that includes material, fittings, devices, appliances, fixtures, apparatuses, and the like used as a part of, or in connection with, an electrical installation.

Equipment Ground. The grounding of exposed conductive materials, such as conduit, switch boxes, or meter frames, that encloses conductors and equipment to prevent the equipment from exceeding ground potential.

Explosion-Proof. Condition of apparatuses enclosed in a case that is capable of withstanding an explosion within the case without igniting flammable materials outside it.

Exposed. Capable of being inadvertently touched or approached nearer than a safe distance. It is applied to parts not suitably guarded, isolated, or insulated.

F

Farad (F). A unit of measurement for capacitance.

Fatal Current. The amount of current needed to kill. Currents between 100 and 100 mA are lethal.

Feeder. All circuit conductors between the service equipment, or the generator switchboard of an isolated plant, and the final branch-circuit overcurrent device.

Feeder Circuit. A circuit that is used to feed others or that takes electrical energy to where it can then be branched off to service other locations.

Filaments. Small coils of resistance wire in light bulbs and vacuum tubes that heat up to glow either red- or white-hot. The filament of a vacuum tube boils off electrons from the cathode. The filament in a light bulb glows to incandescence to produce light.

Fitting. An accessory such as a locknut, bushing, or other part of a wiring system that is intended primarily to perform a mechanical rather than an electrical function.

Fluorescence. The state of glowing or giving off light. Fluorescent lamps produce light when mercury ions collide and strike a fluorescent coating inside the tube.

Fluorescent. Producing light through the action of ultraviolet rays striking a fluorescent material that is coated on the inside of a glass tube.

Forming Shell. A metal structure designed to support a wet-niche lighting fixture assembly and intended for mounting in a pool or fountain structure.

Four-way Switch. A switch used where it is necessary to turn a light or circuit on or off from three or more locations. It needs a minimum of two three-way switches to do its job.

Fuse. A safety device designed to open if excessive current flows through a circuit.

G

Generator. An electromechanical device that converts mechanical energy to electrical energy.

Ground. A conducting connection, whether intentional or accidental, between an electrical circuit or equipment and the earth or to some conducting body that serves in place of the earth.

Ground Fault Circuit Interrupter (GFI or GFCI). A fast-operating circuit breaker that is sensitive to very low levels of current leakage to ground. It is designed to limit electric shock to a current and time duration value below that which can produce serious injury.

Grounded Conductor. A system or circuit conductor that is intentionally grounded.

Grounding. Ensuring that the part to ground is permanent and continuous; it has a low impedance to permit all current-carrying devices on the circuit to work properly.

Grounding Conductor. A conductor used to connect equipment or the grounded circuit of a wiring system to a grounding electrode or electrodes.

Guarded. Covered, shielded, fenced, enclosed, or otherwise protected by means of suitable covers, casings, barriers, rails, screens, mats, or platforms to remove the likelihood of approach or contact by persons or objects to a point of danger.

H

Henry (H). Basic unit of measurement for inductance.

Hertz (Hz). Basic unit of measurement for frequency.

Horsepower (Hp). Unit of measurement of doing work electrically and equating it with the work done by a horse. It means the energy consumed to produce the equivalent work done by a horse lifting 33,000 pounds for a distance of 1 foot in a time period of 1 minute. It takes 746 watts to equal 1 horsepower.

Hydromassage Bathtub. A bathtub equipped with a recirculating piping system, pump, and associated equipment, designed so that it can accept, circulate, and discharge water at each use.

I

Impedance (Z). Total circuit opposition to alternating current, measured in ohms.

Incandescent. Glowing white-hot. The filament in an incandescent lamp glows white-hot to produce heat and light.

Incandescent Lamp. A lamp or bulb that produces light by heating a filament to incandescence (white-hot).

Induced Current. Current produced by electromagnetic induction, usually from a coil or transformer.

Inductance (L). The property of a coil that opposes any change in circuit current, measured in henrys (H).

Induction Heating. The heating of a nominally conductive material due to its own I^2R losses when the material is placed in a varying electromagnetic field.

Inductive Reactance. Opposition presented to alternating (X_L) current by an inductor measured in ohms.

Inerting. Mixing a chemically inert, nonflammable gas with a flammable substance, displacing the oxygen until the percentage of oxygen in the mixture is too low to allow combustion.

Input. Term used to describe the energy applied to a circuit, device, or system.

Insulator. Nonconducting material lacking a sufficient supply of free electrons to allow the movement of electrons without exceptional force or high voltage.

Integrated Electrical System. A unitized segment of an industrial wiring system in which all of the following conditions are met: (1) an orderly shutdown is required to minimize personnel hazard and equipment damage; (2) maintenance and supervision ensure that qualified persons service the system: and (3) effective safeguards, acceptable to the authority having jurisdiction, are established and maintained.

Interlock. A switch so placed that whenever a protective cover is removed, the piece of equipment is safe for human contact.

Interrupting Rating. The highest current at rated voltage that an overcurrent protective device is intended to interrupt under standard test conditions.

Isolated. Not readily accessible unless special means for access are used.

Isolation Transformer. A transformer with a secondary and primary physically separated.

J

Joule. A metric unit of electrical work done by 1 coulomb flowing with a potential difference of 1 volt or 1 watt-second.

Jumper. A short wire, usually with clips on each end, for making temporary connections.

Junction Box. A metal box in which connections of several wires are made; a distribution box.

K

Kilo (k). A prefix that means 1,000.

Kilowatt (kW). 1,000 watts.

Kilowatt Hour (kWh). Measurement of the electrical power (1,000 watts) used by a consumer over a period of 1 month and equated to 1 hour.

Kirchhoff's Law of Voltages. Law that states that the sum of all voltage drops across resistors or loads in a complete loop is equal to the applied voltage.

L

Lighting Outlet. An outlet intended for the direct connection of a lampholder, a lighting fixture, or a pendant cord terminating in a lampholder.

Load. Anything that may draw current from an electrical power source.

Location, Damp. A partially protected location under canopies, marquees, roofed open porches, and like locations, and interior locations subject to moderate degrees of moisture, such as some basements, some barns, and some cold-storage warehouses.

Location, Dry. A location not normally subject to dampness or wetness. A location classified as dry may be temporarily subject to dampness or wetness, as in the case of a building under construction.

Location, Wet. Installations underground or in concrete slabs or masonry in direct contact with the earth, locations subject to saturation with water or other liquids, such as vehicle-washing areas, and locations exposed to weather.

M

Magnet. A device that possesses a magnetic field.

Magnetism. A force produced by an electrical current in a wire or found in nature in certain materials.

Magnet Wire. Copper wire used to wind coils, solenoids, transformers, motors; usually coated with varnish or other insulating material.

Mega. A prefix that means one million (1,000,000).

Meter. An electrical device that is used to measure volts, amps, ohms, watts.

Mica. Insulating material that can withstand high voltages and elevated temperatures used in the manufacture of appliances and capacitors.

Micro. A prefix that means one millionth (0.000001).

Microammeter. A meter limited to measuring microamperes.

Microampere (μA). One millionth of an ampere.

Microvolt (μA). One millionth of a volt.

Milli. A prefix that means one thousandth (0.001).

Milliammeter. A meter limited to measuring milliamperes.

Milliampere (mA). One thousandth of an ampere.

Millivolt (mV). One thousandth of a volt.

Milliwatt (mW). One thousandth of a watt.

Motor. A device used to change electrical to mechanical energy.

Multimeter. A meter capable of measuring (in most instances) volts, ohms, and milliamperes.

Multiplier. A resistor placed in series with a meter movement to handle the extra voltage applied to the movement; and extends the range of the meter movement and serves to extend the range of the meter.

N

Nameplate Rating. A piece of electrical equipment that has the manufacturer's suggested operating voltage, current, phase, and operating temperature information attached to it.

National Electrical Code (NEC). A book of standards, produced by the National Fire Protection Association every three years, that describes the proper installation of various electrical machinery and devices for safe operation.

Neutrons. Tiny particles in the nucleus of an atom that have no electrical charge.

O

Ohmmeter. A device used to measure resistance.

Ohm's Law. Law that states that the current in any circuit is equal to the voltage divided by the resistance.

Omega. The Greek letter (Ω) used to indicate the unit of resistance, the ohm.

OSHA. Occupational Safety and Health Act.

Outlet. A point on the wiring system at which current is taken to supply utilization equipment.

Overcurrent. Any current in excess of the rated current of equipment or the ampacity of a conductor. It may result from overload, short circuit, or ground fault.

Overload. Operation of equipment in excess of normal, full-loading rating or of a conductor in excess of rated ampacity, which, when it persists for a sufficient length of time, causes damage or dangerous overheating. A fault, such as a short circuit or ground fault, is not an overload.

P

Parallel Circuit A circuit where each load (resistance) is connected directly across the voltage source.

Photosensitive. Sensitive to light.

Piezoelectrical Effect. The process of using crystal under pressure to produce electrical energy.

Potentiometer. A variable resistor that has three contact points.

Power (P). Rate of doing work, measured in watts (W).

Proton. Positively charged particle in the nucleus of an atom.

Pulsating Direct Current (PDC). Current produced when AC is changed to DC and left unfiltered; the type of DC generated by a DC generator.

Purging. The use of nonflammable gases to flush flammable gases or vapors from an enclosed space. This process enables the vessel or space involved to be opened or used for fuels without danger of an explosion. Carbon dioxide is used for purging. The amount needed varies with the molecular weight of the substance being purged. Most gases are lighter than carbon dioxide and therefore can be displaced if the purging agent is introduced slowly in gaseous form from the bottom of the container. The flammable gas is vented at the top.

R

Raceway. An enclosed channel designed expressly for holding wires, cables, or busbars, with additional functions as permitted by the NEC.

Rainproof. Constructed, protected, or treated in such a way as to prevent rain from interfering with the successful operation of the apparatus under specified test conditions.

Raintight. Constructed or protected in such a way that exposure to a beating rain will not result in the entrance of water under specified test conditions. Raceways on exterior surfaces of buildings should be made raintight.

Receptacle. A contact device installed at the outlet for the connection of a single attachment plug; an outlet where one or more receptacles is installed.

Rectifier. A device that changes AC to DC, allowing current to flow in only one direction.

Relay. An electromechanical device used for remote switching.

Remote Control. The ability to start or control a device from a location other than that of the device being controlled.

Residential Wiring. The electrical wiring used to supply a home with electrical outlets where needed.

Resistance. Opposition to the movement of electrons, measured in ohms.

Rheostat. A variable resistor, usually with only two points connected into the circuit, used to control voltage by increasing and decreasing resistance.

Root-Mean-Square (rms). A type of reading obtained by using a standard voltmeter or ammeter.

Rotor. A moving part of a motor or generator.

Rural Grid. A network of radial feeders that leave the substation as three-phase circuits with each of the phases fanning out to serve the countryside as single-phase circuits.

S

Series Circuit. A circuit with one resistor or consuming device located in a string or one after another.

Series-Parallel Circuit. A combination of series and parallel circuits where a minimum of three resistors or devices are connected, with at least one in series and at least two in parallel.

Shock. A process whereby an outside source of electricity is such that it overrides the body's normal electrical system and causes the muscles to react involuntarily.

Short Circuit. A circuit that has extremely low resistance.

Shunt. A resistor placed in parallel with a meter movement to handle, or shunt, most of the current around the movement; serves to extend the range of the meter.

Signaling Circuit. Any circuit that energizes signaling equipment.

Slip Ring. A ring of copper mounted on the shaft of a motor or generator through which a brush makes permanent (or constant) contact with the end of the rotor windings; always used in pairs.

Solar Cell. A cell, usually made of silicon, that turns light energy into electrical energy.

Solar Photovoltaic System. The total components and subsystem that in combination convert solar energy into electrical energy suitable for connection to a utilization load.

Solenoid. A coil of wire, an electromagnet, wrapped around a hollow form, usually with some type of core material sucked into the hollow. The movement of the core usually causes a switch or valve to open or close. It can be used to turn electricity, water, oil, or gas on or off.

Splice. A form of electrical connection in which wires are joined directly to one another.

Static Electricity. A form of energy present when there are two charges of opposite polarity in close proximity. Static electricity is generated by friction.

Switchboard. A large single panel, frame, or assembly of panels on which are mounted, on the face or back or both, switches, overcurrent, and other protective devices, buses, and instruments. Switchboards are generally accessible from the rear as well as from the front and are not intended to be installed in cabinets.

System Ground. The grounding of the neutral conductor, or ground leg, of the circuit to prevent lightning or other high voltages from exceeding the design limits of the circuit.

T

Terminal. A connecting point for wires. Terminals are usually present on batteries, cells, switches, relays, motors, and electrical panels.

Thermal Cutout. An overcurrent protective device that contains a heater element in addition to and affecting a renewable fusible member that opens the circuit. It is not designed to interrupt short circuit currents.

Thermocouple. A device made of two different kinds of metals joined at one end. When the junction is heated, an emf is generated across the open ends.

Thermostat. A device that acts as a switch that is operated when heat causes two metals to expand at different rates.

Three-way Switch. A switch used where it is necessary to turn a light or circuit on or off from more than one location.

Toggle Switches. Devices used to turn various circuits on and off or to switch from one device to another. They are made in a number of configurations.

Transformer. A device that can induce electrical energy from one coil to the other by using magnetic lines of force, thereby stepping voltage up or down according to the turns ratio.

Treadle. A foot-operated swiveling device that is used, in some cases, to power simple machines.

U

Underwriters Laboratories, Inc. (UL). A non-profit organization that tests electrical devices, systems, and materials to see whether they meet certain safe operating standards.

Universal Motor. A motor that operates on AC or DC; commonly used in power tools and vacuum cleaners.

Utility. An electric power company that exists to furnish electrical energy to the public.

Utilization Equipment. Equipment that uses electric energy for mechanical, chemical, heating, lighting, or similar purposes.

V

Volt (V). A unit of measurement of electromotive force. Represented by E in Ohm's law journals.

Voltage. Electromotive force (emf) that causes electrons to move along a conductor.

Voltage, Nominal. A nominal value assigned to a circuit or system for the purpose of conveniently designating its voltage class (as 120/240, 480Y/277, 600, etc). The actual voltage at which a circuit operates can vary from the nominal within a range that permits satisfactory operation of equipment.

Voltage-to-Ground. For grounded circuits, the voltage between the given conductor and that point or conductor of the circuit that is grounded; for ungrounded circuits, the greatest voltage between the given conductor and any other conductor of the circuit.

Voltmeter. A meter used to measure voltage.

W

Watt (W). Unit of electrical power.

Watthour-Meter. A meter that measures electrical power consumed over a period of time (usually a month) and equates it to a time period of 1 hour.

Wet-Niche Lighting Fixture. A lighting fixture intended for installation in a metal forming shell mounted in a pool or fountain structure where the fixture will be completely surrounded by water.

X Y Z

Zener. A type of semiconductor diode that breaks down intentionally at a predetermined voltage. Usually used for voltage regulation circuits.

INDEX